ELEMENTARY HUMAN PHYSIOLOGY

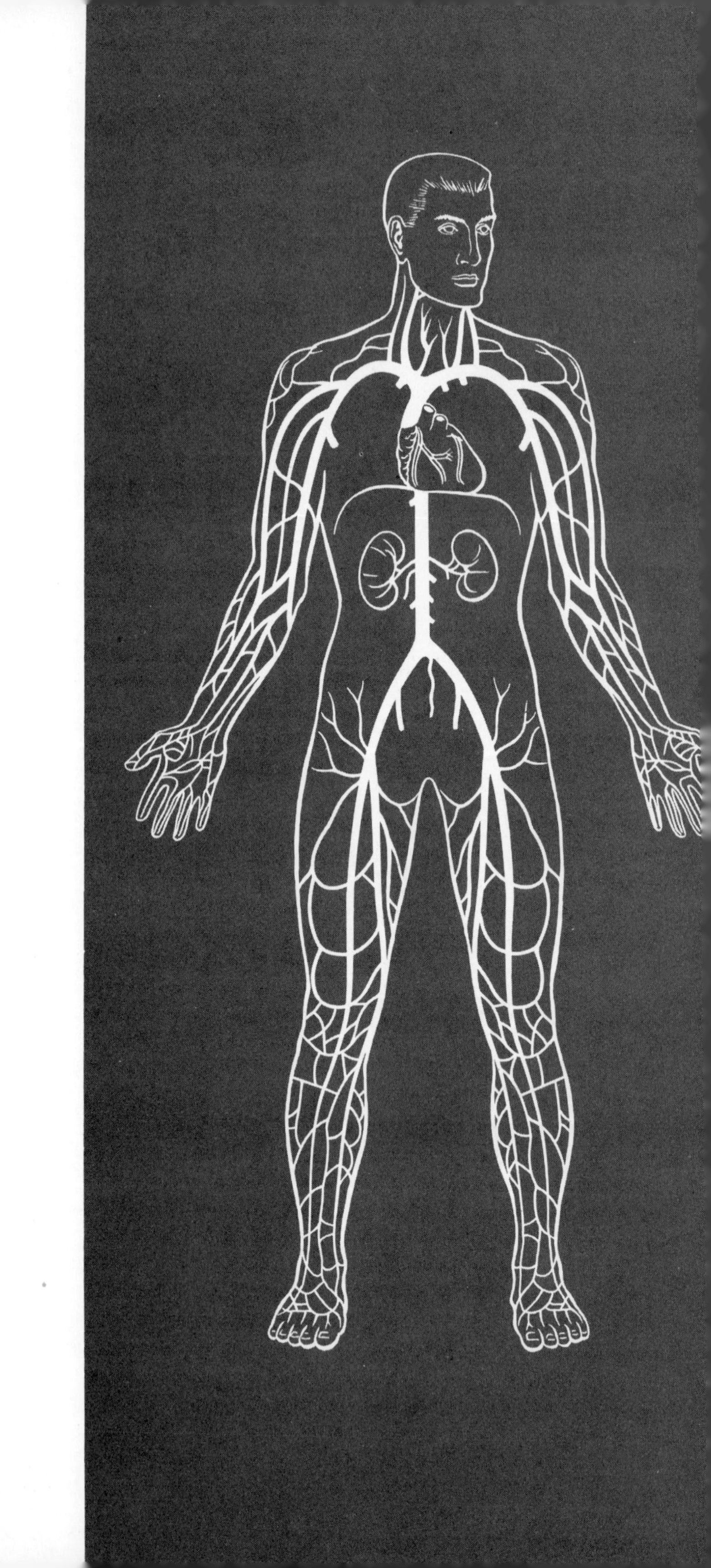

TERENCE A. ROGERS, Ph.D.
Department of Physiology
Stanford University

ELEMENTARY HUMAN PHYSIOLOGY

A TEXT FOR UNDERGRADUATES

JOHN WILEY & SONS, INC., NEW YORK • LONDON

LIBRARY OF CONGRESS CATALOG CARD NUMBER: 61-11245

PRINTED IN THE UNITED STATES OF AMERICA

FOR MY TEACHER, FRANK TREHERN

PREFACE

This is a textbook of human physiology for undergraduates. It is widely agreed that students in nonbiological fields should receive at least some exposure to biological science and I feel, not without bias, that a course in human physiology is entirely suitable. I have written this book therefore, not only for those anticipating careers in the medical sciences or zoology, but also for those students for whom a course in human physiology is their first and last venture in biology.

In order to orient students without a background in elementary biology, the first three chapters describe the general architecture of the mammalian body, its cells and tissues. Many of the processes in the body are best described in chemical terms. Therefore, for the benefit of students without a grounding in freshman college chemistry I have included a brief review of elementary chemistry, and in the remainder of the text I have tried to develop each topic assuming only a reasonable high-school background in this subject. The few sections presenting more sophisticated chemical (and physical) concepts are arranged so that they remain intelligible (if less complete), even without a complete understanding of the quantitative detail.

There is no obvious beginning point in the study of physiology, because the various functions are so interrelated that a full discussion of one must presuppose considerable information about several others. In spite of this difficulty, there is clearly one unifying theme; the maintenance of a suitable composition of the internal environment —or *homeostasis.*

Homeostasis is not only the most important philosophical concept in physiology, it also provides a useful framework for teaching. The functions of all the organs and even their phylogenetic origins can be considered in terms of their dependence on a near-constant internal environment and their contribution to its maintenance. The approach in this book is to introduce the concept of homeostasis very early and to use it as the theme in subsequent discussions. As a result, a somewhat unusual order of presentation is followed, in which muscle and nerve physiology is largely held over to the second part of the book. I feel this avoids some of the difficulties encountered by many beginning students who become hopelessly enmeshed in details of neurophysiology, for example, before they have grasped the overall role of the nervous system. Even with this aim, it has been necessary to introduce an earlier discussion of the nature of excitability and of simple reflexes, so that the nervous control of circulation and respiration can be reasonably presented.

Simple and brief statements about anything as complex as physiology tend to be dogmatic. Wherever reasonable, exceptions from the principles outlined are described, but in the main I have deliberately avoided hedging each statement about with lists of minor exceptions. In particular, it must be borne in mind that since this is a textbook of *human* physiology, any generalizations are to be taken as applicable to mammals. There are inevitably nonmammalian exceptions to these generalizations, but the circumlocutory phrasing or special notes required to take them into account would be tedious and interrupt the main line of discussion. In the same vein, I consider it unnecessary to give chapter and verse from the original literature for material in a beginning textbook. For further detail in any area, the student is referred to the major textbooks of medical physiology, which are all thoroughly documented.

An additional source of supplementary reading is in the many physiological articles which have appeared in the *Scientific American* during the last twelve years. As these semipopular references are not included in the major textbooks, those related to each topic are listed at the ends of the appropriate chapters in this book.

No one can derive the full benefit from a science course without some laboratory experience, but it is difficult for students to learn from experiments which, in their hands, do not work. Laboratory exercises in elementary physiology must be designed to steer a course between a boring over-simplicity and the commoner fault of a complexity that demands too much manipulative skill from the student. The *Laboratory and Demonstration Manual* written by Taylor and Sargent, published by the Burgess Publishing Company, is successful in this respect and is highly suitable for laboratory teaching at the level of this textbook.

Most students are intrigued by the clinical aspects of physiology. While recognizing this interest, I have endeavored to emphasize the importance of an understanding of the normal functions as the groundwork of medicine. The various disease symptoms are not introduced as entities to be followed by physiological *explanations,* but rather they are used to illustrate physiological principles. Physiology as a science has developed to the extent that some of it has the logical orderliness of classical physics. At the same time there are tremendous areas in which our knowledge is fragmentary and full of apparent inconsistencies. This means that even for the beginning student there are great intellectual satisfactions and opportunities for speculation in physiology.

In conclusion I would like to express my gratitude to those colleagues who have helped me in the preparation of this book. They include Professors Max Kleiber, Ronald Grant, and Thomas Algard. The cheerfully acrid comments of my friend Dr. John Lambooy provided even greater assistance and encouragement.

TERENCE A. ROGERS

Stanford, California
March, 1961

CONTENTS

PART 1

INTRODUCTION

OUTLINE OF ANATOMY

1

It is conceivable that the earliest forms of life were blobs of jelly without any structural organization, but most living organisms of the present day comprise one or more *cells.* The cell may be regarded as the unit of living material in much the same way that an individual person is the unit of a society or nation. Furthermore, even the most superficial examination of a dissected animal shows that it is not made of a uniform substance. There is muscle (red meat), fat, intestine, bone, brain, blood, etc. Each of these, including the fluid blood, is made up of one or more *tissues.* A tissue, in turn, comprises a collection of similar cells especially adapted to its functions.

The study of cells, their form and function, is called *cytology;* the study of tissues is *histology.* There is clearly an enormous overlap between these fields, but in general the cytologist is interested in processes *within* the cell and prefers to work with living material whenever possible; the histologist studies the contribution of different cell types to the overall structure of a tissue and usually works with fixed and stained preparations. Histology is an important tool for the *pathologist* who can determine the nature of a disease by microscopic examination of tissues taken at autopsy, or by biopsy.

The *organs,* like the heart, stomach,

ear, etc., are made up of tissues; usually of several different kinds of tissues. Most of the organs are clearly visible to the naked eye and their structure is studied in *gross anatomy.* The study of their *functions* is the major part of *physiology,* the topic of this textbook. We cannot discuss the functions of an organ, however, without some idea of its structure, its location in the body, and the materials of which it is made. This chapter and the next two, therefore, present a brief review of microscopic and gross anatomy.

MAMMALS IN EVOLUTION

The major division among living organisms is between the *plant kingdom* and the *animal kingdom.* Among the simpler forms, it is sometimes difficult to determine whether an organism is a plant or an animal, but in our everyday experience there is little doubt as to which is which.

Animals are classified according to what is presumed to be their evolutionary development. This is indicated diagrammatically in Fig. 1.1. In the present context, it is sufficient to note that the mammals are members of a larger group, the vertebrates, which are members of a still larger group called the chordates. Animals such as insects, crabs, and spiders are not vertebrates and are derived from a separate evolutionary line.

It is well known that the mammals are warm blooded animals; the females bear their young alive and produce milk to nourish them in infancy. It is customary to regard the mammals as the "highest" forms of life, but many of the birds might be considered the equal of some mammals in this respect.

GENERAL ARCHITECTURE OF THE VERTEBRATE BODY

Skeleton. If we hang up the skeletons of different vertebrates, we notice remarkable similarity between such superficially disparate animals as the frog, rat, pigeon, and man (Fig. 1.2). The similarity between the skeletons of relatively closely related mammals is even more striking.

As shown by the example of the rat in Fig. 1.3, the vertebrate skeleton comprises a *vertebral column,* on one end of which the *skull* is articulated. The *pectoral* and *pelvic girdles,* as their names imply, are bones arranged to form flattened loops at what we regard as the upper chest and the hips of the animal. The limbs are articulated on these girdles. The skeleton has joints, or *articula-*

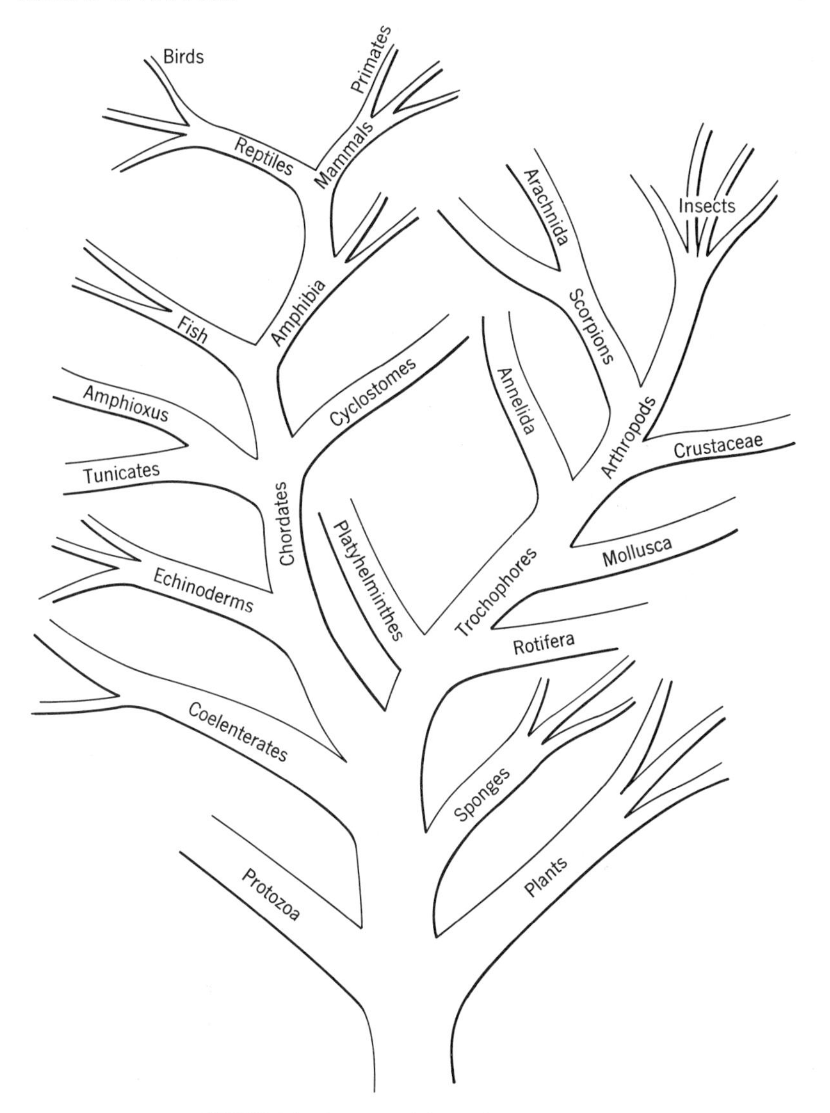

FIGURE 1.1. Diagram of the evolutionary "tree."

tions, by which the limbs are able to move, and these provide various degrees of freedom of movement, depending on their structural arrangement. Ball and socket joints, such as those of the shoulder and hip, permit free rotational movement; hinge-like joints, as in the knee and elbow, provide movement in one plane only; others, like the sacro-iliac joint in the lower back, permit only

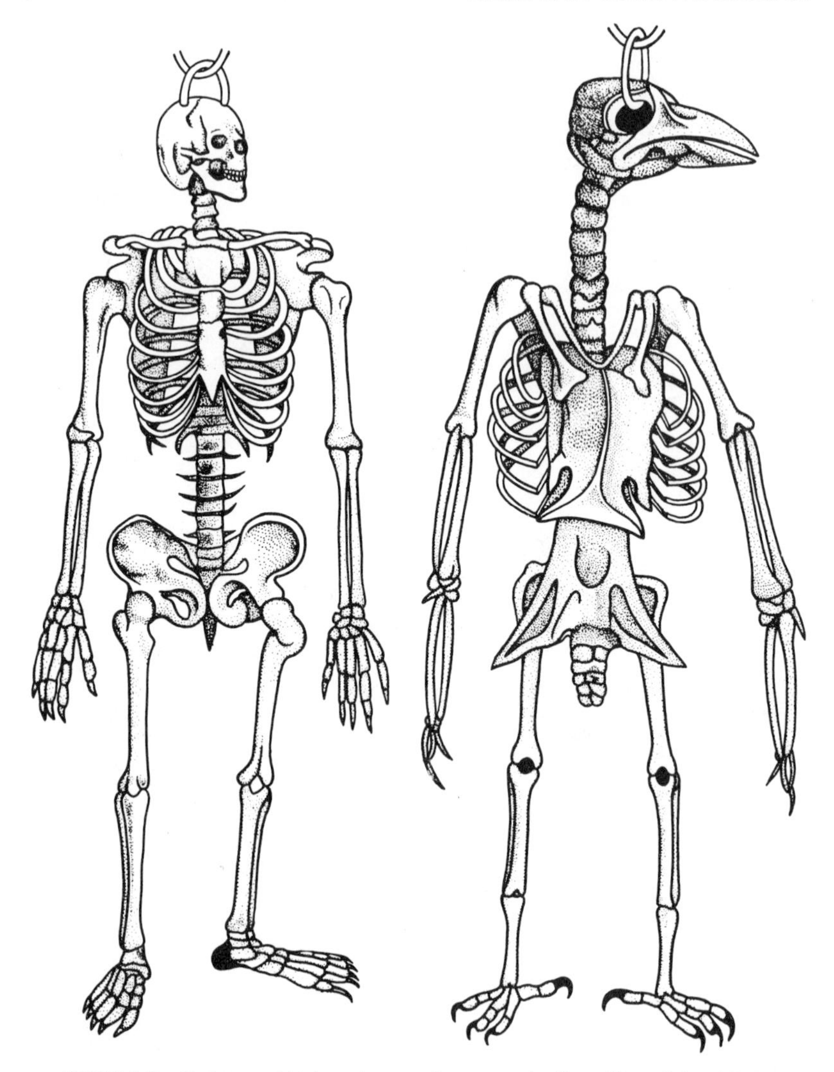

FIGURE 1.2. Skeletons of birds and mammals compared. (From Pierre Belon, 1555.)

limited movement. The junctions of the various bones of the skull are also technically regarded as articulations, but their movement, of course, is practically zero.

The vertebral column is the main axis of the skeleton. In man, it is made up of 24 short segments of bone, the *vertebrae,* and two larger pieces, the *sacrum* and the *coccyx.* Each vertebra is joined to the next by strong ligaments but is separated from it by a pad

around. The muscles are attached to the skeleton in such a manner that their contraction will move the bones and the entire limb.

The muscles comprise about 40% of the body mass as a rule. There are over 400 muscles in the human body, but many of them are arranged in large functional units which can be considered separately. Furthermore, they are arranged in opposing groups, which are especially apparent in the limbs. The muscles that "bend" a joint when they contract are called the *flexor* muscles; they are opposed by the *extensor* muscles, which extend or straighten the joint. In the flexion of a limb, the extensor muscles are normally not merely passive, but their modest contraction

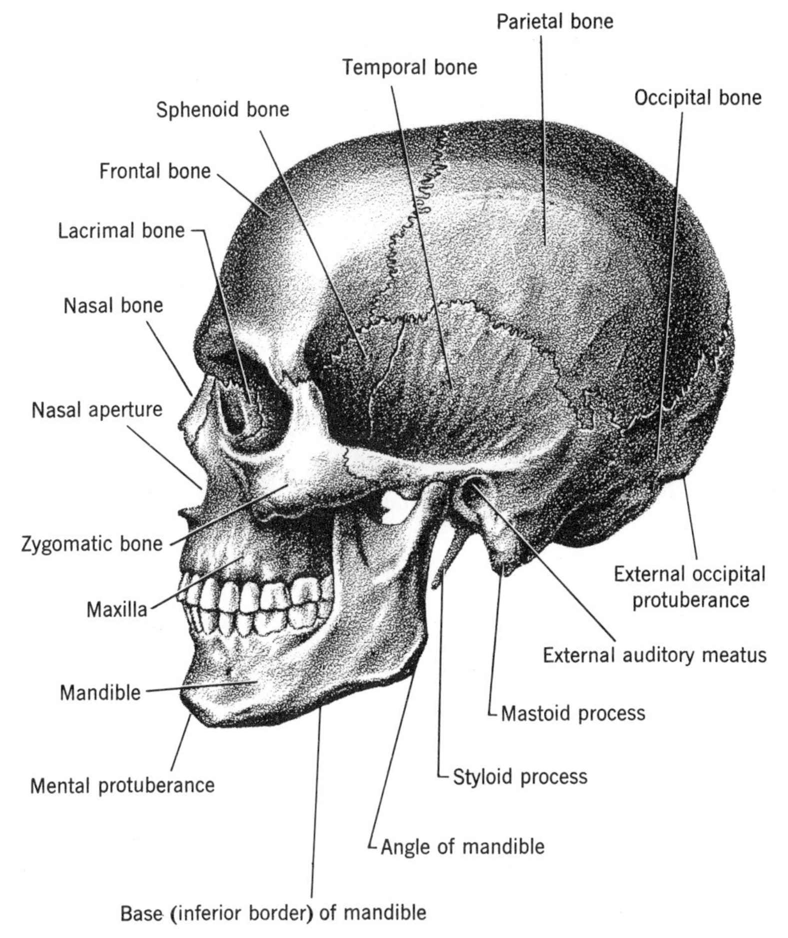

FIGURE 1.6. The skull.

provides an opposition to the flexors, leading to a smoothly controlled flexion.

The musculature, and indeed most other aspects of anatomy, are traditionally studied by regions; the head and neck, the thorax, the abdomen, the perineum, the arms and legs. Because many muscles lie beneath others, the simple examination of a skinned animal does not disclose all of them to view. Therefore, the student should actually dissect an animal to learn the arrangement of the muscles.

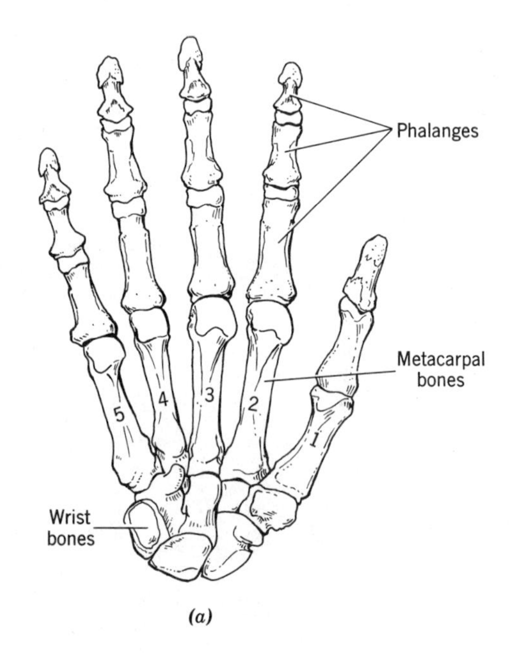

(a)

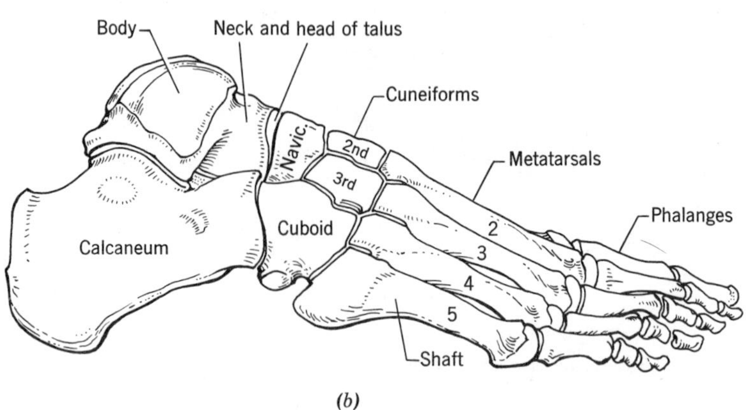

(b)

FIGURE 1.7. (*a*) Bones of the hand. (*b*) Bones of the foot.

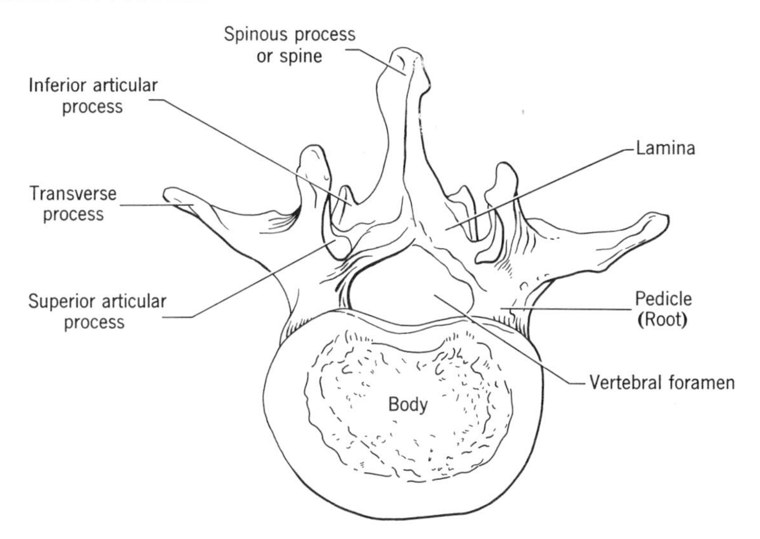

FIGURE 1.8. A vertebra (lumbar).

Even today, the names of many muscles are inherited from the anatomy of the Middle Ages or even earlier, and they frequently refer to the more superficial characteristics of the muscles and not to their real function. The *gastrocnemius* is the large muscle of the calf, and yet its name is derived from the Latin root meaning "*stomach,*" which refers to its shape. The name "*muscle,*" is derived from the Latin root for "*mouse,*" because the early anatomists thought the twitching of muscles in a recently dead animal were mouse-like movements. In the last few years, international conventions of anatomists have straightened up the nomenclature considerably, but the unfortunate student must still settle down to learn an arbitrary and archaic list of names. Figures 1.9 and 1.10 show the superficial muscles of the human body.

Heart muscle differs from skeletal muscle in several important respects, and the type of muscle found in many other internal organs is still more different. These differences in function and structure are fully discussed in a chapter devoted to muscle.

The digestive organs. The digestive tract of an animal, in the simplest form is a tube passing through the animal from mouth to anus. In the mammals and other higher forms, it is extensively convoluted and very long. Food is taken in at the mouth, where it is usually chewed, and then passed down the esophagus to the stomach where digestion begins. From the stomach the food passes in semiliquid form along the intestine where digestive juices act on it, reducing it to forms in which the nutrients can be absorbed

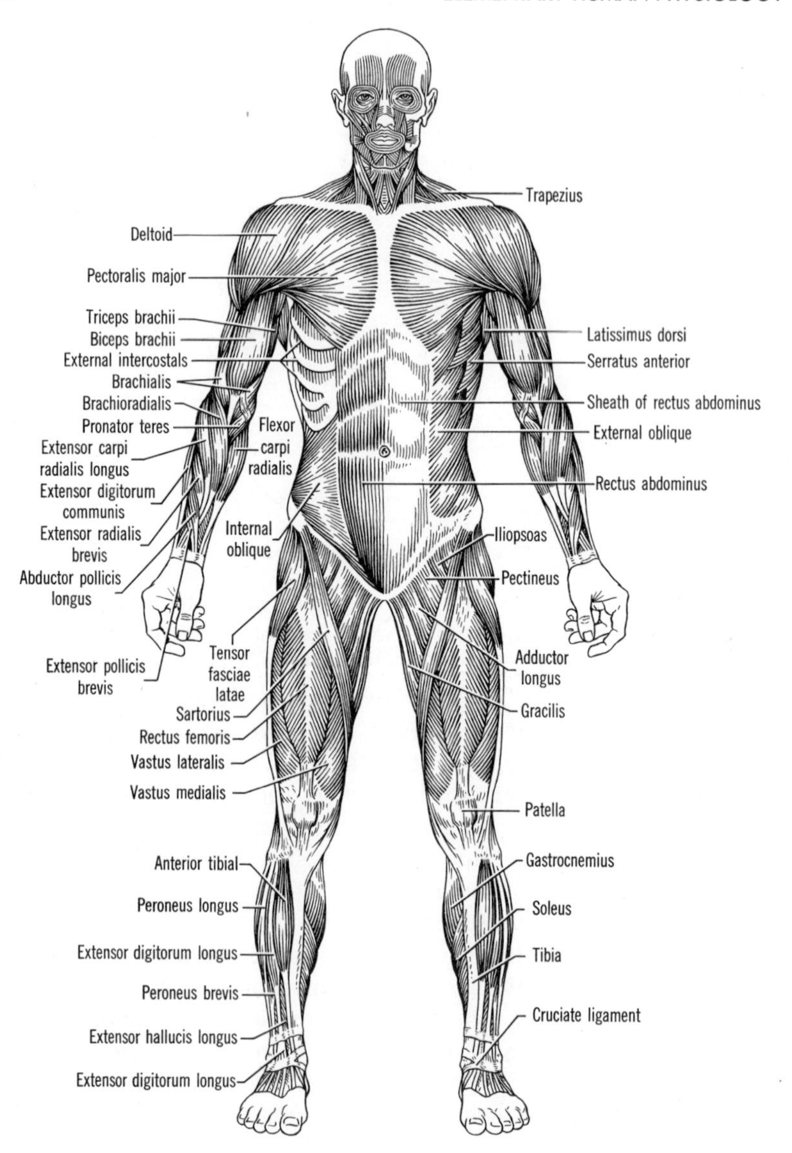

FIGURE 1.9. Muscles of the body, anterior view. The deeper muscles are shown on the right side of the abdomen.

through the intestinal wall into the blood stream. In the lower intestine, water is removed from the residue, which is then excreted as *feces*. The digestive juices are secreted by glands in the walls of the stomach and intestine and also by the separate digestive glands, the *liver* and the *pancreas*.

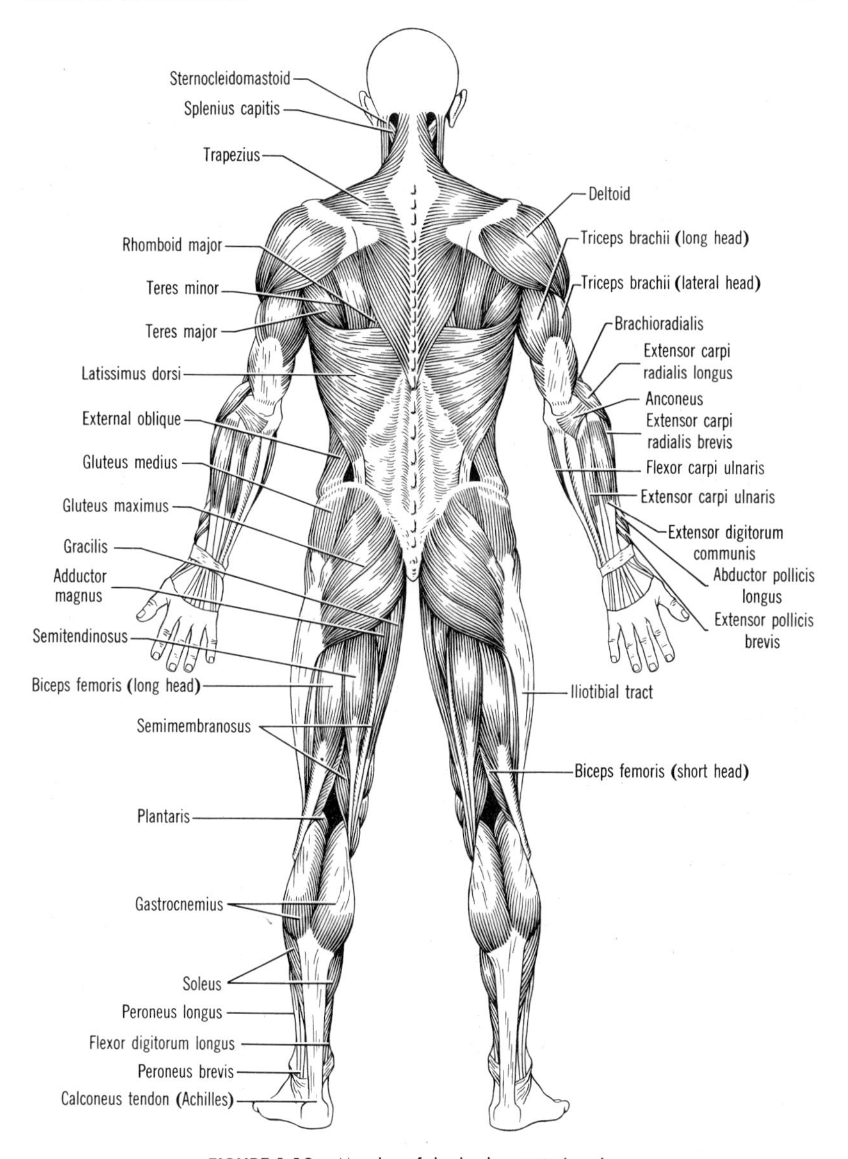

FIGURE 1.10. Muscles of the body, posterior view.

The respiratory system. All animals need oxygen; mammals get theirs by breathing air into the lungs where it is picked up by the blood. (Fishes obtain their oxygen by exchange-diffusion in the gills between blood and the water in which they swim.) The lungs are a pair of sponge-like sacs located in the chest, or *thorax*. They connect to the exterior by two *bronchi* (one for each lung), which join to form the *trachea,* or windpipe, which passes up the neck to the cavity behind the nose and mouth.

The circulatory system. The blood comprises a complex fluid, the *plasma,* in which the *red cells* and a smaller number of *white cells* are suspended. The oxygen from the lungs and the nutrients absorbed across the intestinal walls are distributed to the tissues of the body by the blood. The blood is kept in constant circulation by the pumping of the heart, which is located in the central part of the chest between the lungs. The blood leaves the heart by large elastic vessels called the *arteries,* which subdivide into smaller and smaller vessels supplying the various organs and tissues. The "used blood" from the tissues drains back to the heart in *veins* of increasing diameter approaching the heart. After entering the heart, the venous blood is circulated first through the lungs; then the oxygenated blood returns to the heart and is recirculated to the other tissues.

The excretory system. In the course of their metabolism, the cells produce carbon dioxide. The blood conveys this back to the lungs where, in the simplest possible terms, it is exchanged for oxygen, and the animal breathes out the unwanted carbon dioxide. Part of the blood pumped out by the heart is circulated through the *kidneys* where other kinds of waste products are removed and ultimately discharged in solution into the *bladder.* This solution, the *urine,* accumulates in the bladder and is periodically discharged.

The reproductive system. The essential organs of reproduction are the *testes* in the male and the *ovaries* in the female, which produce the *sperm* and *ova* respectively. The successful union of a sperm and ovum within the female genital tract can lead to the development of a new individual within the *uterus.* The ovaries are in the abdomen of the female, whereas in the male, testes are carried externally in the bag-like *scrotum.* The remainder of the female genitalia comprises the *vagina* (Latin, "sheath") which is a short tube, partly continuous with the uterus in which the young develop. In addition to the testes, the male genitalia comprise the ducts along which the sperm travel, glands contributing to their nutrition, and the *penis,* the external organ through which the sperm are introduced into the vagina of the female. The urine is also discharged from the bladder through the penis.

CELLS

2

The tremendous variety of cells known to biology makes it difficult to generalize about them, but a few characteristics are nearly universal. Typically, a cell has a boundary structure which contains a watery gel of fantastic complexity known as the *cytoplasm,* a denser *nucleus,* and, scattered through the cytoplasm, smaller bodies called *organelles.* Plant cells have rigid cell-walls which contribute to the structural strength of the plant, but animal cells have much frailer boundaries which cannot reasonably be called cell-walls.

The viruses are the smallest living organisms, but they are not strictly cells. In oversimplified terms, a virus is like an isolated cell nucleus, and it can only function and reproduce if it gains entry to another cell. There the virus takes over and uses the cytoplasmic functions of the "host" cell, eventually killing it. Because of this parasitic life habit, the viruses are more likely degenerate forms rather than the evolutionary precursors of cellular organisms. The smallest true cells are those of the bacteria, which are plants. There are many species of rather larger *unicellular* animals and plants, like *Amoeba* and *Chlamydomonas,* which can be studied more easily and are usually regarded as representative of the primitive types from which the higher forms have evolved (Fig. 2.1, *a* and *b*). At this unicellular level, the distinction between animals and plants is not always clear cut. For example, *Euglena* (Fig. 2.1 *c*) is motile and at the same time contains chlorophyll; furthermore, it does not have a rigid cell wall as does *Chlamydomonas*. The motility exhibited by *Euglena* and its frail cell wall are characteristic of animals, whereas the presence of chlorophyll and the photosynthesis that it makes possible are characteristic of some plants.

Animals derive their energy from breaking down complex molecules which they *eat,* whereas the green plants get their energy by breaking down complex molecules which they themselves *synthesize,* using the radiant energy of the sun. For accuracy it should be noted that not all plants are photosynthetic.

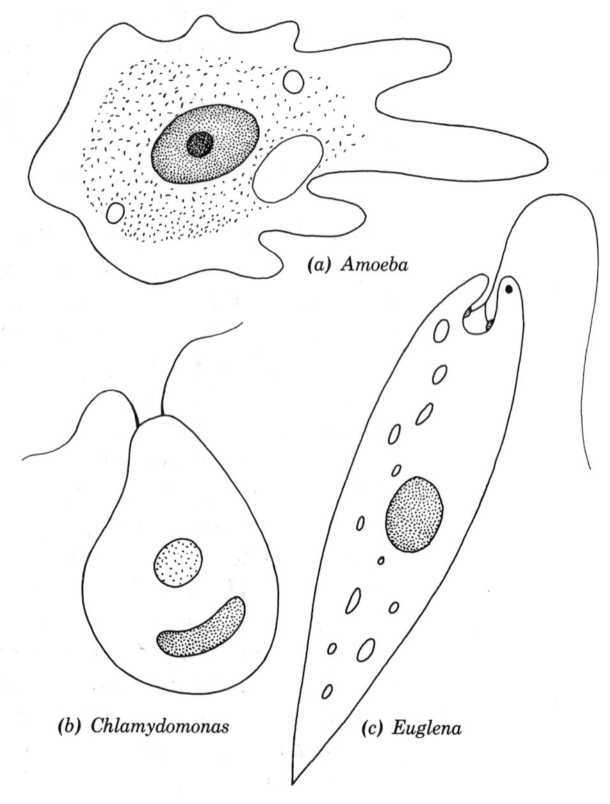

FIGURE 2.1. Three unicellular organisms.

CELLULAR FUNCTION AND SPECIALIZATION

In *Amoeba* (and other forms), the single cell performs all the functions necessary to life. It finds and eats food; it digests it and derives energy from its combustion; it excretes waste products, especially the excess water which continually diffuses in; it respires by the inward diffusion of oxygen and the outward diffusion of carbon dioxide; it moves away from toxic material or toward food; finally, it can reproduce.

Probably the first multicellular animals were actually colonies of similar individuals, each looking after its own affairs as described for *Amoeba.* They enjoyed some advantages of proximity though, especially the sharing of a protective structure. *Volvox* is such a colonial form.

With the evolution of truly multicellular animals, some of the cells specialized in certain functions, with a considerable gain in efficiency. In *Hydra* (Fig. 2.2), the cells of the tentacles are specialized for movement, whereas the cells in the celom specialize in producing digestive enzymes. Some *Hydra* species even have poison organs on the tentacles for killing the prey.

In some of the higher invertebrates, the animal comprises a series of similar segments, each containing most of the types of specialized cells. The earthworm, *Lumbricus,* is an example of this type. The segments are partly independent of each other (as seen when a worm is cut), and yet together they make a more effective whole.

In the vertebrates, the specialization is even more extreme. Every organ has cells of a highly characteristic form and function related to the function of the organ. Muscle cells are specialized for contraction and nerve cells for conduction; although their fundamental properties are similar, they look quite different.

We might compare a unicellular organism with a lone savage who can look after all his requirements himself. He need only find others of his species in order to reproduce. The very simplest multicellular animals, then, are comparable to simple tribal organizations in which each man is a warrior and hunter but with some degree of specialization, if only in rank. A further degree of complexity is similar to that of a frontier village where some men specialize as blacksmiths or storekeepers, but where all of them are familiar with growing crops and the care of animals. Finally, the human body might be likened to a vast industrial city. The specialists in such a community together can make a locomotive or a cyclotron, but

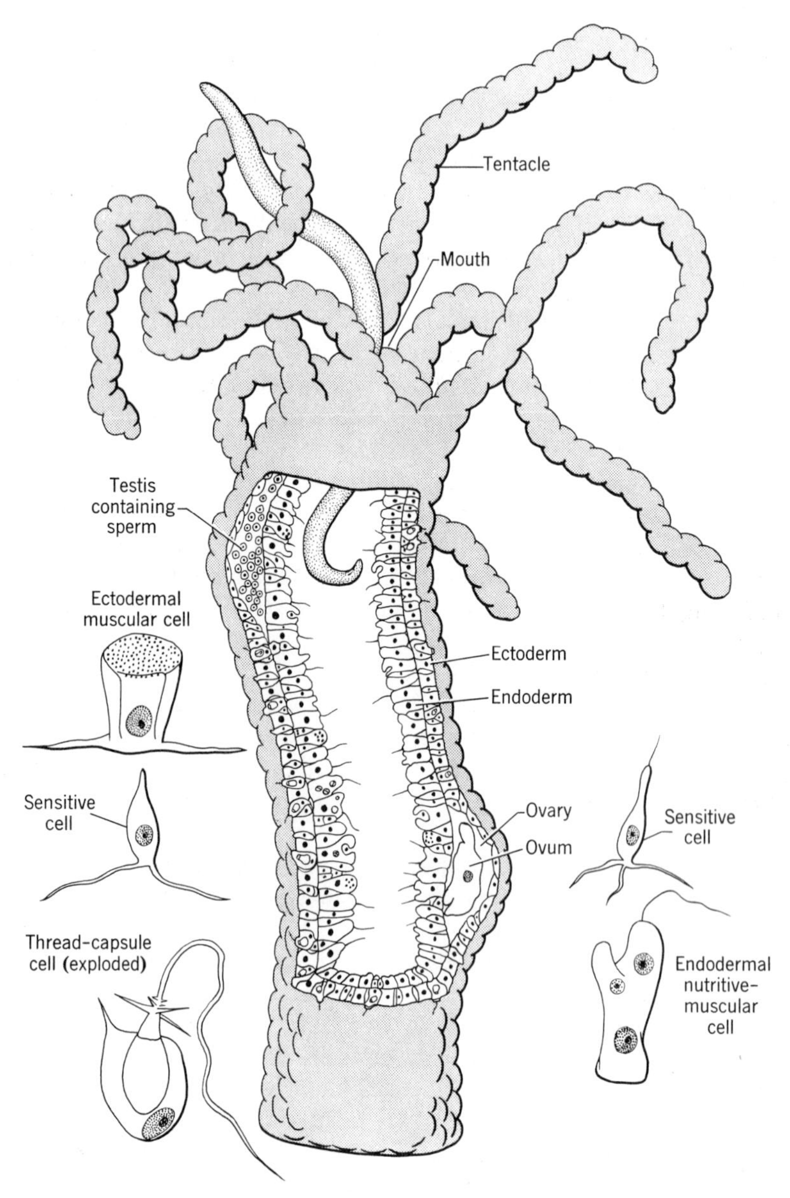

FIGURE 2.2. Hydra. (from Montagna, *Comparative Anatomy*, p. 107, John Wiley)

individually they are unable to butcher a hog or build a log cabin, simple skills taken for granted in a frontier village. As in this analogy, the specialization into nerve cells, muscle cells, blood cells, etc. has been at the expense of flexibility in the capacities of individual cells. Also, the individual cells require oxygen and nutrients just as a savage and a nuclear engineer have the same physical needs.

CELL STRUCTURE

Before discussing the various specialized cells, the basic characteristics of animal cells can be described by reference to a hypothetical "typical" cell shown in Fig. 2.3. The diagram is largely self-explanatory but the following notes on the various structures are pertinent.

The *extracellular fluid* is the environment that bathes the cell and contains the nutrients and oxygen required by the cell.

The *cell membrane* is the structural boundary of the cell.

The *plasma membrane* is the physiological boundary of the cell. The exchange of water and solutes between the cell and the fluid bathing it takes place here as is discussed in Chapter 5. The membrane is *discriminatory* so that the intracellular fluid is maintained with a very different composition to the extracellular fluid.

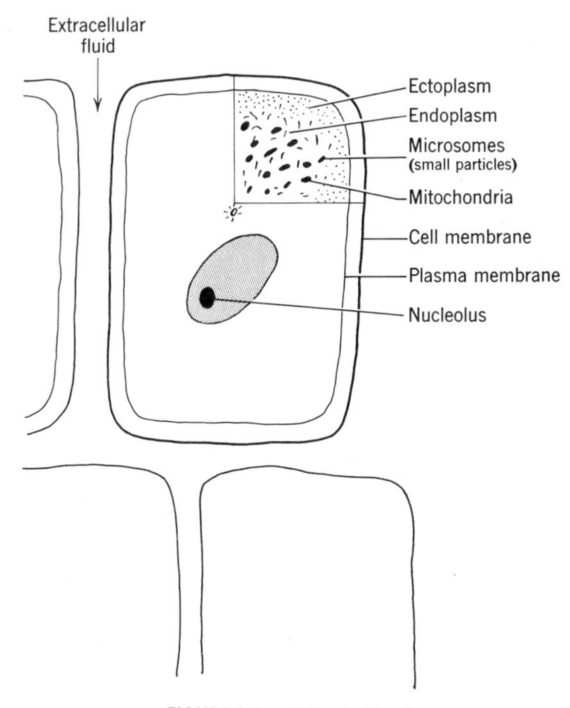

FIGURE 2.3. A "typical" cell.

No one has conclusively shown the plasma membrane in a photomicrograph, and many cytologists doubt its existence. The fact remains, however, that there is a boundary between the intracellular and extracellular fluids because they are so different. It is possible that there is a boundary *zone* rather than a thin membrane, and a *cytoplasmic cortex,* has been proposed. "Cortex" means "bark" as of a tree, and the proposal is that the outer part of the cytoplasm is the discriminatory layer, and that it has a transitional composition.

The *cytoplasm* is the protein-water gel normally making up the bulk of the cell's interior. It contains many *inclusions* as will be described, but the gel itself is not formless either. The *ectoplasm* is clearer and less granular than the *endoplasm,* and microscopic studies with special techniques have shown that the proteins in both are arranged in long strands. Furthermore, electron micrographs have shown a fine *reticulum* throughout the cytoplasm. The reticulum looks like a complicated arrangement of interwoven "membranes," and it has been suggested that convolutions of the plasma membrane may dip down all the way from near the boundary of the cell to every part of the cytoplasm. If this is a fact, the physiological "surface" of the cell is very large indeed, and every part of the cytoplasm is close to the physiological exterior. It has also been pointed out recently that many of the properties of intracellular fluid seem to suggest that the water molecules themselves are oriented with respect to the reticulum, rather than in the random fashion normally found in fluids. This would mean that the intracellular water has almost crystalline properties, as ice does, regardless of the temperature.

The *cytoplasmic inclusions* are many and not all identified. Some are fat droplets or crystals, but others have an organized structure and have an active part in the cell's metabolism. These are the *organelles.*

a. The *mitochondria* are yellowish particles, usually round or sausage shaped and with a discrete structure visible in electron micrographs (Fig. 2.4). They contain large concentrations of the enzymes concerned with the cell's oxidative metabolism.

b. The *microsomes* are smaller, more mysterious particles, probably involved in fat metabolism. Some of the microsomes undoubtedly form part of the plasma membrane, or whatever the discriminatory boundary is.

c. Fat droplets and glycogen granules form the immediate food reserve of the cell.

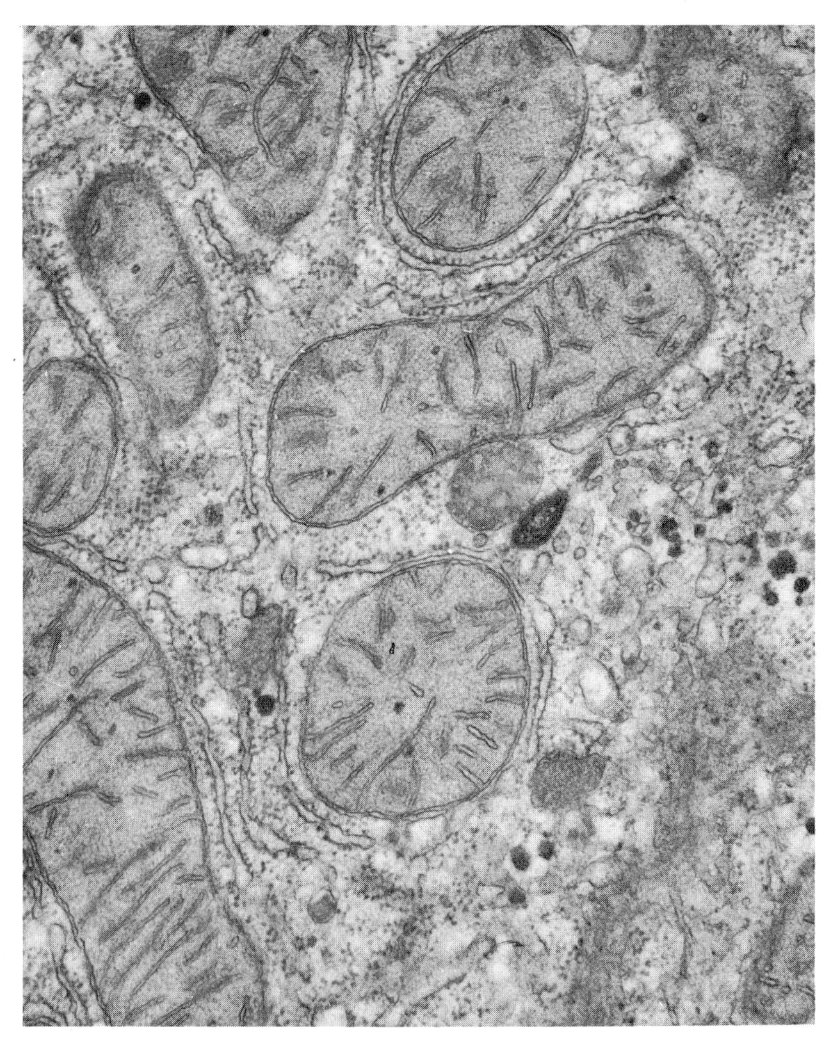

FIGURE 2.4. Hepatic cell mitochrondria—mouse. Magnification × 37,500. (Photo by Dr. H. Parks, Dept. of Anatomy, University of Rochester.)

The *centriole* is a small body of differentiated cytoplasm. It is particularly apparent during cell division when it pairs, and each one is the center of cytoplasmic strands (see p. 23).

The *nucleus* is distinctly different from the surrounding cytoplasm, and its *karyoplasm* is enclosed by a nuclear membrane. It contains a deeply staining material, *chromatin,* which, during cell division, is organized into discrete strands, the *chromosomes.* The nucleus is made of nucleoprotein, and in some way it controls the metabolism of the cell, especially the synthesis of new material. During cell division, the chromosomes also divide, and the duplication of the chromosomes in the new cells ensures that their metabolic patterns are the same as in the mother cell. In sexual reproduction, the offspring get half their chromatin from each parent, as is described in the account of *meiosis*.

The *nucleolus* is a spherical structure of firmer texture than the rest of the nucleus. It is prominent in actively growing cells and is not always visible in mature cells. In consequence, it is reasonably suggested that it is involved with the control of protein synthesis for growth.

CELL DIVISION

An animal grows by making more cells; in only a few instances do the cells themselves increase in size. The reduplication of cells is by *mitosis,* in which an existing cell divides into two *daughter cells.* Mitosis is the basis of growth, but it continues throughout the life-span and does not cease when an animal reaches adult size. Most tissues are subject to continual wear and tear, and the damaged cells are replaced by mitosis. This is seen at its fastest in the skin and intestinal lining which are subject to severe abrasion. The cells of the brain, however, are not replaceable.

A cell in mitosis goes through several well-recognizable phases, which are illustrated in Fig. 2.5. It should be noted that these diagrams "freeze" the action in what is really a smoothly continuous process over a period lasting minutes or hours.

The *interphase* is the "resting" state of a cell not engaged in division.

The *prophase* is the beginning of the mitotic process. The chromatin becomes organized into the chromosomes; the nucleolus and the nuclear membrane disappear. The centriole (in the cytoplasm) divides and the new pair are arranged one on each side of the nucleus.

The *metaphase* is marked by the appearance of rays extending from each centriole to give a spindle-shaped structure. The chromosomes become attached at right angles to the rays at the widest part of the spindle.

In the *anaphase* each chromosome splits longitudinally; the pairs separate and move in opposite directions towards the centrioles. Through this splitting of the chromosomes, the nucleic acids controlling the metabolic patterns of the cell are shared equally (and identically) between the daughter cells.

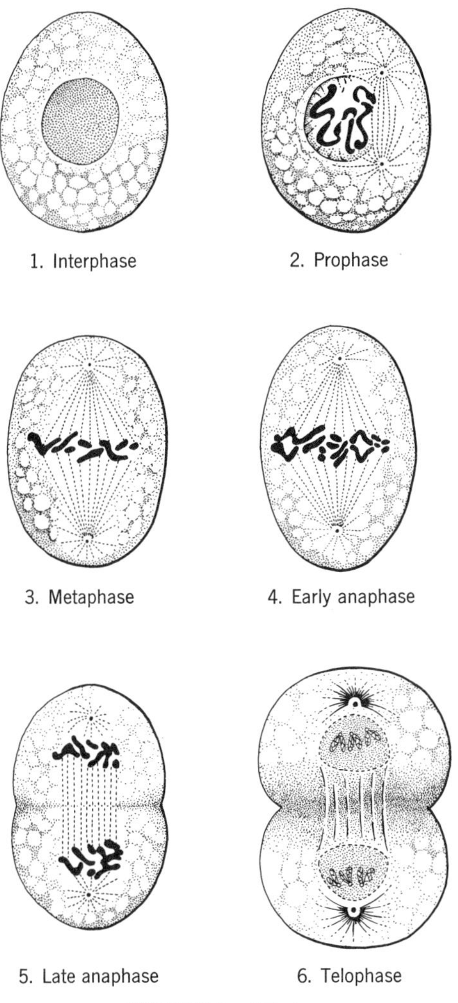

FIGURE 2.5. Mitosis.

In the *telophase* the chromosomes cluster in each half of the still intact cell and revert to chromatin. Nuclear membranes form around each cluster and a nucleolus reappears. For a short time the cell is binuclear, but soon afterwards the cytoplasm divides to complete the formation of the two daughter cells.

MEIOSIS

The cells in the male *testes,* and female *ovaries,* produce sperm and eggs respectively. In mammals, an egg can be fertilized by one of the millions of sperm cells ejaculated into the female's vagina in *coitus*.

Most human cells contain 46 chromosomes (arranged in 23 pairs) and during mitosis, each daughter cell receives all 46. The sperm and egg cells, however, are produced by a special form of cell division in which the chromosome number is halved (giving 23). This is *reduction division* or *meiosis*. When a sperm and egg cell combine, the full complement of 46 is restored and the new individual has chromosome material from both parents.

The chromosomes are strands along which are distributed discrete, submicroscopic units, the *genes*. Each gene seems to control one structural or functional characteristic of the organism. Mitotic division reduplicates cells with identical genetic make-up, whereas meiotic division followed by fusion of the gametes gives a blend of genetic characteristics in almost endless variety. The distribution of the parents' characteristics among the offspring is the essence of *genetics*.

REFERENCES

1. Beadle, George W., "The genes of men and molds," *Sci. American* **179** (3), September 1948.
2. Bonner, John Tyler, "Volvox: a colony of cells," *Sci. American* **182** (5), 52, May 1950.
3. Butler, J. A. V., *Inside the Living Cell,* Basic Books, New York, 1959.
4. Crick, F. H. C., "Nucleic acids," *Sci. American* **197** (3), 188, September 1957.
5. Crick, F. H. C., "The structure of the hereditary material," *Sci. American* **191** (4), 54, October 1954.
6. Dobzhansky, Theodosius, "The genetic basis of evolution," *Sci. American* **182** (1), 32, January 1950.
7. Gay, Helen, "Nuclear control of the cell," *Sci. American* **202** (1), 126, January 1960.
8. Giese, A. C., *Cell Physiology,* Saunders, Philadelphia, 1957.
9. Horowitz, Norman H., "The gene," *Sci. American* **195** (4), 78, October 1956.
10. Mazia, Daniel, "Cell Division," *Sci. American* **189** (2), 53, August 1953.
11. Scheer, Bradley T., *General Physiology,* Wiley, New York, 1953.
12. Siekevitz, Philip, "Powerhouse of the cell," *Sci. American* **197** (1), 131, July 1957.
13. Waddington, C. H., "How do cells differentiate?" *Sci. American* **189** (3), 108, September 1953.

TISSUES

3

In mammals, cells are hardly ever found as isolated units but rather in aggregates ranging from less than a millimeter in size to large areas like the skin. A functional unit made up of similar cells is called a *tissue,* and all of the organs are made of one or more kinds of tissues.

The word "tissue" is frequently used loosely to refer to "kidney tissue," for example, which is actually composed of several kinds of tissue in the more exact sense. This usage is so widespread that it cannot be called incorrect. Nevertheless, the student should be aware of possible confusion.

A systematic classification of the tissues takes into account not only their form and function, but also their embryonic origins. After fertilization of an egg, the cells of the embryo develop in well-defined *germ layers.* The fundamental properties of a tissue and its reaction in disease are frequently more related to the germ layer from which it originated than to its more apparent characteristics in the mature body. Because a discussion of embryology is not appropriate here, and because the topic is treated thoroughly in textbooks of histology, the following notes and diagrams do not take into account this more sophisticated theme. They are based mainly on form and function and are necessarily extremely brief.

Although each category has a wide range of cell structure and function, tissues are of four main kinds. These are:

1. Epithelial tissue
2. Connective tissue
3. Muscle tissue
4. Nerve tissue.

A whole section of this text is devoted to the physiology of muscle and nerve, and the characteristics of these tissues are discussed in some detail in those chapters. The epithelial and connective tissues are described below.

EPITHELIAL TISSUES

Epithelial tissues are frequently found as protective layers over other tissues. The most typical in this respect is the *epidermis,* the outer part of the skin; other epithelial tissues line the respiratory tract, the digestive tract, and the blood vessels. The secretory portions of several glands are also of modified epithelial tissue. The tissue usually comprises one or more layers of cells resting on a *basement membrane.* The cells may be flat (squamous), cuboidal, or columnar in shape. A *simple* epithelium has a single layer of cells, and a *stratified* epithelium has several layers. The various types, together with some examples, are listed below.

Simple squamous epithelium (Fig. 3.1) is made of flattened cells in a pavement-like arrangement. It is found as the inner lining of the blood vessels, the labyrinth of the ear and other cavities. Some histologists place the *endothelial* tissue, which lines blood vessels, in an entirely separate (fifth) tissue category. In appear-

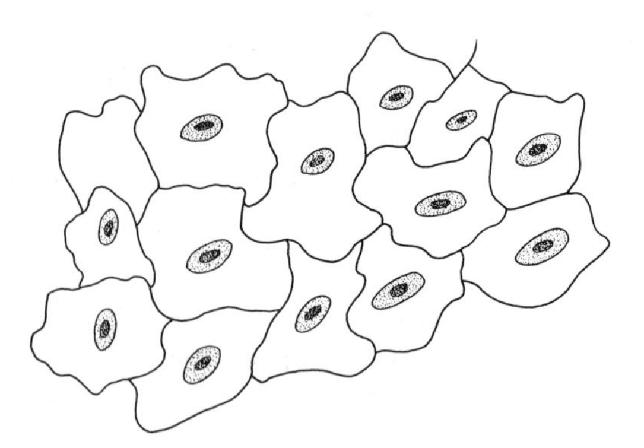

FIGURE 3.1. Simple squamous epithelium.

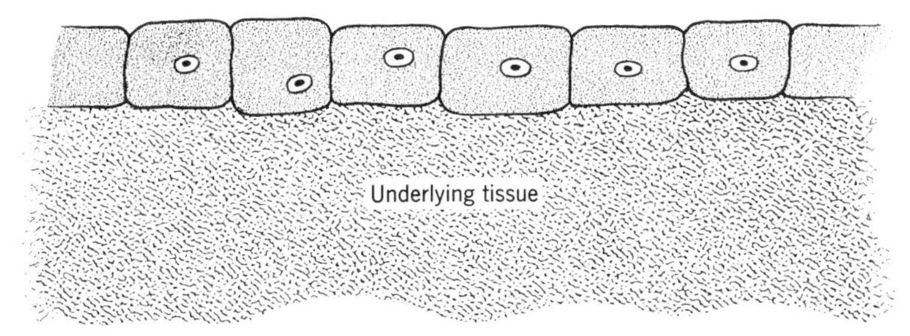

FIGURE 3.2. Simple cuboidal epithelium.

ance, however, it is virtually identical with simple squamous epithelium.

Simple cuboidal epithelium (Fig. 3.2) is found in the thyroid gland (see p. 263) and in the ducts of some exocrine glands.

Simple columnar epithelium, as the name implies, has still taller cells (Fig. 3.3). It is found lining the digestive tract from the stomach to just above the anus. Some of the cells of this type have a secretory function. Secretory products like mucus and digestive juices, for example, are manufactured by columnar epithelial cells and released on to the surface of the epithelium or into the cavity which it lines. In *goblet cells,* mucus accumulates and crowds the cytoplasm and nucleus into one end to give the characteristic appearance shown in Fig. 3.3. Some cells survive the release of a charge and begin to manufacture anew, other cells break down completely after secretion.

Some columnar epithelial cells, like those of the renal tubules, have a *brush border.* Electron microscope studies have shown that this is made up of minute extensions of the cell membrane which give the brush-like appearance. Their presence certainly increases

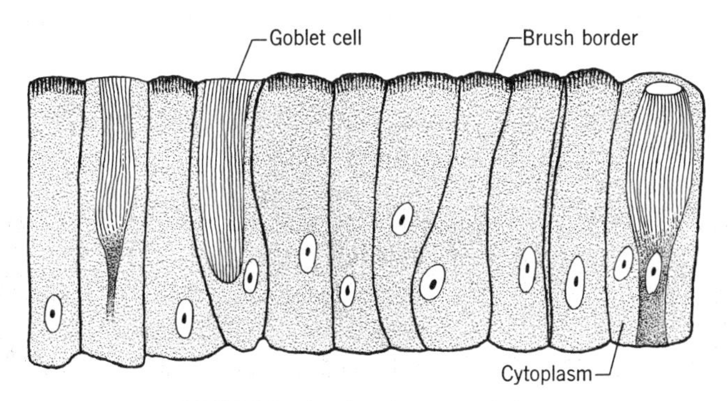

FIGURE 3.3. Simple columnar epithelium.

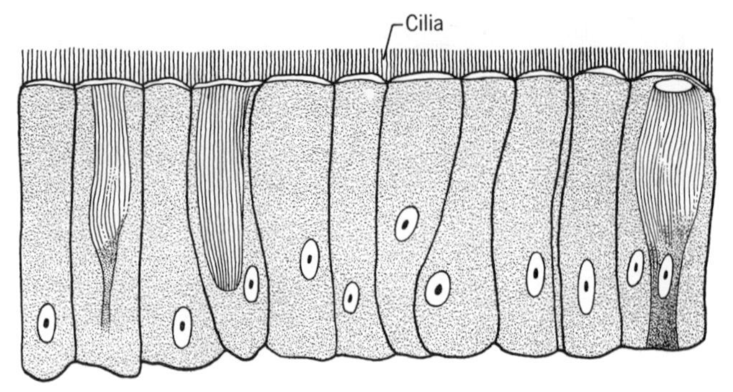

FIGURE 3.4. Simple ciliated columnar epithelium.

the surface area of the cell's luminal surface, and they are probably concerned with absorptive processes.

Simple ciliated columnar epithelium (Fig. 3.4) is similar to the tissue just described and may contain goblet cells and cells with brush borders. In addition, though, the free surface of each cell is covered with hair-like *cilia* (cilium is Latin for eyelash), which are much larger than the brush-border processes, and furthermore they are *motile.* They have a rhythmic motion reminiscent of tall grass in the wind; they beat stiffly in one direction and bend during their slower recovery (see Fig. 3.5). This motion sets up a current in the mucus film covering the epithelium so that small particles falling on it are moved along. The passages of the nose and throat have linings of ciliated epithelium, which not only provide protection for the underlying tissues, but which also transport particles of dust and bacteria towards the mouth.

Stratified squamous epithelium (Fig. 3.6), is a widespread type of tissue, found in the epidermis of the skin, the cornea of the eye, and the linings of the mouth, esophagus, and vagina. The epidermis exemplifies the protective role. New cells are constantly produced by the *Malpighian layer,* which is closest to the underlying connective tissue and blood vessels. The outer and earliest-

FIGURE 3.5. Diagram of ciliary motion.

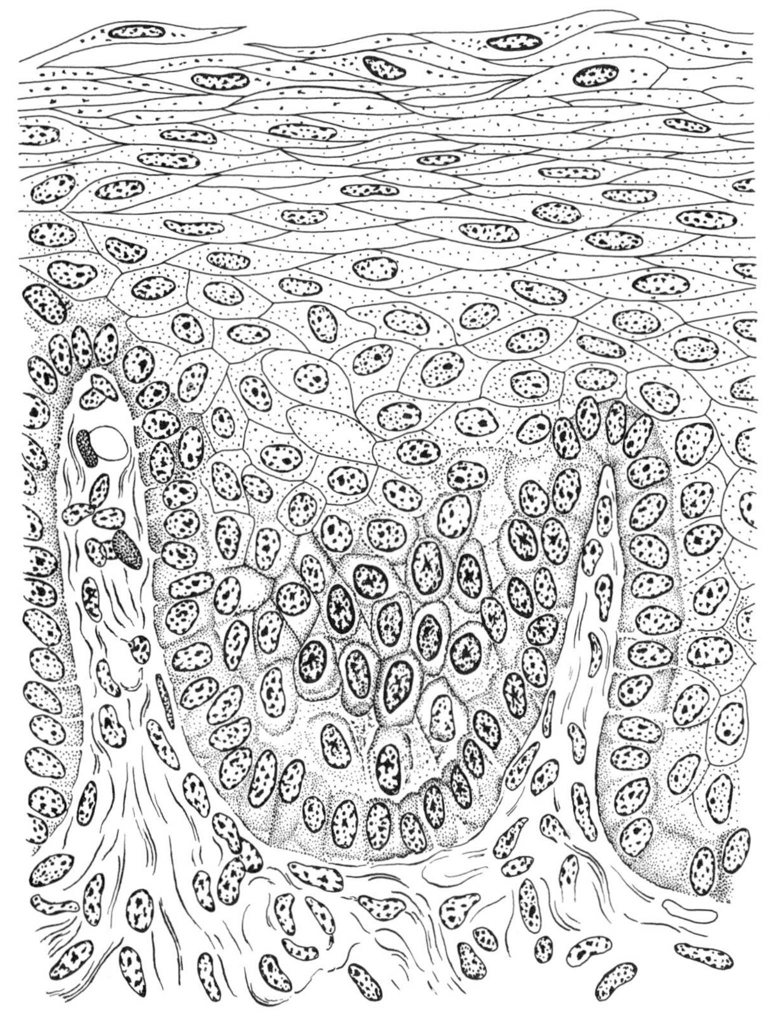

FIGURE 3.6. Stratified squamous epithelium.

produced layers of cells become *cornified,* lose their nuclei, and eventually die. They are continually rubbed off in the wear and tear of daily life, but their production is the most rapid in the regions subject to the greatest wear. On the palms of the hands and soles of the feet, the epidermis can be 1 mm. thick.

Stratified columnar epithelium is rare and found only in the epiglottis, part of the eye, and the urethra.

Pseudostratified columnar epithelium is really a simple columnar epithelium with cells of different heights. Under the microscope, a section looks as if it has several layers because the nuclei are not all at the same level. This type of epithelium is found in the urethra and in salivary glands. It may also be ciliated, as it is in the respiratory passages.

Transitional epithelium is found lining organs subject to stretching such as the bladder. It has several layers of cells when unstretched, but on distension of the organ, it looks somewhat like simple squamous epithelium. Unlike stratified squamous epithelium, the outer layers are not cornified and do not deteriorate.

CONNECTIVE TISSUE

Connective tissue is found in every part of the body. Superficially, at least, the types of connective tissue are as widely different as their functions, which fall into six main categories.

Packing. The spaces around blood vessels and between other more organized tissues are filled with a white and sticky connective tissue.

Binding. The muscles and other organs, such as the spleen, are enclosed in tough capsules partly composed of a fibrous connective tissue. Tendons are the connections between muscles and bone which move the joints. These are formed of even tougher fibrous connective tissue.

Support. Cartilage and bone form the rigid skeleton. These connective tissues have very sparse cellular elements.

Food Storage. *Adipose tissue* has cells which can become filled with fat, serving as a food reserve.

Blood cell manufacture (*Hemopoiesis*). The bone marrow which produces both red and white blood cells, and the lymphoid tissues producing only white cells, are connective tissues. Since the red and white cells are derived from these, blood may be also regarded as a type of connective tissue, with the amorphous *plasma* as the intercellular ground substance. As with endothelium, some authorities regard blood as a separate category of tissue.

Destruction of invading organisms (*Phagocytosis*). The lymphoid tissue which produces white cells also contains cells which literally *eat* cell debris and invading micro-organisms. The spleen and liver contain similar cells, which are also regarded as connective tissue. All the tissues participating in this function form the *reticulo-endothelial system.*

Intercellular Substance

In each of the first three functional categories above, the connective tissue cells are usually scattered sparsely through an intercellular ground substance produced by them. In the embryo, the spaces between the developing organs contain the primitive *mesenchymal* cells. These produce an amorphous jelly-like material which fills the intercellular spaces. In mature connective tissues, the intercellular substance comprises an amorphous ground substance or matrix, in which are embedded fibers of a type and density that vary with the tissue and its function.

There are three kinds of fibers—*reticular, collagen,* and *elastic.* Reticular fibers are very small and are usually invisible in ordinary connective tissue preparations. Collagen fibers are white and wavy, elastic fibers are yellow and straight. Partly because of the difficulty of observing them, very little is known of reticular fibers and their function. Collagen fibers are especially plentiful in the white connective tissue holding the skin on the body. Elastic fibers are the main constituents of yellow ligaments like the *ligamentum-nuchae* of the neck. Other ligaments are mainly of collagen but with elastic components.

The *cellular elements* of connective tissue produce the intercellular substance. Their embryonic forerunners are the mesenchymal cells, but these are never found after birth. The basic type of connective tissue cell is the *fibroblast,* which is associated with the production of the fibers in the fibrous tissues. *Chrondroblasts* and *osteoblasts* produce cartilage and bone respectively, but they may be regarded as specialized fibroblasts. Histologists have described many different kinds of cells in those connective tissues which have only very little intercellular substance, but in the interests of brevity, only three will be referred to here. There are the *fat cells* in adipose tissue, the *macrophages* in the tissues in the reticulo-endothelial system, and the large *mast cells* which are also found in most connective tissues; the latter are believed to produce *heparin,* a compound which prevents the indiscriminate clotting of the blood in the vessels.

Types of connective tissue. The principal kinds of connective tissues are listed below in the same order as their functions are discussed.

Areolar tissue (Fig. 3.7) is the "packing" connective tissue referred to above. It contains a few elastic fibers, but has mainly bundles of reticular and collagen fibers running in all directions.

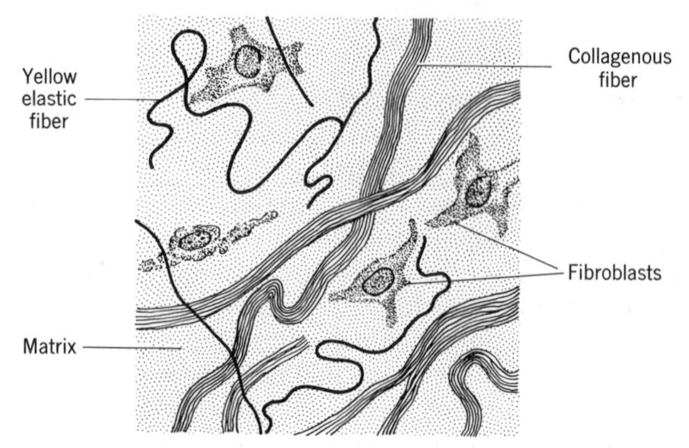

FIGURE 3.7. Aroelar tissue. Reproduced by permission from Millard, King, and Showers, *Human Anatomy and Physiology*, W. B. Saunders Co., 1956.

Fibrous tissue has a denser arrangement of both collagen and elastic fibers. It is found in the body with varying degrees of density of the fibers and varying degrees of the regularity of their arrangement.

i. Irregular fibrous tissue has the fibers running in all directions. The deep *fascia* enclosing muscles in tough sheaths and the scar tissue which repairs a wound are both examples of this type.

ii. Regular fibrous tissue has the fibers arranged longitudinally, giving the tissue a white, glistening appearance. Tendons are typical examples and have the great strength characteristic of this tissue.

Cartilage is found in three forms, and the different functions are reflected in the fibrosity of their intercellular substance

i. *Hyaline cartilage* (Fig. 3.8) is extremely strong and flexible, with a translucent glassy-appearing intercellular substance. It actually contains many fibers but these are not visible in slides prepared by the usual methods. This kind of cartilage is found in the nose, the flexible ends of the ribs (costal cartilage), the surfaces of the bones where they come together in the joints, and, until after puberty, in the epiphyseal cartilages (to be discussed). In the embryo, all the long bones of the skeleton are first formed of hyaline cartilage. This begins to be replaced by bone at the second month of embryonic life and the process is completed at birth with the noted exceptions.

ii. *Fibrocartilage* has a dense content of collagen fibers in the matrix. It is very strong and forms the capsules of joints, some ligaments, and the intervertebral disks.

iii. *Elastic cartilage* contains all three kinds of fibers and, as would be expected, is highly flexible. It is found in the external part of the ear (pinna), in the larynx (voice box) and in the epiglottis.

Bone is the most important supporting tissue. The fibrous intercellular substance becomes hard through its impregnation with crystals of *hydroxy-apatite,* a complex salt of calcium, magnesium, phosphorous, carbonate, and some fluoride. The pieces of bone usually examined in the laboratory are dried specimens, in which only the mineral, nonliving parts remain. Such samples give a wrong impression since bone in the body has an active metabolism and a considerable blood supply. Microscopic examination of fresh bone shows that cells and blood vessels are scattered through even the densest portions. Even the mineral constituents are not laid down once and for all. Studies with radioactive isotopes show that their molecules are in a constant state of turnover, with the removal of old material and the deposition of new.

Examination of a skeleton (Fig. 1.5) shows that there are two kinds of bone; flat bones like those of the skull and shoulder blade (scapula) and long bones like those of the arms and legs.

Many flat bones appear first in the embryo as membranes of fibrous connective tissue, which turn to bone when the fibroblasts become *osteoblasts.* They deposit bone salts within the membrane; this kind of flat bone is therefore known as *intramembranous bone.*

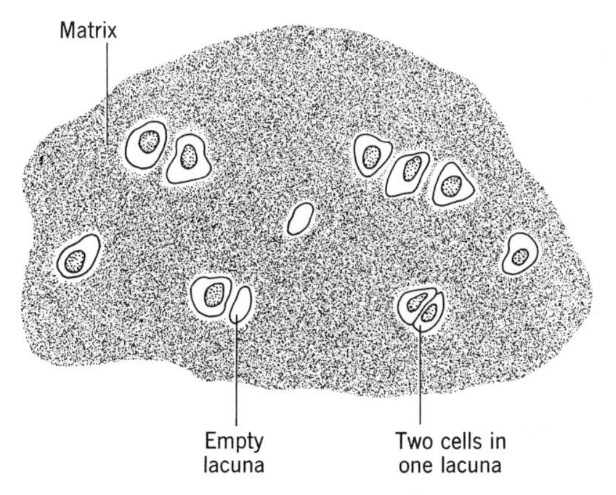

FIGURE 3.8. Hyaline cartilage.

The mineral is deposited as interconnecting spicules with plenty of space between them; this is *trabecular, cancellous,* or *spongy* bone. On the outer part of the bone, the osteoblasts are regularly arranged in a nutritive membrane, the *periosteum.* Here, they deposit the mineral in dense layers as *lamellar* or *compact* bone. The mature bone then has an inner and an outer surface of smooth, dense bone and an interior of sponge-like (but hard) bone with marrow in the interstices.

The long bones are those which first appear as cartilage models of their mature shape. The mineralization of the cartilage proceeds from *centers of ossification,* both inside and outside the bone. The deposition of bone salts within the cartilage leads this type to be called *endochondral* bones.

The centers of ossification are in the shaft (*diaphysis*) and heads (*epiphyses*) of a bone (Fig. 3.9), and the first mineralization is from the periosteum, on the outside. At the same time, small blood vessels and associated osteoblasts grow inward from the periosteum, apparently by removing the calcified intercellular substance in their path. Bone is deposited around these channels to give spicules similar to those found in intramembranous bones. Thus, the outer part of the bone is made of continually thickening lamellar bone, and the interior of trabecular bone. As the bone matures, the trabeculae in the diaphysis are reabsorbed so that only a thick tube of lamellar bone is left. This contains marrow with many fat cells and a yellow color. In the epiphyses, the trabecular interior persists, and its interstices are filled with *red* marrow, which is the tissue producing new red blood cells.

Most of the ossification of the original cartilage is completed *in utero;* at birth there are only two wide bands of cartilage left, one in each epiphysis. These *epiphyseal cartilages* persist until maturity, and the lengthening of the long bones proceeds from their edges. Once they have "closed," there can be no further growth in bone length, and the individual has reached his maximum height. Malnutrition and some diseases can lead to early closure of the epiphyses, thus stunting the height. A shortage of vitamin D, calcium, or phosphorus in the diet can lead to excessively wide epiphyseal cartilages and inadequate mineralization of the shafts. The bones then do not grow straight and give the stunted, bowlegged look associated with *rickets.* (See Chapter 11)

Bone growth is subject to *hormonal* control by the *parathyroid* gland, the *adenohypophysis,* and the *gonads.* These endocrine effects are described in Chapter 16.

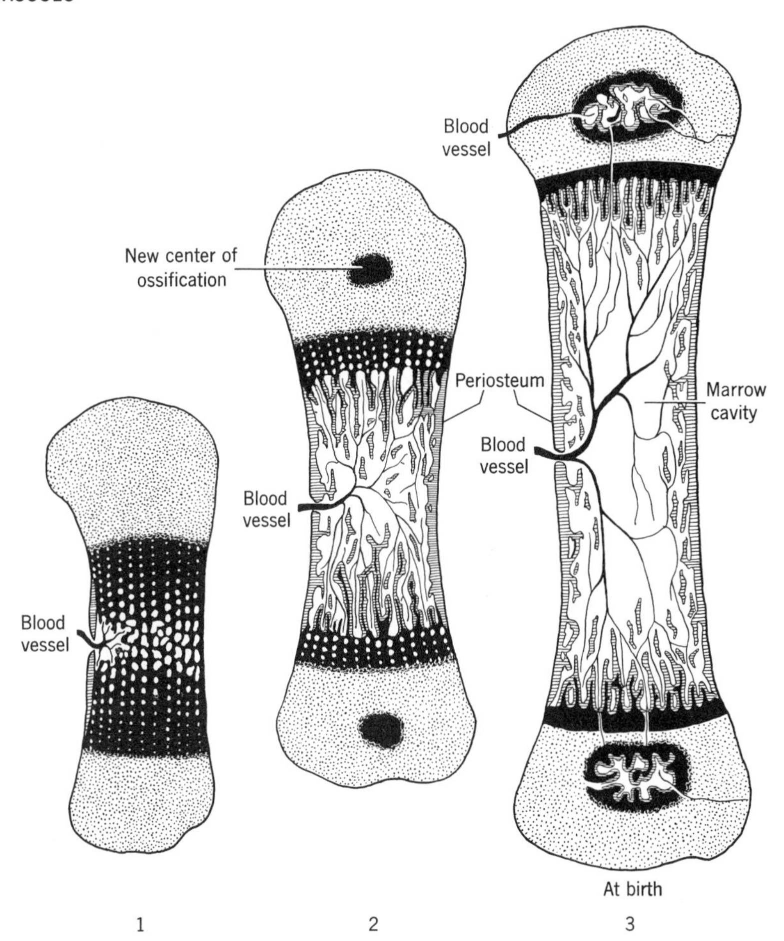

FIGURE 3.9. Centers of ossification. Reproduced by permission from Nonidez and Windle, *Textbook of Histology,* McGraw-Hill, 1949.

If we compare the size of a child's femur at the ages of seven and fourteen, for example, it is clear that it grows in *all* its dimensions. The growth is not by the mere accretion of new material as in the growth of a crystal, because the marrow *cavities* also enlarge. The growth of bone involves, therefore, the continual removal of old material as well as the deposition of new; the removal is performed by specialized cells, the *osteoclasts.*

The microscopic structure of compact bone is shown in Fig. 3.10. The unit of structure is a *Haversian system,* comprising concentric lamellae of bone mineral and the diminished remains of the osteo-

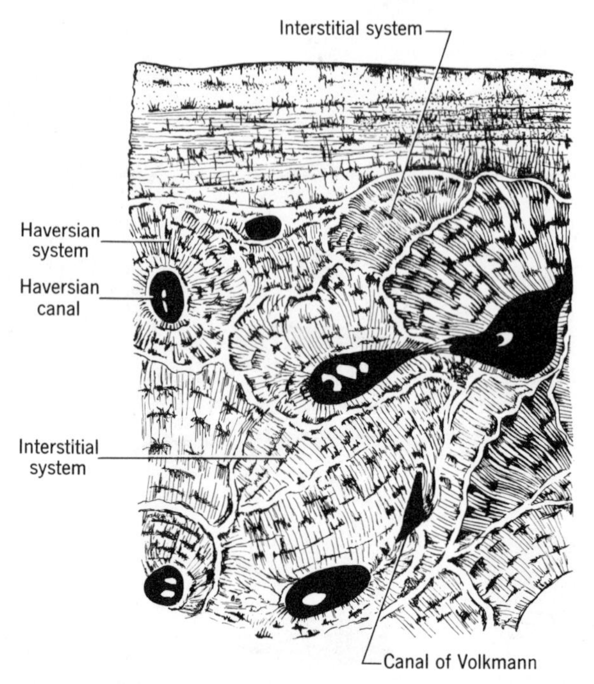

FIGURE 3.10. Bone section showing lamellar bone and Haversian systems. (After Schaffer)

blasts that laid it down, all centered on a canal containing blood vessels. When actively depositing bone, osteoblasts have long protoplasmic extensions, but in mature bone these deteriorate and leave fine channels through the mineral. These *canaliculi* intercommunicate with similar channels from other cells and provide a route for the exchange of fluid and nutrients between the vessels in the Haversian canal and the cells. The shrunken osteoblasts are called *osteocytes.* They occur singly or in small groups in cavities in the mineral called *lacunae.*

Adipose tissue is found in lobular arrangements in the *fat depots* of the body. Each cell is intrinsically small but accumulates droplets of fat which coalesce into a globule much larger than the original cell. When full, the cell's nucleus and cytoplasm are pushed to one side so that the cell has the signet ring appearance seen in Fig. 3.11. The fat content is depleted in starvation. Isotope studies have shown that fat, like bone mineral, does not remain inertly in the tissue. Even when the weight of fat in the body is constant, the individual fat molecules in the depots are in a constant state of turnover.

Hemopoietic tissue is found in the bone marrow which produces red cells and some white cells and in the lymphatic tissue which produces only white cells. The *myeloid* tissue of red bone marrow is very complex. The red cells (erythrocytes) are produced by the mitotic division of cells called *myeloblasts,* but the subsequent development is through several stages, which are all represented at the same time in a prepared marrow sample. Figure 3.12 shows some of the stages. Mature erythrocytes are nonnucleated, and in their last stage of development the nucleus is fragmented just prior to dissolution. A cell at this stage is a *reticulocyte;* some of them are

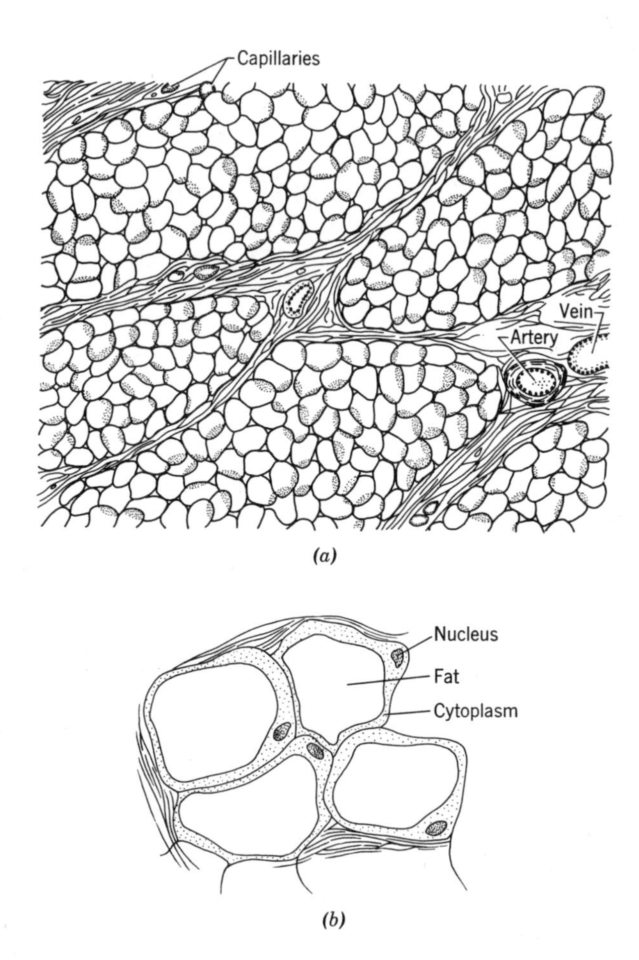

FIGURE 3.11. (*a*) Adipose tissue under low magnification. (*b*) Adipose tissue cells showing "signet-ring" appearance. Reproduced by permission from Millard, King, and Showers, *Human Anatomy and Physiology,* W. B. Saunders, 1956.

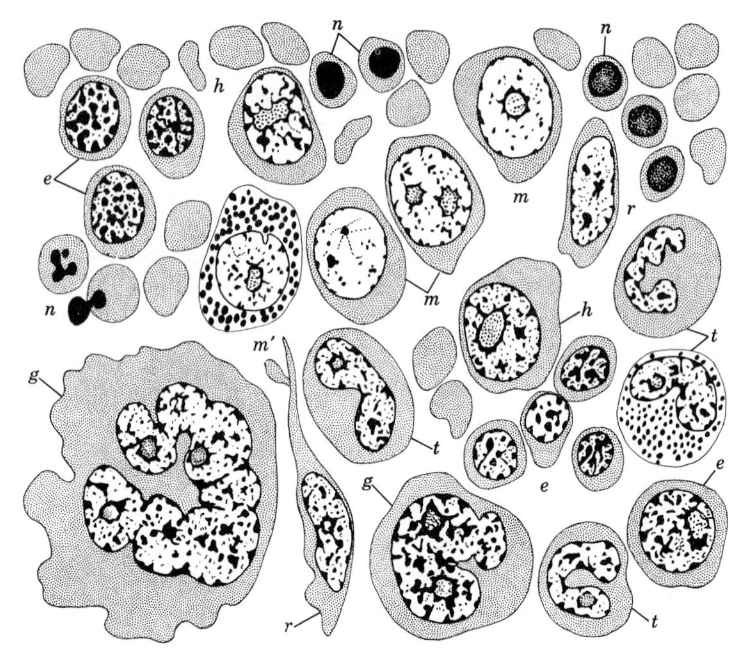

FIGURE 3.12. Marrow cells of a newborn kitten, 900 ×. Reproduced by permission from Nonidez and Windle, *Textbook of Histology*, McGraw-Hill, 1949.

released into the blood stream and complete their development while circulating. In a shortage of erythrocytes, however, the demands on the hemopoietic tissue for new cells are so great, that more and more immature cells appear in the circulation. Therefore, an increased reticulocyte count in the blood is an indication of some kinds of *anemia* (see Chapter 6).

The myeloblasts also produce the *granular leukocytes* (white cells), which are described in the section on blood (p. 83). Myeloid tissue accordingly contains cells which are in the various stages of development of the leukocytes as well.

The nongranular leukocytes are produced in certain centers of the lymphatic tissue. The primary cells in this case are called *stem cells,* or *lymphoblasts.*

Phagocytic cells are distributed all over the body as the active units of the reticulo-endothelial system. They are present in the lymphatic tissues and also in the bone marrow, spleen and liver. The unit cells are *macrophages* or *histiocytes;* they engulf and destroy particles which come into contact with them. In this way, invading microorganisms are dealt with, but the macrophages also have a normal scavenging function in removing the debris of the animal's own dead cells. Wornout erythrocytes in particular are removed in this fashion.

REFERENCES

1. Clark, W. E. LeGros, *The Tissues of the Body,* Oxford, 1952.
2. Cowdry, F. E., *Textbook of Histology,* Lea and Febiger, Philadelphia, 1950.
3. Maximow and Bloom, *A Textbook of Histology,* 7th ed. Saunders, Philadelphia, 1957.
4. McLean, Franklin C., "Bone," *Sci. American* **192** (2), 84, February 1955.
5. Sognnaes, Reidar F., "The skin of your teeth," *Sci. American* **188** (6), 38, June 1953.

CHEMISTRY, A REVIEW

4

ATOMS AND FUNDAMENTAL PARTICLES

The smallest particles in which matter can exist, in a familiar form at least, are the *atoms*. The atoms are made up of even more fundamental particles common to matter throughout the universe. The physicists are constantly discovering more and more of the fundamental particles, but for our purposes we need consider only three, *neutrons, protons,* and *electrons*.

The atom has a *nucleus* which usually contains several protons and neutrons; this is the heavy part of the atom. The protons each have a positive electrical charge, and the neutrons are neutral. Around the nucleus, in well defined *shells,* or *orbitals,* are the electrons, which have virtually no weight but have a negative charge each. There are as many electrons as protons, so that the net charge is zero.

ELEMENTS

The elements, such as iron, copper, carbon, hydrogen, silicon, and nitrogen, owe their individual characteristics to the structure of their atoms. The elements can be arranged in advancing order of *atomic weights,* which depend on the number of neutrons and protons in their nuclei.

The electron shells around the nucleus can each accommodate just so many electrons—as if they were buses with so many seats. In an atom with the inner shell full, any extra electrons are in the next outer shell, which again has a specific capacity but not the same as the first shell. The most stable state for an atom is when the outer shell has its full complement. When the number of electrons in the outer shell (in use) is not the full complement, that atom may be regarded as "avid" for electrons to fill the shell and thereby attain a more stable state. The extra electrons can come from another atom, and so the "avidity" of the atom for electrons (or the ease with which it will surrender electrons) is the basis of its chemical *reactivity.*

The elements can be arranged in a *Periodic Series* (See Table 4.1), in which elements with surprisingly similar characteristics recur periodically as one ascends the series of increasing atomic weights. This periodicity is broadly due to the recurrence of similar patterns of incompleteness in the outer electron shells.

The elements react together atom by atom and in any single reaction the ratio of the weights of the reactants is always the same. This is a fundamental law of chemistry.

MOLECULES

The elements may combine to yield *compounds* with distinct properties of their own:

$$(\underset{\text{Hydrogen}}{2H_2} + \underset{\text{oxygen}}{O_2} \rightarrow \underset{\text{water}}{2H_2O})$$

The smallest units in which the compounds can exist without being broken down into the atoms of the constituent elements, are called *molecules.* Because a molecule of any particular compound is always made of the same number of atoms, the sum of the atomic weights is the *molecular weight.* Many elements such as oxygen and hydrogen do not ordinarily exist as single atoms but rather join up in twos or more. The resulting units, such as H_2 and O_2, are also molecules.

The atomic weight of *hydrogen,* the lightest element, was originally taken as 1.0. On this basis, the atomic weight of oxygen was about 16.0. The atomic weight of oxygen is now taken as 16.0 exactly and that of hydrogen is not quite 1.0. The other atomic weights are expressed as ratios of oxygen; those of the commoner elements are given in Table 4.2. A ratio is simply a number, it has

TABLE 4.1 The Periodic System of the Elements

H 1	He 2

0	I	II	III	IV	V	VI	VII	0
He 2	Li 3	Be 4	B 5	C 6	N 7	O 8	F 9	Ne 10
Ne 10	Na 11	Mg 12	Al 13	Si 14	P 15	S 16	Cl 17	A 18

0	Ia	IIa	IIIa	IVa	Va	VIa	VIIa	VIII			Ib	IIb	IIIb	IVb	Vb	VIb	VIIb	0
A 18	K 19	Ca 20	Sc 21	Ti 22	V 23	Cr 24	Mn 25	Fe 26	Co 27	Ni 28	Cu 29	Zn 30	Ga 31	Ge 32	As 33	Se 34	Br 35	Kr 36
Kr 36	Rb 37	Sr 38	Y 39	Zr 40	Cb 41	Mo 42	Tc 43	Ru 44	Rh 45	Pd 46	Ag 47	Cd 48	In 49	Sn 50	Sb 51	Te 52	I 53	Xe 54
Xe 54	Cs 55	Ba 56	La 57 *	Hf 72	Ta 73	W 74	Re 75	Os 76	Ir 77	Pt 78	Au 79	Hg 80	Tl 81	Pb 82	Bi 83	Po 84	At 85	Rn 86
Rn 86	Fa 87	Ra 88	Ac 89 †	Th 90	Pa 91	U 92	Np 93	Pu 94	Am 95	Cm 96								

*Rare earth metals	Ce 58	Pr 59	Nd 60	Pm 61	Sm 62	Eu 63	Gd 64	Tb 65	Dy 66	Ho 67	Er 68	Tm 69	Yb 70	Lu 71
†Uranium metals	Th 90	Pa 91	U 92	Np 93	Pu 94	Am 95	Cm 96							

no dimensions (like grams, inches, or gallons). The same number of *grams* of an element as the atomic weight is called the *gram atomic weight.* This also applies to molecular weights, and a gram molecular weight of a compound is called 1 *mole.* When 1 mole is dissolved in water and made up to 1 liter, the resulting solution is called a *molar* solution. A millimole is 1/1000 of a mole.

A mole of any compound always contains the same number of molecules, because the weight of material in the mole depends on the ratio of the molecule's weight to that of the hydrogen atom. This astronomic figure is called *Avogadro's number,* and is equal to 6.0235×10^{23}, which is about 6 with 23 zeros after it.

VALENCE

An atom can be part of different kinds of molecules insofar that it can form *bonds* with other atoms. The number of such possible bonds is called the *valence* of the atom and depends on the relative "emptiness" of its outer electron shell. A few elements have outer shells that are already full and they can form no bonds at all. Their valences, therefore, are zero. These are the inert gases, which already have a stable electron configuration and cannot combine with anything, even themselves. They are helium, neon, argon, crypton, etc. Next are elements like sodium and potassium with

TABLE 4.2 Atomic Weights

Aluminum	Al	26.97	Lithium	Li	6.94
Antimony	Sb	121.76	Magnesium	Mg	24.32
Arsenic	As	74.91	Manganese	Mn	54.93
Barium	Ba	137.36	Mercury	Hg	200.61
Bromine	Br	79.92	Nickel	Ni	58.69
Calcium	Ca	40.08	Nitrogen	N	14.008
Carbon	C	12.01	Oxygen	O	16.00
Chlorine	Cl	35.46	Phosphorus	P	30.98
Chromium	Cr	52.01	Platinum	Pt	195.09
Copper	Cu	63.57	Potassium	K	39.10
Fluorine	F	19.00	Selenium	Se	78.96
Hydrogen	H	1.008	Silver	Ag	107.88
Iodine	I	126.92	Sodium	Na	23.00
Iron	Fe	55.85	Sulfur	S	32.01
Lead	Pb	207.21	Tin	Sn	118.70
			Zinc	Zn	65.38

valences of 1, then those like magnesium and calcium with 2, and so on.

EQUIVALENTS

Atoms and molecules react in constant ratios, sometimes 1 for 1, but not necessarily so, depending on their valences. For example, 1 molecule (or 1 mole) of hydrochloric acid will combine with 1 of sodium hydroxide:

$$HCl + NaOH \rightarrow NaCl + H_2O$$

The amount of HCl *equivalent* to 1 mole of sodium hydroxide was also 1 mole. The *equivalent weight* here is therefore the same as the molecular weight. If we now consider sodium hydroxide and *sulfuric* acid:

$$2NaOH + H_2SO_4 \rightarrow Na_2SO_4 + 2H_2O$$

In this reaction, 1 mole of H_2SO_4 combined with (or was *equivalent to*) 2 moles of sodium hydroxide. The equivalent weight of H_2SO_4 is therefore *half* the molecular weight.

In much of chemistry, and certainly in physiology, we are more interested in the *equivalence* of the various components of a mixed solution than in the actual weights dissolved. Therefore, solution strengths are the most conveniently expressed as *equivalents per liter,* or because biological fluids are usually so dilute, in *milliequivalents per liter.*

Thus, a sodium chloride (NaCl) solution with 140 milliequivalents per liter of sodium has 140 millimoles per liter of sodium. As the atomic weight of *sodium* (*not* sodium chloride) is 23, the actual weight of sodium dissolved is $140 \times \frac{23}{1000}$ grams = 3.220 grams per liter. On the other hand, a magnesium chloride ($MgCl_2$) solution with 140 milliequivalents per liter of magnesium has only 70 millimoles per liter. The magnesium atom is divalent, and so each atomic weight provides 2 equivalents. The solution contains therefor $70 \times \frac{24}{1000}$ = 1.680 grams per liter of *magnesium.*

STATES OF MATTER

Matter can ordinarily exist as gas, liquid, or solid. In gases, the molecules are free and hurtle about with a velocity that depends on the temperature. Through this random motion, a gas immediately fills all of whatever space encloses it, and the millions of impacts of the gas molecules against the walls of the container exert what we recognize as the *pressure* of the gas. In liquids, the molecules move around more slowly and their mutual attraction holds them to a certain volume, although a liquid will adapt its *shape* to its container. The random motion of the liquid molecules enables them to escape through the surface occasionally; this we call *evaporation.* The random motion is speeded up as the temperature rises and so the rate of evaporation increases.

In solids there is very little molecular motion. *Crystals* such as ice or sodium chloride are solids in which the molecules are arranged in the definite pattern that gives the crystal its characteristic shape. *Amorphous* solids are those without crystalline structure, like lampblack. A few solids, such as glass, are really super-cooled liquids.

SOLUTION AND SOLVATION

A solid, liquid, or gas may be dissolved in a liquid. That which is dissolved is the *solute,* the liquid in which it is dissolved is the *solvent.* The molecules of the solute are dispersed through the solvent much as if they were gas molecules filling a space of that volume. The behavior of molecules in weak solution is very like that of gases in many respects. When, as so often happens, the solute reacts chemically with the solvent as well as dissolving in it, we call this *solvation.* Carbon dioxide, a gas we will discuss extensively, provides an example; it dissolves in water but also reacts with the water to yield *carbonic acid,* viz.:

$$CO_2 + H_2O \overset{\text{Equilibrium}}{\rightleftharpoons} H_2CO_3$$

IONIZATION

We have indicated that some atoms can achieve a stable configuration of the outer shell of electrons (like that of the inert gases) by acquiring or surrendering electrons. Sodium chloride is formed by the sodium atom's giving up an electron which is assumed by the chlorine atom. The products, with more stable configurations, are called the *ions* of the elements, and even in the solid sodium chloride crystal, they exist as distinct units. The loss of an electron means that the sodium ion has a net positive charge. The ions may therefore be represented as Na^+ and Cl^-. When a salt such as sodium chloride is dissolved in water, the constituent ions exist as separate particles in the solution, and we say the salt is completely *dissociated.*

The sodium *ion* is clearly quite different from the sodium *atom.* The ion has a stable electron configuration and is relatively unreactive, whereas the unstable configuration of the sodium atom causes it to react violently with water.

$$2Na + 2H_2O \rightarrow NaOH + H_2$$

The word "ion" is from the Greek and means "wanderer." The positively charged ions are called *cations* and the negative ones are called *anions,* because in electrolysis they migrate to the *cathode* and *anode* respectively.

Other ionizing molecules like sodium chloride are listed below:

$$KCl \rightarrow K^+ + Cl^-$$
$$MgSO_4 \rightarrow Mg^{++} + SO_4^=$$
$$Na_2SO_4 \rightarrow Na^+ + Na^+ + SO_4^=$$
$$CaCl_2 \rightarrow Ca^{++} + Cl^- + Cl^-$$

The ionization of these molecules permits them to carry an electrical current in solution. The electrochemical laws are too complex for the present discussion, but because of this property, such molecules are called *electrolytes.* They are extremely important in physiology. Some salts do not completely dissociate in solution, and are said to be weakly ionized, or *weak electrolytes.*

IONS IN PHYSIOLOGY

In the body fluids, the metals sodium, calcium, magnesium, and potassium occur almost exclusively in their ionic form. These elements are therefore frequently referred to by physiologists as "the cations." Similarly, chloride, phosphate, sulfate, and bicarbonate are called "the anions." The whole range of them are called the "electrolytes."

ACIDS AND BASES

An acid is any molecule that will dissociate to yield one or more *hydrogen ions.* The hydrogen ion, like the sodium ion just discussed, has a positive charge.

$$HCl \rightarrow H^+ + Cl^-$$

The free hydrogen ions give acids the properties we associate with them.

Bases are molecules dissociating to yield hydroxyl ions, which are negatively charged.

$$NaOH \rightarrow Na^+ + OH^-$$

Some acids, such as hydrochloric acid, dissociate virtually completely; others, particularly the *organic acids* and carbonic acid (referred to previously), are weakly dissociated and the lesser hydrogen ion concentration makes them weaker acids.

$$H_2CO_3 \xrightleftharpoons{\text{Equilibrium}} H^+ + HCO_3^-$$

pH

Pure water is a neutral molecule and it can be thought of as dissociating weakly to give an equal number of hydrogen ions and hydroxyl ions:

$$H_2O \rightleftharpoons H^+ + OH^-$$

The actual hydrogen ion concentration in water, expressed as grams per liter, is very small. In order to make the numbers more manageable and because of certain theoretical advantages, the hydrogen ion concentration is expressed as the logarithm of its reciprocal. It is clearly inappropriate to discuss the nature of logarithms in this context; it must suffice to say that this logarithmic expression of the hydrogen ion concentration is called the pH. The pH of water is 7.0, which is neutrality. The relationship of pH to the hydrogen ion concentration is such that as the latter *increases,* the pH gets smaller. Thus, an acid solution has a pH of less than 7.0, and a basic (or alkaline) solution has one greater (up to 14.0).

It should be noted that a logarithmic difference relates to the *percent* change rather than the *absolute* change. Therefore a pH of 5.0 is ten times more acid than pH 6.0, and a pH of 4.0 is ten times more acid again, or 100 times as acid as pH 6.0. The pH of the human body fluids is 7.4, just on the alkaline side. The pH of the stomach is often about 2.0 which is more than 100,000 times the hydrogen ion concentration of the blood.

The reaction of equivalent amounts of acid and base *neutralize* each other and yield a *salt:*

$$NaOH + HCl \rightarrow NaCl + H_2O$$

ELEMENTS IN LIVING MATERIAL

A very large number of the elements are found in living tissue. Some of them, present in only minute amounts, serve vital roles, but many are there by "accident." With a few exceptions that are not important in the present context, the plentiful elements in living tissue are:

Carbon	Sulfur
Hydrogen	Phosphorus
Nitrogen	Potassium
Oxygen	Sodium
Iron	Magnesium
Calcium	Chlorine

The less plentiful but equally vital *trace elements* are:

Copper
Manganese
Zinc
Iodine
Cobalt (in some species)

Carbon is the element probably the most characteristic of life. Indeed, we might say that it is the properties of this element that make life, as we know it, possible. The carbon atom is *quadrivalent* and forms *covalent* bonds. These bonds are different from the ionic bonds we have discussed so far. Carbon atoms react with other atoms, typically hydrogens, to form a product with a stable electron configuration. In this reaction, there is no exchange of electrons, but rather a *sharing* to make *electron pairs.* We may think of the outer shells of the two atoms being completed by each other. The outer shell of electrons of the unreacted, isolated carbon atom can be diagrammatically shown as

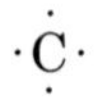

and the hydrogen atom as H.. Four hydrogens combine with one carbon to give a stable gas, methane

$$\begin{array}{c} H \\ H:\ddot{\underset{..}{C}}:H \\ H \end{array}$$

In this fashion the carbon atom's outer shell gets the total of 8 electrons it needs for stability and each hydrogen atom gets the 2 it needs. These covalent bonds are much closer than ionic bonds and there is no dissociation in solution.

A carbon atom can also combine with 3 hydrogens and another similarly equipped carbon to give ethane,

$$\begin{array}{c} \text{H}\ \ \text{H} \\ \text{H}:\ddot{\text{C}}:\ddot{\text{C}}:\text{H} \\ \ddot{\text{H}}\ \ \ddot{\text{H}} \end{array}$$

This molecule can be added to in a similar fashion and, within certain limits, the newly added radical can be almost anything. The hydrogens on the individual carbons can be replaced by oxygens, amine groups ($-NH_2$), sulfur, or an almost endless variety of other carbon-containing radicals. The carbon atoms can also combine with each other in ring-shaped structures, also of endless possible combinations. This is enough to show that the carbon chemistry is almost infinite. The branch of chemistry dealing with these endless, but rather systematic compounds is called *organic chemistry,* because in earlier days it related almost entirely to compounds of organic or living origin.

IMPORTANT ORGANIC COMPOUNDS

Animal tissues are made of proteins, fats, carbohydrates, and the mineral parts of the bones and teeth. The chemistry of these materials is briefly discussed in appropriate sections throughout the book, but a few points may be profitably noted at this stage.

Fatty acids. Fatty acids are made of chains of carbon units with a *carboxyl* group at the end (—COOH). This group, or *radical,* can dissociate to yield a hydrogen ion, thereby giving the acid character, for example acetic acid.

$$CH_3COOH \rightleftharpoons CH_3COO^- + H^+$$

This is a strong acid as organic acids go, but it is still much weaker than hydrochloric acid for example.

Propionic and butyric acids are similar, but with one or two more carbons respectively:

$$CH_3CH_2COOH \text{ and } CH_3CH_2CH_2COOH$$

There is a *homologous series* of these fatty acids, increasing in length in this fashion. Those which make up the plant oils and animal fats have between 14 and 20 carbons each (always an even number).

Amino acids. The chemical units of the proteins are α-amino acids; they are like the short-chain fatty acids above but with an *amine group* ($—NH_2$) replacing one of the hydrogens on the carbon next to the carboxyl group, the α-carbon. Thus:

$$\begin{array}{l} CH_3\ \underset{\displaystyle NH_2}{\underset{|}{CH}}\ COOH\ \text{(Alanine)} \end{array}$$

Beyond the α-carbon, the chain can be much more complex than we have indicated; some amino acids have ring structures and others have sulfur atoms attached. The amino acids are joined together from amine group to carboxyl group, in chains to form the proteins.

Carbohydrates. Carbohydrates are typified by the sugar glucose. The structure of this *monosaccharide* may be written thus:

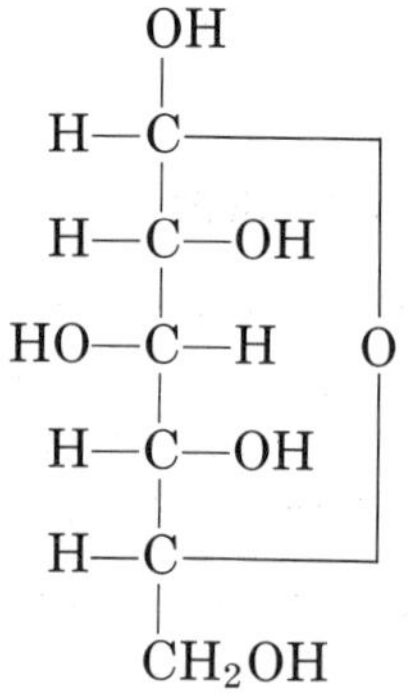

```
        OH
        |
    H—C————————┐
        |              |
    H—C—OH         |
        |              |
  HO—C—H          O
        |              |
    H—C—OH         |
        |              |
    H—C————————┘
        |
        CH2OH
```

or more representatively of its probable molecular arrangement as

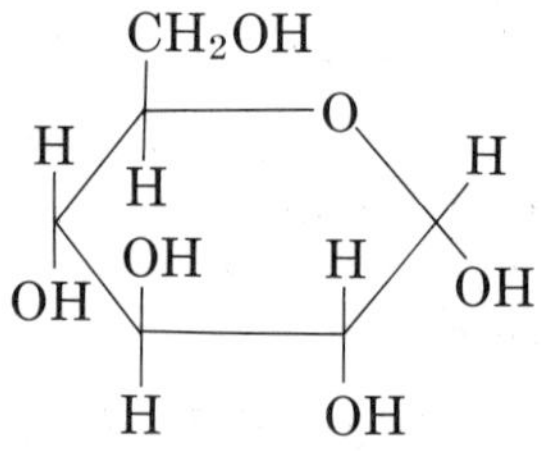

```
          CH2OH
           |
      H   /|————————O
      |  / H          \      H
      | /              \   /
      |\  OH      H     \ /
  OH   \ |        |     / \
        \|________|/      OH
         |        |
         H        OH
```

There are several monosaccharides resembling this, and some of them are discussed in the chapter on nutrition. *Disaccharides* are made up of two units of the same, or dissimilar, monosaccharides. *Polysaccharides,* like starch or glycogen, are made of a great many such units.

Catalysis. Some chemical reactions will proceed much faster in the presence of another material which does not enter into the new compound and *remains unchanged* from beginning to end. Such a material is called a *catalyst.* A simple example is in the production of oxygen by heating potassium chlorate.

$$2KClO_3 \rightarrow 2KCl + 3O_2$$

The reaction goes faster with a small amount of manganese dioxide (MnO_2) present as a catalyst. Some reactions do not go at all without a suitable catalyst.

Modern theories of catalysis propose that one of the reactants combines briefly with the catalyst to yield an *active complex,* which combines with the other very readily, whereupon the catalyst splits off again unchanged. Consistent with this is the observation that most catalysts are more effective in finely divided form. The total surface area of a gram of platinum catalyst as a powder is immensely greater than that of a 1 gram cube and offers more opportunity to form active complexes.

Enzymes. The catalysts in the body that permit relatively violent chemical rearrangements to proceed smoothly and quietly are the enzymes. They are extremely complex proteins, usually with a simpler molecule attached, called the *prosthetic group.* A good laboratory example of enzyme action is the splitting of starch. Starch is made of a great many glucose units which can be split off in ones and twos by boiling starch with hydrochloric acid for an hour or so. The addition of a few cubic centimeters of saliva to some starch solution achieves the same result in a few minutes and at room temperature. The enzyme *ptyalin* in the saliva catalyzes the hydrolysis of the starch.

The glands of the digestive tract secrete powerful enzymes to break the food down to simple soluble or finely divided materials which can be absorbed. Within the cells, equally complex enzymes engage in further breakdown processes to liberate energy, and other enzymes build simple units (like amino acids) into complicated tissue proteins.

OXIDATION AND REDUCTION

When charcoal burns, it combines with oxygen to give carbon dioxide, releasing large amounts of heat.

$$C + O_2 \rightarrow CO_2$$

Similarly, hydrogen burns in oxygen to give water, again with the evolution of heat.

$$2H_2 + O_2 \rightarrow 2H_2O$$

We may also burn finely divided iron in an oxygen atmosphere to give iron oxide.

$$4Fe + 3O_2 \rightarrow 2Fe_2O_3$$

This last reaction may also proceed very slowly in air, and this we recognize as "rusting." These reactions are all *oxidations,* that is, an oxygen atom is added. In terms of our discussion of the electron shells, oxidations all involve the removal of an electron from the substance *oxidized.* The reactant which accepts the electron is an *oxidizing agent,* and is itself *reduced,* the opposite of being oxidized. Oxidation, then, is the removal of an electron; reduction is the gain of an electron.

If we define oxidation and reduction in the general terms of electron transfer, some reactions not involving the element oxygen at all are technically "oxidations," viz:

$$2Na + Cl_2 \rightarrow 2NaCl \rightarrow 2Na^+ + 2Cl^-$$

In biological systems, most oxidations are of the kind involving the addition of oxygen or the removal of hydrogen atoms (which then combine with oxygen to give water). Glucose can be completely oxidized to carbon dioxide and water by burning it, and the heat evolved is apparent in the flame.

$$C_6H_{12}O_6 + 6O_2 \rightarrow 6CO_2 + 6H_2O$$

The same end result is reached in the body cells by a series of step-wise reactions, catalyzed by enzymes and releasing the energy gently and usefully instead of as a brief, fierce flame. The total amount of energy is the same however, regardless of how many step reactions there are in achieving the same end result.

The energy release from oxidation is due to the rearrangement of the atoms in a more stable, and therefore more "likely" form. The conversion of an atom to an unstable electron configuration, or the arrangement of atoms in complex molecules, both require energy. This energy is released when the atoms or molecules return to their lower energy state, which is a more "likely" state than the unstable or complex state. In the oxidation of carbon, its electron configuration in the carbon dioxide molecule is more stable than it is as charcoal, and energy is released. In the oxidation of a complex carbon-containing molecule (like glucose), this same energy is available from the oxidation of the carbon atoms, plus some of the energy previously locked in the carbon-to-carbon bonds.

As discussed in the introduction to the chapter on metabolism (p. 221), not all the energy from such a reaction is available, and not all of the energy invested in synthesizing glucose can be realized by oxidizing it.

DIFFUSION AND OSMOSIS

We have mentioned that molecules in solution behave similarly to gas molecules, but their movements are not as rapid, and in a large volume of solvent their complete distribution may take some time. We may take as an example the dissolving of a crystal of potassium permanganate in a beaker of water (Fig. 4.1). At first, the concentration of permanganate solution in the immediate vicinity of the still dissolving crystal is high, as shown by the color, and in distant parts of the beaker the concentration is negligible. Within the space of a few hours, however, the uniform color of the solution shows that the permanganate concentration is the same throughout the beaker, even without stirring. The movement of a dissolved material which leads to its uniform distribution throughout the solvent is known as *diffusion.* Dissolved gases behave in a similar fashion.

Nutrients and oxygen are available to living cells in dissolved form and they enter the cells by diffusion. In doing so they must cross one or more *membranes,* which can present barriers to diffusion. The properties of living membranes are exceedingly complex and so it is convenient to consider first diffusion across simpler, artificial membranes. One such membrane is the plastic tubing now used for sausage casings in place of the intestines of pigs used in more robust times.

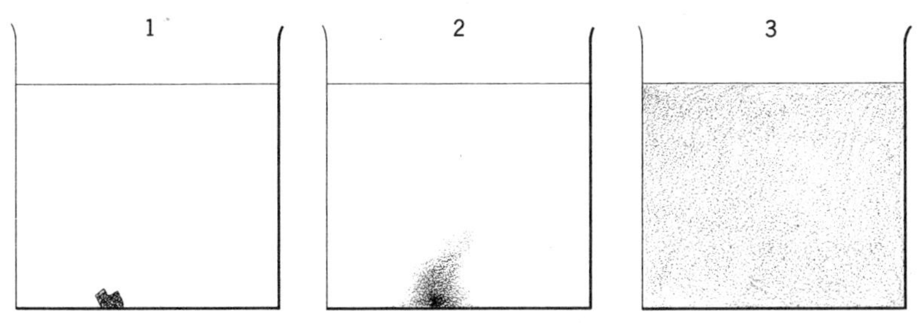

FIGURE 4.1. Dissolving of a crystal of permanganate and diffusion of the solution.

We can think of the solvent and solute molecules diffusing through pores in the membrane. Water and salt molecules are small enough to pass through the membrane as if it were literally not there, but larger molecules are partially or completely prevented from passing through. If a bag made of the plastic tubing is filled with a concentrated solution of sucrose and then suspended in a beaker of water, the bag will swell. This is because the sucrose cannot diffuse into the water to make a uniform concentration of sucrose, but water from the beaker *can* diffuse into the sucrose solution, tending to dilute it. The same observation can be made if the beaker contains a sucrose solution more dilute than that in the bag, or if the beaker contains a solution of any material with small molecules which can pass readily across the membrane. If the bag is tied to the bottom of a tube as shown in Fig. 4.2, the water diffusing into the bag causes the liquid to rise in the tube. This continues until the column of diluted sucrose solution exerts a hydrostatic pressure sufficient to oppose the incoming water from the beaker. The pressure to produce this equilibrium depends on the concentration of the sucrose solution; it is in fact directly proportional to it over a wide range. This pressure is the *osmotic pressure.* Subject to some variation, a solute exerts an osmotic pressure equal to the pressure it would exert if it were in gaseous form and contained in a volume equal to that of the solvent in which it is dissolved. Thus, 1 mole of a gas exerts 1 atmosphere of pressure when contained in 22.4 liters. Similarly, 1 mole of sucrose exerts 1 atmosphere of osmotic pressure when dissolved in 22.4 liters of water and the osmotic pressure of 1 mole in a liter is theoretically equal to 22.4 atmospheres.

Other materials with large molecules show osmotic properties at artificial membranes. These include glucose, egg white, gelatin, etc. Each of these materials also exerts an osmotic pressure at cell membranes. For reasons discussed in a later section, cell membranes are *functionally* impermeable to extracellular sodium ions and therefore, although the sodium ion is small, it exerts an osmotic pressure at the cell membrane. This pressure is normally equalled by the opposite osmotic effect of the intracellular salts and proteins. If the concentration of the extracellular sodium chloride is increased, water diffuses out of the cells, as if to make the total *osmolarity* the same on each side of the membrane. Conversely, if the extracellular fluid's osmolarity is reduced (by adding water) then water diffuses into the cells causing them to swell. It should be noted that water diffuses across the membranes in each direction all the time, but in an osmotic disequilibrium there is a *net* movement to the more concentrated side.

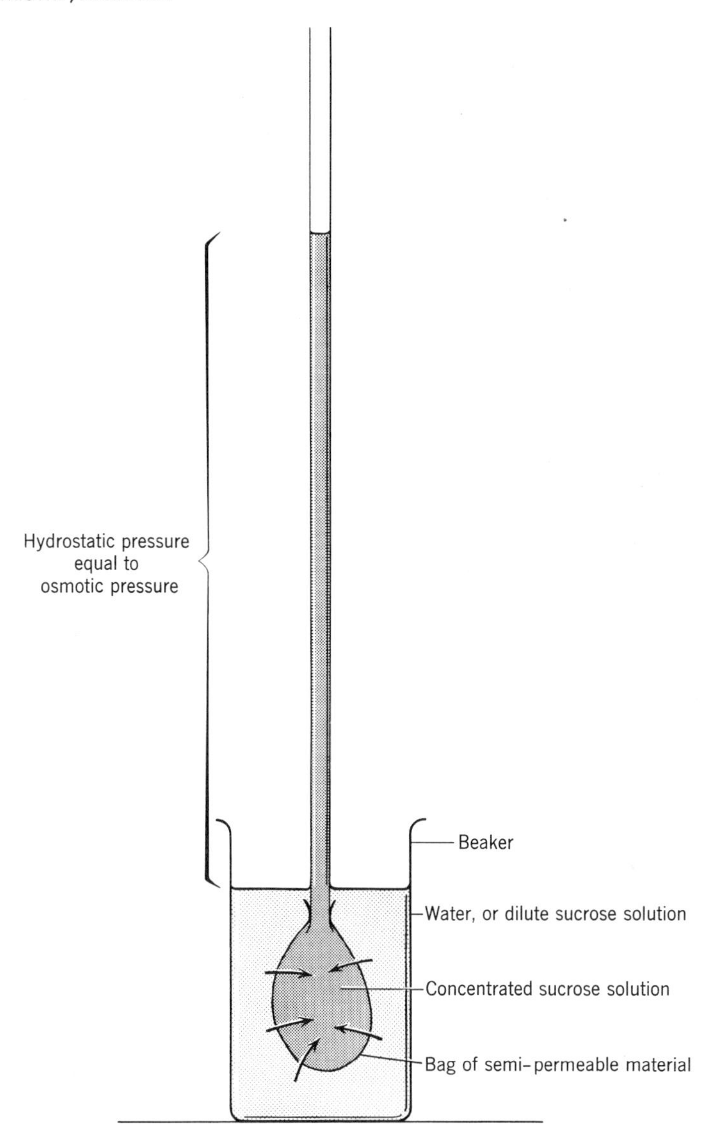

FIGURE 4.2. Osmotic pressure.

Some aspects of osmosis are conveniently demonstrated with red blood cells. If they are suspended in distilled water there is a rapid movement of water into the cells causing them to swell and rupture. The suspension then becomes a clear red solution of hemoglobin. If they are suspended in $0.15M$ NaCl they remain intact. At concentrations of NaCl intermediate between zero and $0.15M$, some or all of the red cells rupture, depending on their individual fragility. At concentrations above 0.15*M* NaCl, water diffuses out of the red cells causing them to collapse in a wrinkled state called *crenation.* The 0.15*M* solution of NaCl is *isosmolar* with the red cell contents, and it is said to be *isotonic* with respect to these cells. Weaker solutions are *hypotonic* and stronger solutions are *hypertonic.* Because NaCl dissociates in solution to give Na^+ and Cl^- ions, a solution contains twice as many *particles* as it does *molecules* of NaCl. A 0.15*M* solution of NaCl contains 0.3 *osmole.* Compounds such as glucose and sucrose do not dissociate in solution and so solutions of these materials must be 0.3*M* to be isotonic with red cell contents.

It will be recalled that a solute only exerts an osmotic pressure if its diffusion across a membrane is impeded. Therefore, it is possible to make 0.3 osmolar solutions which are *not* isotonic to red cells. Among these solutions is urea, a waste product of the body which can pass freely across membranes and is evenly distributed between intracellular and extracellular water. If red cells are suspended in 0.3*M* urea they rupture as rapidly as if they were in distilled water.

Not all cells are as susceptible to osmotic variations as are red blood cells, but serious consequences can result from the shrinkage or swelling of brain cells in disturbances of osmolarity in dehydration or water intoxication.

REFERENCES

1. Daniels, Farrington, *Outlines of Physical Chemistry,* Wiley, New York, 1948.
2. MacInnes, Duncan A., "pH," *Sci. American* **184** (1), 40, January 1951.
3. Noller, Carl R., *Chemistry of Organic Compounds,* Saunders, Philadelphia, 1951.
4. Pauling, Linus, *General Chemistry,* Freeman, San Francisco, 1949.
5. Scheer, Bradley T., *General Physiology,* Wiley, New York, 1953.

PART 2

THE INTERNAL ENVIRONMENT

BODY FLUIDS

5

The life of a cell, an individual, or even of a civilization depends on an adequate supply of water. It is not only the most plentiful constituent of living tissue, but it is also the very medium in which life proceeds. The flux of chemical reactions which we call life is carried on entirely in the water inside the cells. Similarly, the water outside the cells contains nutrients and oxygen which must diffuse into the cells for the continuance of the vital processes. All of the body fluids are water solutions; without water there can be no life.

The earliest forms of life developed in the sea, and the first terrestrial animals were probably amphibians similar to our modern frogs and newts, which must pass their life span within easy reach of water. The higher animals are not so dependent on water that they must pass part of their time in it, but with very few exceptions, they still have to drink water frequently to maintain their watery internal environment. Their individual cells are just as much aquatic organisms as are the primordial unicellular animals and plants.

THE FLUID COMPARTMENTS

The body fluids can be considered in two broad categories: those inside the cell (*intracellular*) and those outside (*extracellular*). The water molecules inside and outside the cells can exchange readily, but the exchange of the dissolved constituents is limited by various mechanisms so that their concentrations inside and outside the cells are maintained at different levels.

In man, 60% of the body weight is water. Two-thirds of this, or 40% of the body weight, is in the *intracellular fluid* and the remainder is in the *extracellular fluid.* The subdivisions of the body fluids are shown in the diagram of Fig. 5.1. The extracellular fluid is mainly the fluid in the spaces around the cells, the *interstitial fluid.* The plasma of the blood and the lymph, however, are also parts of the extracellular fluid, and, as we shall see, they are in rapid equilibration with the interstitial fluid. The blood itself is a fluid in the usual sense, but the red cells which comprise about one-half its volume are discrete particles suspended in it. For this reason we consider only the plasma rather than the whole blood as part of the extracellular fluid. The fluid secretions in the intestines are indeed outside the cells, but they are also outside the body in the strictest sense and so are not part of the extracellular fluid.

The intracellular fluid is the total fluid content of all the cells; it is not as homogeneous as the extracellular fluid. The fluid inside liver cells, for example, is different from the fluid inside kidney cells, which is different again from that of brain. Even within a single cell, the nucleus and mitochondria contain less water than the cytoplasm, and even the cytoplasm is not of uniform consistency throughout the cell. Some cells contain more water than others, but all of them are at least half composed of water. The nonliving parts of bones and teeth are very low in water, but the

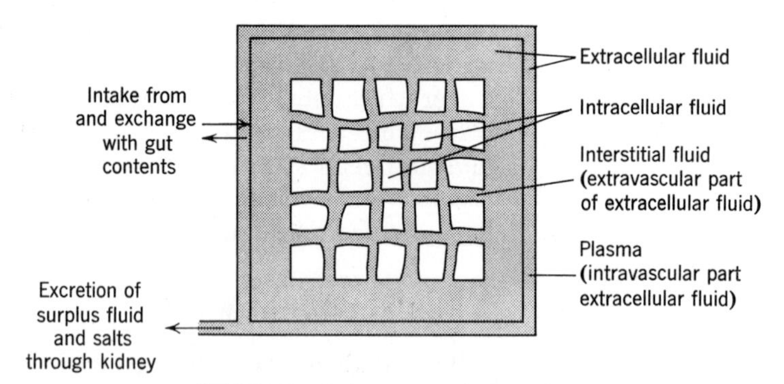

FIGURE 5.1. Subdivisions of the body fluids.

living cells that secrete the mineral structure are similar to the other tissues of the body in their high water content.

The extracellular fluid. The interstitial fluid is the immediate environment of the animal's cells; it bathes the cells in much the same way that the ocean or pond water bathes unicellular organisms. For this reason, Claude Bernard, a great French physiologist called it "the internal environment." In order to maintain optimal conditions in the internal environment, particularly with respect to the oxygen supply, the fluid actually in contact with the cell surfaces must be continually renewed. In very simple animals, the tissue fluids are moved around adequately by the squeezing and mixing of the locomotory movements of the animal, but in higher animals, and especially with increased size, the actively metabolizing cells require a more vigorous renewal of the fluid around them. This is accomplished by the *circulation of the blood.*

As described in the introductory chapter of this book, the blood comprises chiefly the plasma and red cells (erythrocytes). The latter are individual cells (although without nuclei in the mammals) which float in the plasma and are responsible for the carriage of oxygen and carbon dioxide to and from the tissues. The truly liquid plasma is part of the extracellular fluid. Since it is contained within the blood vessels it is called the *intravascular* part of the extracellular fluid, whereas the interstitial fluid is the *extravascular* portion.

The blood is pumped by the heart through vessels to all parts of the body. In the tissues, the blood vessels subdivide into very small, thin-walled *capillaries.* The oxygen carried by the red cells is able to diffuse through these walls into the interstitial fluid, from which it diffuses into the cells. The carbon dioxide follows a similar path in the opposite direction. The walls of the capillaries are permeable to plasma water and to the smaller molecules dissolved in it. Therefore, because the blood is under pressure in the capillaries, there is constant leakage of water and solutes from the plasma into the interstitial fluid. As is described in the chapter on circulation, a nearly equal volume of interstitial fluid diffuses back into the capillaries at the same time and balances the loss from the plasma. By this rapid exchange, the internal environment is constantly renewed by fresh fluid from the plasma. The advantage of the exchange is that during each circulation of the blood, a large proportion of the plasma and red cells pass through *regulatory organs* which add to, or subtract from it, so that the concentration of nutrients, oxygen, and waste products is constantly adjusted to an optimum. Since the plasma is in rapid exchange with the interstitial fluid, it follows that the composition of all the extracellular fluid is kept constant within very narrow limits.

In addition to the rapid exchange between the interstitial fluid and the plasma, the contents of the intracellular fluids are similarly in exchange with the interstitial fluids. This exchange is across the cell membranes, which are more discriminatory in many respects than the walls of the capillaries. In health, the respective volumes of the intracellular and extracellular fluids do not vary much, and we therefore speak of the intracellular and extracellular, "compartments." This term is useful but possibly misleading, since individual molecules are in constant movement between the compartments.

Intracellular fluid. The cell membranes are similar in many ways to the artificial semipermeable membranes described in the discussion of osmosis in Chap. 4. The intracellular volume is largely determined by the concentration of solutes on each side of the cell membrane. Any increase in the osmolarity of the extracellular fluid causes a shift of water out of the cells and a relative dehydration of the cells' contents. This is seen in thirst, where water loss is initially from the extracellular fluid. Conversely, any decrease in the osmolarity of the extracellular fluid will cause a shift of water *into* the cells, frequently with results just as disastrous as the dehydration of extreme thirst. The volume of each of the main compartments, therefore, depends chiefly on the regulation of the osmolarity of the extracellular fluid. This is one of the principal roles of the *kidney*.

A few of the solutes in the body fluids, the waste product urea, for example, are uniformly dissolved in all the body water and so are distributed as if the cell membranes were indeed simple semipermeable membranes. Most of the other solutes have more complex distributions which at first sight seem to defy the laws of diffusion.

THE EQUILIBRIUM ACROSS THE CELL MEMBRANE

In unicellular aquatic organisms the cell contents are all derived from the water in which they live, but the amounts and proportions within the cell are different from those found outside. The incorporation of material into the cell must therefore be *selective*. An example of this is seen in unicellular animals, which select and engulf food particles which can be broken down and the components used for energy or synthetic purposes. Less obvious, but equally important, is the difference in the concentrations of the electrolytes in the intracellular fluid and the environment. Most organisms maintain intracellular electrolyte concentrations quite different from the medium in which they live; this is especially true of the cells of higher animals. The extracellular fluid (the internal environment) is typically high in sodium, calcium, and chloride and low in potassium, magnesium, and phosphate. The intracellular fluid has the *reverse* characteristics (see Fig. 5.2).

The extracellular fluid of mammals bears some resemblance to sea water in its composition, although it is much less salty. It has been suggested that perhaps it was originally derived from the sea water which was "trapped" when the primitive animals first closed off an internal environment, making them more independent of the exterior. Since the oceans have become steadily more salty throughout geological time, the lower salt concentration of extracellular fluid was thought to reflect that of the ocean at that time in the remote past when the vertebrate ancestors first differentiated. This attractive theory is not borne out by paleochemical studies. It is now believed that the sea at that time already had a higher salt concentration than does mammalian blood, and that

Intracellular fluid

Ion	meq/L.H_2O
K^+	150
Na^+	10
Mg^{++}	40
Ca^{++}	Trace
Cl	Trace
HCO_3^-	8
HPO_4^-	135
Protein	

Interstitial fluid

Ion	meq/L.H_2O
K^+	5
Na^+	140
Mg^{++}	2
Ca^{++}	5
Cl	100
HCO_3^-	27
HPO_4^-	2

FIGURE 5.2. A comparison of the composition of intracellular and extracellular fluid.

the concentration of magnesium in particular was already much too high. Even if this "primitive ocean" theory is incorrect, it is still a useful concept for envisaging the role of the extracellular fluid.

The concentration gradient of potassium and sodium. One of the biological problems under the most active investigation is the mechanism by which the cells maintain their high concentration of potassium and magnesium and at the same time exclude sodium and calcium. The unequal distribution of these electrolytes across the cell membrane is highly characteristic of living tissue. It is responsible for the electrical properties of cells, which are exemplified by the conduction of impulses in nerve and muscle tissues. These two tissues are special examples of the property of *irritability* (responsiveness to stimuli) which is displayed by living cells.

For reasons beyond the scope of this text, the unequal distribution of sodium and potassium gives rise to an electrical potential across the cell membrane. That is, the inside and outside of the membrane are like the poles of a battery, across which a voltage, or *potential* exists. A stimulus acts by disrupting the equilibrium at the membrane so that the potential suddenly decreases and actually reverses. This is called *depolarization.* The change in potential at one point affects the adjacent areas on the membrane and causes them to depolarize too. In consequence, a wave of depolarization spreads in each direction from the original point of stimulus. After the passage of a wave of depolarization, the normal electrolyte concentrations are rapidly restored and the membrane potential returns to its resting level.

It is apparent, even from this brief summary, that it is vital to cells to maintain this unequal distribution of electrolytes across the membrane and that furthermore, their capacity to do so is a fundamental characteristic of life itself. The way in which the cells manage this remains a mystery, but several mechanisms have been proposed. Some of them are reviewed.

For a long time it was assumed that cell membranes were impermeable to sodium and potassium and that the sodium stayed in the extracellular fluid simply because it could not enter the cells. This simple theory did not explain, however, the presence of potassium in the cells in the first place. There would need to be some local concentration of potassium whenever a new cell was made.

Another early theory got around this difficulty in part by taking into account the different sizes of the ions concerned. The sodium ion is actually smaller than the potassium ion, but when in solution, these positively charged particles attract around them a number of water molecules forming a *shell of hydration.* The sodium ion attracts more water molecules than does potassium, so that when hydrated, it is actually *larger* than the hydrated potassium ion. The theory postulated that there are pores in the cell membranes which are large enough to admit the hydrated ions of potassium, but not those of sodium.

Both theories have been rejected on the evidence of studies with radioactive isotopes of sodium and potassium. It has been shown that both ions can pass freely across the cell membrane, and that the normal distribution is a "steady state" of a system in constant flux in each direction. A theory widely accepted at present is that sodium can enter the cells but is immediately extruded by a "pump" mechanism. If all the sodium is removed as fast as it enters, the potassium is able to accumulate more or less passively within the cell.

This theory has several weaknesses, but it is consistent with many of our observations on the maintenance of the potassium gradient across the cell membrane. For example, if cells are subjected to a shortage of oxygen, or if their metabolism is inhibited by various poisons, the normal distribution of sodium and potassium is no longer maintained. The potassium inside the cell leaks out and is replaced by sodium until the distribution approaches that which would be expected if there were no selection at the membrane. The membrane potential declines and the irritability is lost. The metabolic poisons and a shortage of oxygen interfere with the production of energy within the cell, and the prompt influx of sodium under these conditions suggests that sodium is usually kept out of the cell by some energy-using system. A "pump" of course requires energy, and so in this respect the theory is consistent with experimental facts. The pump is envisaged by many physiologists as a "carrier" molecule which shuttles to and fro across the membrane carrying sodium in the outward direction and returning unloaded. Such a carrier would have to discriminate between sodium and potassium, which are chemically very similar. No compound with suitable characteristics has ever been isolated from cells, but this is, of course, no evidence that one does not exist.

Still another current theory suggests that because of the smaller size of the hydrated potassium ion, it can be bound more closely by the negatively charged sites on the proteins within the cells than can the larger sodium ion. According to this theory, the loss of potassium under unfavorable metabolic conditions is due to the chemical changes in the intracellular proteins, which reduce the number and availability of sites suitable for binding potassium.

Each of these theories has something to offer, and time may show that all of the suggested mechanisms are involved to some extent. It is interesting to note that the concentration of magnesium within the cells and the relative exclusion of calcium have not received so much attention. In addition to the unequal distribution of cations —that is, the positively charged ions—the anions, chloride, and phosphate, are largely confined to the extracellular and intracellular fluids respectively.

The internal environment also contains organic materials in solution. Chief among these is glucose, which is the principle source of energy for the cells. There is good evidence that glucose does not simply diffuse into the cells, but that there is rather an *active transport* process for its passage of the membrane. Such a process is analogous to that proposed for the extrusion of sodium, but of course in the opposite direction. The transport of glucose, sodium, or of any other material in the direction opposite to that expected from the concentration gradient must involve the expenditure of energy for the active transport. Part of the cell's *resting metabolism* is certainly devoted to that end, although it is not known how the energy from the combustion of nutrients is coupled to the transport process.

Determination of body water. In research and in clinical medicine it is often of interest to determine the water content of an animal and to determine the distribution of water between the "compartments" of the body fluids.

The water content of a dead animal is most easily determined by weighing the animal at the time of death. The minced carcass is dried in an oven at 110°C and then reweighed. The difference between the weights is clearly that of the water driven off by the high temperature. This method has obvious disadvantages in human medicine and several less direct, and on the whole kinder, methods are used.

Concept of Lean-Body Mass

The body water content varies widely among individuals, and numerous analyses of cadavers have been made to characterize this. Analysis has shown that the muscles and visceral organs (*the lean-body mass*) comprise 73% water with remarkable constancy. Two-thirds of this is intracellular and the remainder is extracellular. The intact body, however, consists also of the skeleton and the fat deposits. In health, the relative weights (and water contents) of the extracellular fluid and the skeleton are quite constant, but the fat is subject to great variations among individuals, and indeed, in the same individual from one year to another. The fat deposits (adipose tissue) contain relatively little water, so that the obese individual has a relatively smaller water content than a lean person. This is the main source of the variation in total body water among individuals (see Fig. 5.3).

On the basis of extensive analyses, the water content of an individual can be predicted quite accurately if the fat content is known. The fat content can be determined from the specific gravity of the body, since the lower density of fat tissues reduces the specific gravity of the whole body in a consistent fashion. Then, knowing the fat content, the lean-body mass can be calculated. The technique is not widely used, however, because in order to determine the specific gravity of the subject, he has to be immersed in water to find his body volume. This is cumbersome at the best of times and quite out of the question with sick patients. For most purposes, the water content can be estimated from the weight of the subject, making allowance for the apparent fatness. The experienced clinician also considers the sex of the subject, since females have a distinctly lower relative water content. This sex difference does not apply in children and only becomes apparent after puberty. The difference between adult males and females is partly due to the greater fat content of the average female, but lean females still have less body water per kilogram body weight than do lean males.

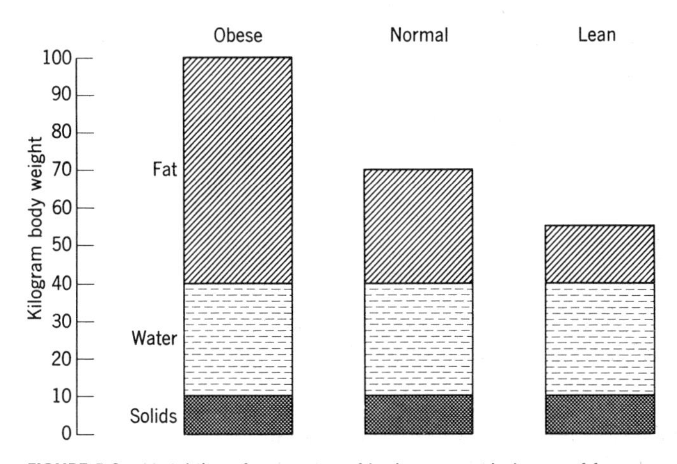

FIGURE 5.3. Variability of proportion of body water with degree of fatness.

Newborn infants have a greater relative body water than adults and this decreases through childhood.

Use of heavy water. When it is necessary to determine total body water with fair precision, it is now customary to use the heavy-water *dilution technique.* Water is composed of two atoms of hydrogen and one of oxygen according to the formula H_2O. There is a heavy isotope of hydrogen called *deuterium,* which has almost identical chemical properties, but each atom is twice as heavy as an ordinary hydrogen atom. Deuterium oxide, D_2O, is therefore the same as water chemically, but it is heavier and so known as "heavy water." By various techniques, it is possible to determine the amount of heavy water present in a sample of ordinary water.

If a known amount of heavy water is injected into a subject, it disperses through all the body water (since all cell membranes are permeable to water). After about three hours, it reaches an equilibrium at which the concentration of heavy water is uniform in all the body fluids. A sample is then taken from the body and the concentration of heavy water is determined. The sample may be of plasma water or even of the condensate from the breath against a cold surface. It is evident that the dilution of the originally injected heavy water depends on the amount of ordinary water in the body, and this is of course what we need to know.

$$\text{Amount } D_2O \text{ injected} = \text{Conc. in injected soln.} \times \text{Vol. injected}$$

and

$$\text{Amount } D_2O \text{ remaining} = \text{Conc. in body water} \times \text{Vol. of body water}$$

If we ignore (or make corrections for) the loss of D_2O by excretion and evaporation, the amount injected is the same as the amount remaining at equilibrium. Therefore:

$$\text{Conc. injected} \times \text{Vol. injected} = \\ \text{Conc. in body water} \times \text{Vol. of body water}$$

and

$$\frac{\text{Conc. injected} \times \text{Vol. injected}}{\text{Conc. in body water}} = \text{Volume of body water}$$

This elegant method is an example of the *isotope dilution* technique, which has many applications in modern biology and chemistry.

Determination of extracellular fluid volume. The heavy water dilution technique for total body-water determination depends on the complete distribution of the injected isotope through all the body water. If a material is injected which cannot penetrate the cells, it can be used in a similar fashion to measure the extracellular fluid volume, since the injected dose is diluted by the extracellular fluid only. Two such materials are inulin and sodium thiocyanate. Despite its toxic sounding name, sodium thiocyanate is quite harmless and can be injected without any ill effects. Inulin is a starch-like compound obtained from the roots of dahlias, Jerusalem artichokes, and similar plants. It is not to be confused with the hormone *insulin* which is instrumental in controlling the sugar metabolism of the body. Inulin and thiocyanate are used in roughly the same way as the heavy water, except that the concentration in the body after equilibration is determined by chemical analysis of the blood plasma. As before:

$$\frac{\text{Conc. injected} \times \text{Vol. injected}}{\text{Conc. in plasma}} = \text{Extracellular fluid vol.}$$

These materials are injected into a vein in the subject, but their molecules are small enough to leave the blood vessels in the capillaries in the same way as described for water and salts so that they distribute themselves in all of the extracellular fluid.

Determination of plasma volume. By exactly analagous methods, the volume of the plasma may also be measured. This requires a material which will not leave the blood stream after injection. A suitable compound is one with a molecule too large to pass out through the capillary walls into the interstitial fluid. A synthetic plasma expander called dextran is used in this fashion and its dilution by the plasma is measured chemically. Probably more widely used is the dye, Evans Blue or T 1824. This dye does not have large molecules in itself, but forms a complex with the plasma proteins that is not easily broken. Since the plasma protein molecules are too large to pass out through the walls of the capillaries, the T 1824 is retained in the blood vessels as well. It will be noted that this method determines the *plasma* volume only and not the blood volume. In order to find the blood volume, a blood sample is centrifuged and the respective fractions of plasma and cells are determined. The blood volume can then be calculated by multiplying the plasma volume by the appropriate factor.

Determination of electrolyte concentrations. Because of the profound effects of changes in osmolarity and relative ion concentrations, the electrolyte composition of the plasma and other body fluids is of great interest to the physiologist and clinician. In particular, a great deal of information can be gleaned from the sodium, potassium, and calcium concentrations in the plasma. Until recently the quantitative chemical determinations of these ions was too tedious for routine usage in the clinical laboratory, but the invention of the *flame photometer* has greatly changed this situation in the last few years.

Flame Photometer

This instrument has made possible so many important strides in physiology and medicine that it merits a brief description. It will be recalled from elementary chemistry, that a flame of characteristic color is seen when a loop of platinum wire is dipped into a sodium chloride solution and then held in the flame of the bunsen burner. The flames resulting from the presence of potassium or calcium are of different colors from that of sodium; that is, the wavelengths of their emitted light are different. In the flame photometer, a solution (such as dilute plasma) is aspirated into a small gas flame and the resulting light is focused by a lens through a filter on to a photoelectric cell. The filter allows light of only one wavelength to pass, so that only the light due to sodium, for example, reaches the photoelectric cell even if other cations are present in the flame. The current developed in the photoelectric cell depends on the intensity of the light at that wavelength, and the intensity is proportional to the amount of sodium in the solution aspirated into the flame. The response of the photoelectric cell is indicated on a galvanometer on the outside of the instrument. If the galvanometer readings are standardized by introducing solutions of known sodium content into the flame, the sodium content of unknown solutions can be calculated. The concentrations of potassium and calcium are measured in the same way by using filters which select only the appropriate wavelengths of light.

In this fashion it is possible for a skilled laboratory technician to determine the electrolyte composition of up to one hundred samples of plasma in a day. For research purposes, the electrolyte composition of the solid tissues may be measured by digesting a sample in a hot nitric acid which can afterwards be diluted to a known volume and run through the flame photometer.

Other chemical determinations in body fluids. The electrolytes referred to in the previous section are important constituents of the body fluids, but others are of equal interest. The anions chloride, phosphate, and sulfate are measured by various chemical methods, as are the concentrations of organic compounds like glucose, urea, creatinine, and many others. By the newer techniques of electrophoresis and the ultracentrifuge the amount and kinds of plasma proteins can be characterized.

By these methods it is literally possible to take the body fluids apart, which is of great value in our understanding of the regulation of the internal environment and of the consequences of changes. Furthermore, departures from the range of normal in the concentrations of most constituents are frequently associated with disease. The change in concentration is usually a *result* of a diseased state, but the visible symptoms of the disease are often consequences of the subtle change in the internal environment.

A common situation is for a disease to affect one of the regulatory organs, for example, a kidney infection. The diseased kidney is less able to fulfill its role in the regulation of the internal environment, with the result that the waste-product concentration in the plasma reaches toxic proportions and upsets the metabolism of all the cells in the body. Among those disturbed are the brain cells, and the patient becomes confused and eventually comatose. In this oversimplified case, the mental symptoms are caused by the change in the internal environment, which was caused by the loss of effectiveness of the kidney subsequent to bacterial infection.

THE REGULATION OF THE EXTRACELLULAR FLUID COMPOSITION

Whatever may be the nature of the mechanisms at the membrane which control the intracellular composition, there is overwhelming evidence that the composition of the extracellular fluid exerts a critical influence on the intracellular fluid and on the whole metabolism of the cell. The cells of higher animals are particularly sensitive to changes in the extracellular fluid, which is of course their environment. The primitive, aquatic, unicellular organisms are at the mercy of changes in their environment, but they are still able to tolerate a fair range of variation in its composition, and indeed of their own composition. It appears as if the cells of higher animals have lost some adaptability as a price paid for their independence of the external environment and the specialization that this has made possible. In particular, the irritability of the cells is drastically altered by changes of calcium and potassium concentrations in the extracellular fluid, even if those changes are so small as to cause no measureable change in the intracellular fluid. Such changes which occur in disease in man may affect the irritability of the nerve cells, leading to paralysis or convulsions depending on the nature and direction of the changes.

Because of the dependence of cells on the composition of the bathing fluid, tissues removed from an animal must be kept in special *physiological* solutions if they are to retain their irritability and other characteristics for experimental purposes. Physiological solutions must closely approximate normal interstitial fluid in osmotic pressure, pH, nutrient content, oxygenation, and in the concentration of the individual ions of sodium, potassium, calcium, magnesium, chloride, and phosphate. The heart can be removed from an animal and will continue to beat if perfused with an appropriate solution, but if the calcium or potassium concentrations are altered, the heart stops immediately. The composition and usefulness of several physiological solutions is given in Table 5.1.

TABLE 5.1 Some Useful Physiological Solutions

Salt	For Frog Tissue	For Mammalian Tissues	
		Tyrode for Rabbit Gut	Krebs-Ringer for Rat Tissues
NaCl	6.5*	8.0	8.25
KCl	0.1	0.2	0.42
$CaCl_2$	0.2	0.2	0.34
$MgCl_2$	—	0.1	—
Glucose	2.0	1.0	1.0
Buffer	Phosphate	Phosphate	Bicarbonate—CO_2
pH	7.1	7.4	7.4
KH_2PO_4	—	—	0.19
$MgSO_4 7H_2O$	—	—	0.35

* Numbers are grams per liter.

Therefore, in addition to the regulation at the cell membrane, another set of regulatory mechanisms controls the composition of the extracellular fluid. As we have seen, the interstitial fluid is derived from the plasma, and the regulation of all the extracellular fluid is accomplished by controlling the composition of the plasma.

The composition of the plasma at any time is quite evidently the net result of the processes adding material to it and those taking material away. A suitable balance of these processes results in the plasma composition (and therefore the interstitial fluid composition) remaining at a stable optimum. This is the very essence of *homeostasis* and the study of these regulations comprises the largest part of the study of physiology.

Before discussing the regulatory mechanisms themselves, it would be profitable to review the factors which need to be regulated and the consequences if they were not so regulated.

Nutrients and oxygen. The metabolism of the tissues is continually depleting the interstitial fluid of oxygen and nutrients. Unless these are replaced the cells rapidly die.

Waste products, including carbon dioxide. The metabolism is similarly producing end products which diffuse out into the interstitial fluid and would reach toxic levels if they were not removed.

Water. There is a constant evaporation of water from the skin and particularly from the lung surfaces (our breath is moist). If the water is not replaced, the end result is dehydration and death.

Salts and osmolarity. Salts and water must be maintained in the right proportions to give the optimal osmotic pressure, ion balance, and acid-base balance.

These incipient changes are dealt with continuously by regulatory mechanisms which operate simultaneously and in a coordinated fashion, as do the instruments of an orchestra. For simplicity, however, they must be described singly. Furthermore, this review is for orientation only, as each of the mechanisms to be described is dealt with more fully in a later chapter.

Lungs. The blood is continually circulated through the lungs before being pumped to the other tissues. In the lungs, the blood equilibrates with fresh air so that the carbon dioxide it has brought from the tissues is released and a fresh supply of oxygen is taken up. This ensures that the oxygen and carbon dioxide contents of the interstitial fluid remain at nearly constant levels. The efficiency of the gas exchange in the lungs depends on the rate and depth of breathing, and on the blood flow through the lungs. These can be varied according to the amount of oxygen used and the amount of carbon dioxide produced in the tissues. In exercise, the demand for oxygen is increased and we *automatically* breathe faster and more deeply and the blood flow is increased in the lungs and other tissues. If the subject is in good health and the exercise is not excessive, the concentrations of oxygen and carbon dioxide do not vary much from those during rest. The automaticity of the respiratory control is through nervous *reflexes* stimulated by the changes in the gas composition of the plasma.

Intake and storage of nutrients. It is obvious that the composition of an animal must, in the final analysis, depend on the nature of the material eaten. The food and drink intakes are subject to long-term regulation by the appetite and thirst mechanisms, but these are subject to wide variations among individuals and from day to day in the same individual. Over a period of about a month, we eat and drink fairly exactly what we require, but from meal to meal we may eat much more or less than is needed.

If the nutrient level in the plasma depended solely on the rate of absorption from the gut, there would be an excess of nutrients immediately following a meal, followed by a serious shortage several hours later when all the food had been absorbed. The blood which absorbs the nutrients in the intestinal walls returns to the heart by way of the hepatic portal system. That is, it drains through the liver first, where the excess nutrients are removed and stored temporarily. Then, during a subsequent time of fasting, the stored nutrients (glucose and amino acids) are released piecemeal to maintain the extracellular fluid concentrations at their optimum.

The uptake of glucose by the liver (and other tissues) is under further fine control by hormone systems which will be described in later sections. If the intake of foodstuffs is so large that the storage capacity of the liver is exceeded, the excess is stored in the form of fat at various locations in the body. At a subsequent time when the food intake is inadequate, the reserves of fat are drawn on as a source of energy. The deposition of fat in many wild animals serves as a reserve against the food shortage they encounter during the winter months.

The release of stored nutrients. The removal of nutrients and oxygen from the extracellular fluid depends on the level of activity in the cells. For example, rapidly exercising muscle cells require more fuel as well as more oxygen in order to put out more energy. Accordingly, they rapidly use up the glucose in the interstitial fluid and this is immediately reflected by a lowered plasma glucose concentration. The liver responds to this situation by releasing more stored glucose until the normal concentration can be maintained at that level of muscular activity. The response of the liver is so prompt that frequently the blood glucose level actually *rises* during exercise.

Elimination of waste products. The metabolism of the tissues and their breakdown and repair result in a continuous production of waste materials. Some of these are merely useless, but others are positively toxic, especially if allowed to accumulate. The toxic materials are frequently rendered harmless by combination with other materials, mainly in the liver, but all have to be excreted from the body finally. The removal of these materials from the plasma is one of the prime functions of the kidney, which incorporates them in the urine so that they may be excreted.

Control of osmotic pressure. The osmotic pressure of the body fluids is regulated by the relative intake and excretion of water and salts. As mentioned previously, the intake is subject only to a rather loose and slow-acting control, so that the fine adjustment is necessarily made by discriminatory excretion of water and salts by the kidney. Excess fluid is excreted by the kidneys, and it is a common observation that drinking large amounts of water or other more attractive fluids is followed by a copious urine output. When the water intake is low, the kidney responds by excreting a more concentrated urine, thereby conserving water. The regulation of the total salt content, and of the relative concentrations of individual electrolytes, is also accomplished by suitable excretion or retention by the kidney. In this, and in water excretion, the kidney is under hormonal control which is activated by the composition of the interstitial fluid.

If the food and drink of an animal produces excess acid (as normally happens with our type of diet), the kidney corrects this by excreting a more acid urine. Conversely, the presence of excess base (as in grass-eating animals) is adjusted by the excretion of an alkaline urine. The regulation of acid-base balance is also mediated in part by the respiratory system.

Blood pressure and blood flow. However efficiently the regulatory organs work, the maintenance of the interstitial fluid composition depends on the rate of exchange between it and the plasma. Therefore, when any of the foregoing mechanisms are called on for extra effort, they can be aided by the more rapid circulation of the blood. The passage of the plasma through the regulatory organs is speeded up and so is the presentation of new plasma in the capillaries. The circulation of the blood is increased by the heart's contracting more rapidly and more strongly, and this is under both hormonal and nervous reflex control. The blood flow through an organ can be increased by the dilation of the small blood vessels, especially if there is a simultaneous constriction of blood vessels in other organs.

From this brief discussion of the regulatory mechanisms it is clear that the body could not survive if deprived of any of them. However, the organ displaying the most exacting regulatory activity is the kidney. Homer Smith has stated that the composition of an animal is determined not so much by what it eats, but by what the kidney keeps.

REFERENCES

1. Bland, John H., *Disturbances of Body Fluids,* 2nd ed. Saunders, Philadelphia, 1956.
2. Fenn, Wallace O., "Potassium," *Sci. American* **181** (2), 16, August 1949.
3. Heibrumn, L. V., "Calcium and life," *Sci. American* **184** (6), 60, June 1951.
4. Wolf, A. V., "Body water," *Sci. American* **199** (5), 125, November 1958.

BLOOD

6

The internal environment is continually renewed and restored to its "ideal" composition by the circulation of the blood. From the capillaries, water and crystalloid components of the plasma enter the intercellular spaces by ultrafiltration, and a similar volume of interstitial fluid returns to the plasma at the distal ends of the capillaries by osmosis. Meanwhile, the blood composition is regulated by the lungs, kidneys and liver, etc. The circulation and exchange in the capillaries is so effective that the regulation of the relatively small volume of plasma is sufficient to maintain the entire internal environment at a near-constant composition.

Everything required by the tissues for energy or synthesis is carried in the blood, and all the waste products derived from the tissues are removed by the blood. As a result, the blood is extremely complex, but certain major characteristics may be described.

Physical characteristics of whole blood. Blood is a sticky, opaque fluid about $4\frac{1}{2}$ times as viscous as water and with a specific gravity of 1.040–1.060. It has a characteristic taste and odor. The blood is usually about 9% of the body weight but this relative amount may be changed in disease and is normally increased in hot climates. The blood consists of the *plasma* and *formed elements,* each of which make about half of the total blood volume. The formed elements are chiefly the red cells (erythrocytes) and much smaller volumes of white cells (leukocytes) and platelets (thrombocytes).

Plasma. Plasma is a true liquid about $1\frac{1}{2}$ times as viscous as water and usually of a pale straw color. Its normal composition is given in Table 6.1. Except for the 7% protein in the plasma, the composition of interstitial fluid is closely similar, which is to be expected since the interstitial fluid is an ultrafiltrate of plasma. The role of the plasma in maintaining homeostasis has already been described. Other important functions include the *clotting* or *coagulation* process and the *immune reactions.* The role of the plasma buffers in acid-base balance is taken up in a later chapter.

TABLE 6.1 Some Constituents of Human Plasma

(The values given are typical rather than universal.)

Constituent	Concentration
Calcium	4.8 milliequivalents per liter
Chloride	102.0 milliequivalents per liter
Magnesium	1.8 milliequivalents per liter
Phosphate	3.4 milliequivalents per liter
Potassium	3.6 milliequivalents per liter
Sodium	140.0 milliequivalents per liter
Protein	70.0 grams per liter
Lipid	1.0 grams per liter
Lactic acid	0.4 grams per liter

Erythrocytes. The structure of red cells is shown in Fig. 6.1. They are biconcave discs suspended in the blood plasma in fantastic numbers. There are 5,000,000 per cubic millimeter, and their presence gives both the red color and opacity to blood. The color, which is actually greenish under the microscope, is due to the protein, *hemoglobin,* contained in the structural protein (*stroma*) of the erythrocytes. Hemoglobin is a protein conjugated with four molecules of heme, each of which contains an atom of iron. The large resulting molecule has phenomenal powers of loosely binding with either oxygen or carbon dioxide, so that the blood is able to carry much larger quantities of these gases than it can in simple solution. The transport of the respiratory gases is the main function of the erythrocytes and this is covered in detail in the section on respiration.

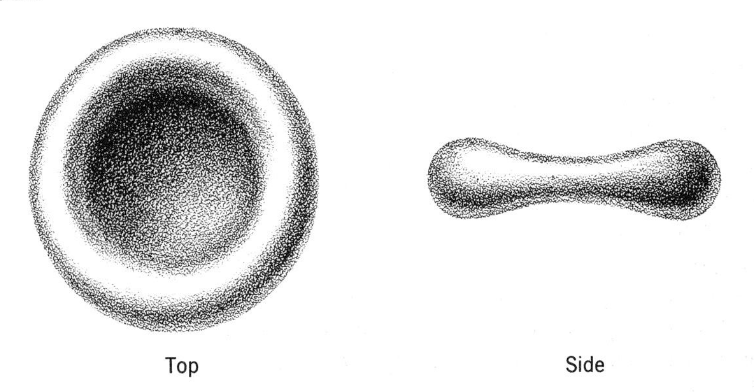

FIGURE 6.1. Structure of erythrocyte.

BLOOD CELLS

Origin and life span of erythrocytes. The manufacture of red cells is called *hemopoiesis.* In the unborn child it is carried on in special tissues of the liver, spleen, and red bone marrow, but after birth it is restricted to the red bone marrow. The stages in the development of a mature erythrocyte are shown in Fig. 3.12. They are originally nucleated but lose their nuclei while still in the bone marrow and before their release into the blood stream, but in certain *anemias* the release of cells is speeded up as if to make good the defect, and immature, nucleated red cells appear in the circulating blood.

Worn-out red blood cells become more and more fragile and eventually rupture. The debris is removed by *phagocytic* cells of the *reticulo-endothelial* system of the liver and spleen. Studies with radioactive erythrocytes have shown that their life span is about 120 days. Since the body normally contain 20,000,000,000,000 erythrocytes, this life span suggests that they must be destroyed (and produced) at the rate of 115,000,000 per *minute.*

Anemias. A deficiency of hemoglobin in the blood is called anemia, and it results in a reduced oxygen-transport capacity. Anemia occurs in several forms. It may be primarily due to inadequate hemoglobin synthesis, to a shortage of cell-stroma material which contains the hemoglobin, or both. In some types of anemia, the red cells are larger than normal but contain subnormal amounts of hemoglobin (*macrocytic hypochromic anemia*). Conversely the red cells may be small but contain extra hemoglobin (*microcytic *hyperchromic anemia*), or there may be any other combinations of deficiency.

Probably the simplest type of anemia is the *acute traumatic* anemia, which is seen after a severe hemorrhage. If an animal loses blood through a wound, it inevitably loses red cells, and these are not replaced as rapidly as are the fluid components of the blood. This type of anemia is readily determined by measuring the *hematocrit*. If some blood is centrifuged at high speed in a narrow tube closed at one end, the red cells (and the much fewer white cells) become packed in the lower part of the tube. Then, the fraction of the total volume made up by the packed cells can be determined by measuring the total length of the cells and plasma and that of the cells alone (see Fig. 6.2). In men the hematocrit is about 45%, or we may say the *hematocrit ratio* is 0.45. In women the hematocrit is normally about 40%. Figures substantially below these values are indicative of anemia.

Most forms of anemia are associated with a low hematocrit. This does not necessarily follow, however, because there may be normal numbers of cells with a subnormal hemoglobin content. For this reason, a determination of the actual hemoglobin content of the blood may be more informative than a hematocrit measurement. The intensity of the color due to the hemoglobin is compared with standards of known hemoglobin content. In more refined tests, the hemoglobin is first subjected to chemical reactions which yield a more stable colored compound which is then meas-

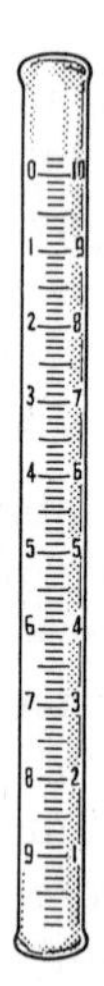

FIGURE 6.2. Winthrobe tube for hematocrit determinations.

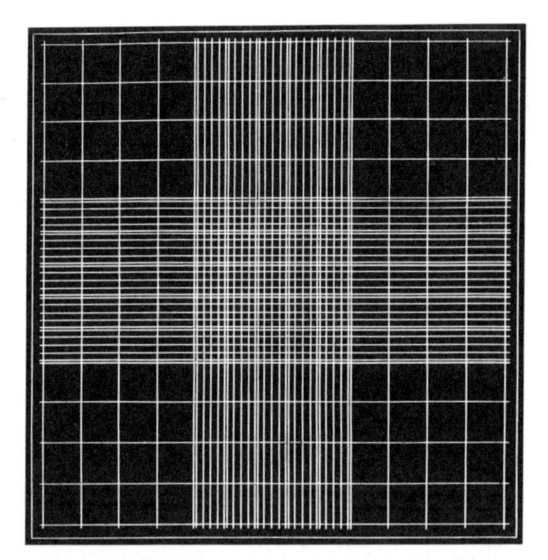

FIGURE 6.3. A hemocytometer counting chamber enlarged. Red cells are counted in the small squares in the center.

ured. The normal hemoglobin in men is 16 grams per 100 milliliters and 14 grams in women.

BLOOD COUNT

A determination of the actual number of corpuscles per unit volume is a further useful index of the state of a patient's blood. A small amount of blood is obtained in a pipette from a stab wound in the finger of the patient. The blood is diluted in standard fashion, and the dilute suspension of blood cells is placed in a special microscope slide which has a square depression of a standard volume. The slide has ruled lines which are visible under the microscope, and the number of blood cells found in the squares between the ruled lines can be related to the number of cells per cubic millimeter (Fig. 6.3). A similar technique, using a lesser dilution of the blood, is used for counting the number of white cells. The significance of the white cell count is discussed later.

Anemia can also result from the excessive destruction of red cells, but this is unusual. Far more common is an impaired hemopoiesis. Although much of the iron and protein from destroyed erythrocytes is re-used, there has to be a good supply of raw material to maintain the numbers and quality in the blood. Since the red cells are composed of protein and iron compounds, it follows that a deficiency of either in the diet will interfere with normal hemopoiesis. Milk is low in iron, and an iron-deficiency anemia is occasionally encountered among children given a largely milk diet after infancy. Other factors are equally essential, especially the vitamins B_{12} and folic acid. When one or both of these vitamins are lacking, the red cells do not mature properly and the blood contains not only immature cells but also large, irregularly shaped and fragile cells known as *macrocytes*. A dietary deficiency of vitamin B_{12} can be induced in experimental animals, but it is virtually unknown in man since the intestinal bacteria synthesize the vitamin, even if there is none in the diet. It is now known, however, that *pernicious anemia* is caused by a defective *absorption* of B_{12}. The vitamin itself is known as the *extrinsic factor*, which is only absorbed in the presence of the yet unidentified *intrinsic factor* normally present in the gastric and duodenal mucosa. Some individuals lack the intrinsic factor and so become anemic. The disease is treated by *injecting* vitamin B_{12} so that the absorption stage is by-passed.

Hemopoiesis is impaired in a number of diseases and especially by radiation damage. The hemopoietic activity in the bone marrow can be estimated by removing a small quantity by suction from a puncture in the *sternum* or *iliac crest*. The red bone marrow tends to be progressively invaded by fat cells in the adult years; the amount of hemopoietic tissue is therefore reduced, and in old age there is usually a mild anemia.

In severe anemia the hematocrit may be as low as 15%. This reduces the oxygen-carrying capacity of the blood and the tissues become chronically short of oxygen (hypoxic). The anemic patient's breathing and circulation rates are consequently increased as if in an endeavor to deliver as much oxygen as possible to the tissues. Over long periods this causes strain and enlargement of the heart.

Polycythemia is a relatively unusual condition in which the hematocrit is above normal. People who live at high altitudes (say about 12,000 feet) develop a physiological polycythemia to compensate for the lower pressure of oxygen at that altitude. Polycythemia also occurs in a disease in which hemopoiesis runs wild in a manner similar to the uncontrolled growth of tumors in other tissues. In polycythemia the hematocrit may be as high as 75%, and the blood is consequently very viscous and sluggish in circulation.

Hemolysis. As described in the section on osmosis, red cells are sensitive to osmotic changes and they will rupture in hypotonic solutions as a result of the excessive influx of water. The rupturing of red cells and the consequent release of hemoglobin is called *hemolysis.* The red cells are not uniformly sensitive to rupture in hypotonic solutions. A sodium chloride solution of 0.9% strength is *isotonic* to red cells and they remain intact in such a solution; on the other hand, they will all hemolyze in distilled water. If some blood cells are added to a series of sodium chloride solutions at intermediate concentrations of 0.1%, 0.2%, 0.3%, etc., up to 0.9%, it is found that there is a gradation in the proportion of cells hemolyzed, because some cells are more *fragile* than others. All of them are hemolyzed at the lowest concentrations and none of them at the highest. Excessive fragility is associated with some diseases and so this forms the basis of a simple diagnostic test.

Hemolysis may happen from a variety of causes, and, as might be expected, osmotic hemolysis is unusual in the living animal since the osmotic pressure of the plasma is so rigidly controlled. It can happen to some extent if water or hypotonic solutions are injected, however, and for this reason glucose and other solutions injected for therapeutic reasons are always in concentrations isotonic to the cells. A more usual cause of intravascular hemolysis is the chemical destruction of the red cell membranes. Compounds which disrupt these lipoid membranes are known as *lysins,* or specifically, as *hemolysins.* They are found in toxins liberated by some bacteria and in some snake venoms. The hemolysis is then one of the evil sequels to the bacterial infection or the snake bite. In addition, an animal's serum may contain a compound which does not affect its own red cells but will cause hemolysis if injected into another animal. Blood transfusions from one species to another, therefore are seldom successful, and even within a species, one individual's serum may be incompatible with another's red cells. This is the reason for the typing of blood for human transfusions.

Leukocytes. The white cells of the blood are leukocytes, which are concerned with the body's defense against infection. There are five types of white cells in normal blood: *polymorphonuclear basophils, polymorphonuclear eosinophils, polymorphonuclear neutrophils, monocytes,* and *lymphocytes.* The *platelets* (*thrombocytes*) are also white cells, but their functions are so different that they will be discussed separately. The five types listed are illustrated in Figure 6.4. The polymorphonuclear types are so called because of their irregularly shaped nuclei and are distinguished from each other by their staining reactions. Their cytoplasm is granular and they are therefore referred to as *granulocytes.*

The granulocytes and monocytes are *phagocytes.* That is, they combat invading pathogens by eating them. The lymphocytes' role is the production of the *immune bodies* which combat invading organisms in a different manner, to be described in a later section. This function of the lymphocytes is shared by the *lymph glands* in which the lymphocytes are produced. The granulocytes and monocytes originate in the bone marrow, as do red cells.

Normally the adult has about 7,000 white cells per cubic millimeter of blood and these are normally in the following proportions in each 100 cells:

Polymorphonuclear neutrophils	63.0
Polymorphonuclear eosinophils	1.6
Polymorphonuclear basophils	0.4
Monocytes	5.0
Lymphocytes	30.0

Children have a higher proportion of lymphocytes as well as a higher total white cell count. In bacterial infections the numbers increase to as high as 30,000 per cubic millimeter with most of the increase in the neutrophils. In more chronic infections the lymphocyte numbers increase. Therefore, it is useful to make a *differential count* of the white cells when infections are suspected and so to determine not only the total number but also the composition of the leukocyte population.

The granulocytes and monocytes do not stay in the blood stream but can actually squeeze through pores in the blood vessels into the intercellular spaces where they move around by amoeboid motion. This enables them to attack invading pathogens wherever they may enter the tissues.

Inflammation. Damage to tissues, whether by bacteria or a physical blow, causes the death of some cells. These cells release a compound which makes the local blood capillaries more permeable, so that plasma leaks into the interstitial fluid. The plasma clots and blocks up the intersteces between the cells. At the same time, the release of another compound attracts leukocytes to the scene and they phagocytize the invading organisms and the debris from the dead cells.

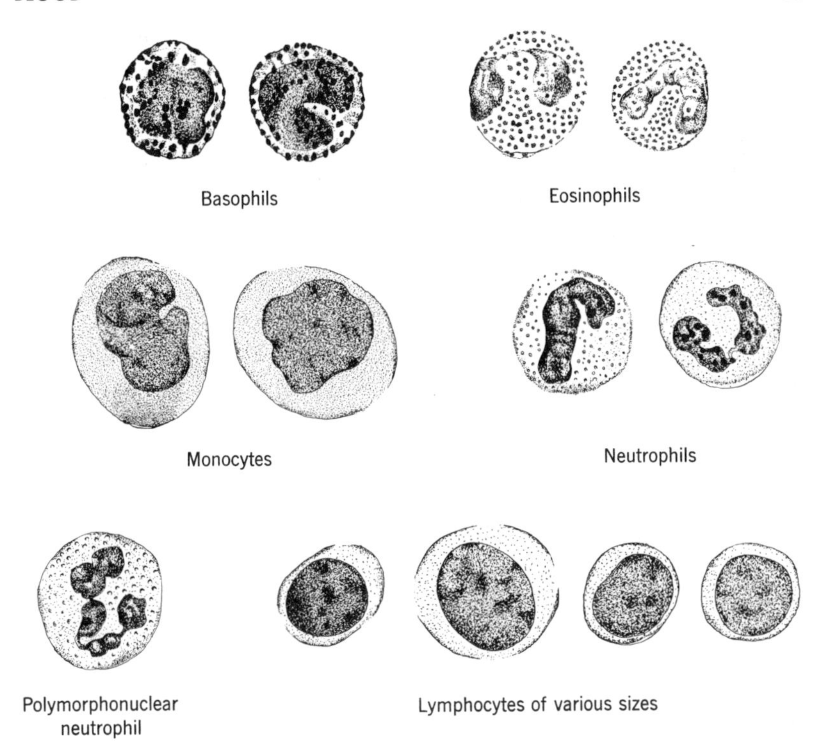

FIGURE 6.4. Leukocytes (Wright's stain).

The net effect of the interstitial clotting and the congregation of leukocytes is to "wall off" the site of infection and to impede the transport of bacteria away from their point of entry. This is one of the reasons why it is inadvisable to squeeze pimples, as the mechanical pressure spreads the infective organisms more widely into the surrounding tissue fluids. In the course of phagocytizing bacteria, the leukocytes themselves are killed and the accumulation of dead, liquefied leukocytes is the *pus* found in infected tissues.

Agranulocytosis. Serious consequences can arise from a failure in leukocyte production, because the body is then unprotected against the bacterial infections to which it is continually exposed. Granulocyte production in the bone marrow is impaired by exposure to several poisonous industrial chemicals containing benzene or anthracene, or even by some otherwise useful drugs. Radiation damage from an atom bomb or overexposure to X-rays will also cause this condition of agranulocytosis with nearly always fatal results.

Leukemia is a disease in which the total white cell count may rise as high as 800,000 per cubic millimeter. This enormous production is due to tumors in the bone marrow or lymphoid tissues in which the leukocytes are variously produced. Among laymen this disease is aptly called "cancer of the blood." It is nearly always fatal within a few months.

PLASMA

Important functions of the plasma, independent of its contributing to the interstitial fluid, are those concerned with the proteins that remain in the vascular system. About 7% of the plasma is protein, of which 3.5–5.0% is *albumin* and the remainder *globulin.*

Albumin. This protein is synthesized in the liver. Its great importance is in its contribution to the plasma osmotic pressure. Since it cannot pass out of the capillaries (in any quantity), the concentrations of the diffusible ions in the plasma are slightly less than those in the interstitial fluid when in osmotic equilibrium. At the proximal end of the capillary (that nearer the heart), the blood pressure is sufficient to cause an ultrafiltration of plasma into the intersteces of the tissues, despite the osmotic equality. At the distal end of the capillary, the blood pressure is lower and the plasma now has a higher osmotic pressure because some of the water and salts have been lost and the relative protein content is greater. This causes an influx of water and salts from the interstitial fluid to restore osmotic equilibrium (see Chapter 4). Furthermore, the ultrafiltration of fluid into the interstitial spaces sets up a "tissue pressure," which helps to force fluid back into the distal ends of the capillaries.

Solutions of materials with large particles are called *colloidal solutions,* and protein solutions like plasma are in this category. The difference of osmotic pressure across the capillary membrane at the distal end is due to the impermeability of the membrane to the large, colloidal protein particles. This difference is therefore called the *colloid osmotic* pressure as distinct from the *crystalloid osmotic pressure* (due to the dissolved salts, etc.). The colloid osmotic pressure is now usually referred to as the *oncotic pressure.*

Figure 6.5 shows the quantitative relationships in the exchange across the capillary walls. At the arterial end of the capillary, the oncotic pressure of the plasma is 36 millimeters of H_2O, and the hydrostatic pressure due to the heart's action may be taken as 45 millimeters of H_2O. The difference, 9 millimeters of H_2O, is the

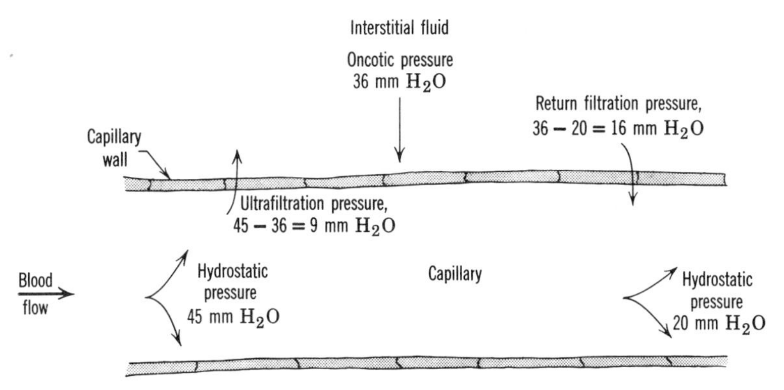

FIGURE 6.5. Diagram of Starling Principle.

ultrafiltration pressure. At the venous end of the capillary, the hydrostatic pressure drops to 20 millimeters of H_2O or less, and the oncotic pressure remains the same (or even slightly higher). Therefore, the pressure for fluid *return* is $36 - 20 = 16$ millimeters of H_2O.

The albumin therefore serves a vital role in the exchange and restoration of the internal environment. In some conditions the albumin concentration is lowered, and the return of fluid from the intersteces is reduced. As a result, the interstitial fluid volume is increased and the tissues become *edematous*. This occurs in a chronic deficiency of dietary protein or in renal diseases which cause loss of protein in the urine.

Globulin. By chemical methods, by ultracentrifugation, and most of all by electrophoresis, the individual components of the albumin and globulin of the plasma may be separated. These methods yield several globulins, among which are the alpha-1, alpha-2, beta, gamma, and fibrinogen. With the exception of fibrinogen, the globulins are the proteins involved in the immune reactions in the body's defenses against infection. Fibrinogen, although strictly a globulin, is concerned with the coagulation of the blood and so may be considered separately.

Coagulation of the Blood

If there were no blood clotting, even the smallest wound would permit a continual leakage of blood with fatal results. In normal individuals, minor cuts bleed for only a few minutes and then a clot closes off the end of the cut vessel. The clot is formed when the soluble fibrinogen is converted to insoluble *fibrin*. In this reaction the fibrinogen molecules join together to make long complex chains which are less soluble; this is called *polymerization*. The polymerization of fibrinogen is catalyzed by the enzyme *thrombin*. Ordinarily the concentration of thrombin in the blood is very low, or otherwise there would be indiscriminate clotting in the circulating blood. It is present mainly in the form of an inactive precursor called *prothrombin,* and the conversion (or *activation*) occurs only when certain conditions are met. Studies with isolated prothrombin have shown that its activation (and therefore the coagulation of blood) needs the presence of calcium ions, a "plasma factor," a "platelet factor," and *thromboplastin.*

There is no coagulation if the calcium ions are removed from blood, and this can be accomplished by adding oxalate, forming insoluble calcium oxalate which ties up the calcium ions. The addition of citrate has a similar effect except that the calcium citrate remains in solution. Since it does not dissociate in solution, however, there are no free calcium *ions*. This is the method used to prevent clotting in blood collected from a donor for subsequent transfusions.

The platelet and plasma factors are ill-defined compounds still under intensive study. Calcium ions, the plasma factor, and the platelet factor are normally always present in the blood. It is the thromboplastin production, therefore, which is the trigger mechanism in blood clotting. When cells are ruptured or damaged, the phospholipids in the membranes are released. Thromboplastin is made up of some of these phospholipids. Damage to any cells will cause the release of thromboplastin, but the thrombocytes (platelets) are particularly susceptible to rupture. And so, although they do not contain more potential thromboplastin than other cells, their ease of rupture makes them the most important source. The platelets are negatively charged and are normally held away from the smooth endothelial lining of blood vessels by like charges. However, if the wall of the blood vessel is damaged (as by a cut) the repelling charge fails and the platelets stick to the wall causing their immediate rupture and the release of thromboplastin. Then, if the other factors are present, coagulation ensues. These events are summarized in Fig. 6.6.

Formation of a clot. At first, the fibrin appears as a network of stringy fibers in which red cells and platelets are entrapped. The platelets release more thromboplastin, which causes more fibrin formation, and the clot gets bigger. After about half an hour, the fibrin of the clot retracts and squeezes out the *defibrinated plasma,*

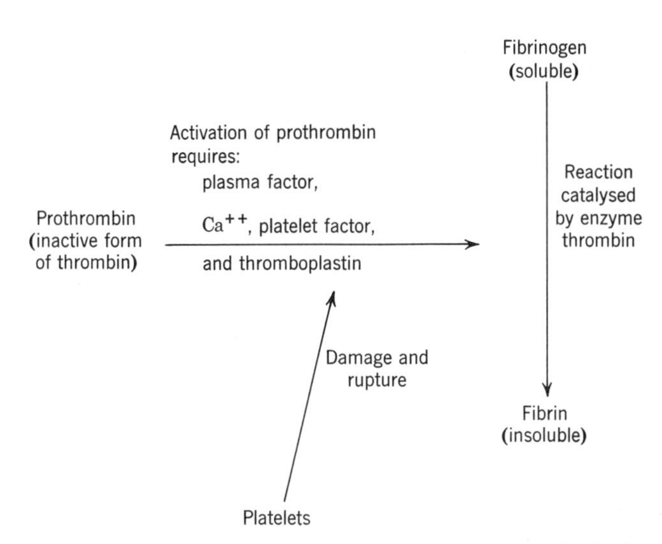

FIGURE 6.6. Blood clotting. Production of thromboplastin subsequent to platelet damage is the trigger for activation of prothrombin to thrombin, which catalyses conversion of soluble fibrinogen to insoluble fibrin.

which is known as *serum*. As the fibrin strands are attached to the walls of the blocked vessel, their contraction constricts the open end.

Inhibition of coagulation. Blood samples removed from the body may be prevented from clotting by removal of the calcium ions. The body itself has other mechanisms to inhibit clotting; not only to prevent unwanted intravascular clotting but also to prevent a desirable and useful clot from extending indefinitely. Plasma contains an enzyme, *antithrombin,* which destroys the small amount of thrombin which is always being produced. *Heparin,* a compound found in most cells, is another safety measure since it can block the transformation of prothrombin to thrombin. Heparin is extracted from the lungs of slaughtered animals and used to prevent clotting in situations where oxalate or citrate are not suitable. For example, it is injected into experimental animals in which operative procedures are likely to cause clotting, and into patients who are subject to the intravascular clots known as *thrombi* (plural of *thrombus*).

Intravascular clotting can happen if the platelets encounter rough linings in the blood vessels. This is seen even outside the body, where blood clots much more readily in a rough or dirty test tube than in a very clean one, or in one lined with a layer of oil or silicone. The inside of blood vessels may be rough where there are deposits of fatty material and calcified plaques, as in *arteriosclerosis*. In patients with this condition, the dangers of their partially occluded blood vessels can be multiplied by the formation of clots which can block them off altogether. The blockage of an important vessel in a limb can lead to *gangrene* (death of the tissue) and loss of the limb. If the thrombus forms in one of the blood vessels supplying the heart muscle itself, we speak of a *coronary thrombosis,* which is a common form of "heart attack." If the thrombus forms in the brain, we have one form of cerebro-vascular accident or "stroke."

Slowly moving blood is much more likely to clot than blood in rapid circulation. This can happen if a limb is held still in a doubled-up position or pressed tightly against the edge of a chair in such a way that the circulation is obstructed for some time. Large clots have formed in the legs of otherwise normal people who have sat with their legs doubled beneath them while enthralled with a television program. A long period of driving without moving the legs is another potentially dangerous situation in our modern life. This condition has been aptly named a "thruway thrombosis."

The danger of thrombi in limbs lies not only in the chances of gangrene but in the possibility that the clot will break loose before it is dissolved. A clot which breaks loose from its site of formation is called an *embolus*. A large clot originating in the veins of the thigh moves to the heart when it breaks loose and is pumped out again into the pulmonary circulation. If the clot is large enough, it can block both pulmonary arteries with immediately fatal results. Even smaller clots will block arteries when they reach narrow enough vessels, and the subsequent occlusion of blood flow causes the death of some of the lung tissue. Small emboli can also lodge in the coronary or cerebral blood vessels with the same disastrous effects as the formation of thrombi.

IMMUNITY

The body is continually invaded by microorganisms which enter through the respiratory passages and through small wounds in the skin. In most cases the invasions are harmless since they are immediately dealt with by the phagocytic leukocytes and by the *immune reactions*.

Invading organisms and the toxins they produce are made of proteins. The danger of a foreign protein lies presumably in the reactive chemical groups on its surface which can combine with groups in the body and interfere with normal processes. This can be prevented if the invading protein is "smothered" by another protein's combining with it. Such a combination is the essence of the immune reaction. A foreign protein that will evoke the immune reaction is called an *antigen,* and the native protein that reacts with it is an *antibody.* The antibodies are part of the globulin fraction of the blood and are produced by the lymphoid tissues and lymphocytes. Each antigen has a specific antibody which can react with it alone. The specificity of the antibodies is probably due to electrostatic charges on their surfaces which are geometrically arranged to match charges of the opposite signs on the surfaces of the antigens. By the electrostatic attraction, the antibodies are firmly attached to the antigens, rendering them harmless, and the combination is then destroyed by the reticulo-endothelial system or by lysis of the invading organism.

We have immunity to some organisms because at birth our blood already contains suitable antibodies. This is known as *natural immunity.* For other organisms, even though there is no native antibody, an exposure to them will cause the production of antibodies leading to an *acquired immunity.* This is why one infection with measles, for example, usually conveys a life-long immunity to that organism.

Antibody production and consequent immunity to an organism can be induced artificially as a protective measure. For example, an infection with the harmless cow-pox (*vaccinia*) causes the production of antibodies which will also protect against the virulent small-pox virus, presumably because of a similarity in the electrostatic configuration in the two viruses. The term "vaccination" is of course derived from this practice. Other *live virus vaccines* are available to induce immunity against yellow fever and other diseases. Because the protein constituent of the invading organism is involved in the immune reaction, antibody formation can be induced by the intact, dead organism as well as the live. Therefore *killed virus vaccines* are successfully employed to protect against several diseases. Similarly, the toxins produced by the disease organism will also cause the production of specific antibodies even without the presence of the organism itself. The separated toxins may be subjected to a mild chemical treatment which makes them harmless but leaves them sufficiently intact to induce the appropriate immunity. Such a *toxoid* is used against diphtheria, which used to be a widespread disease.

The antibody production reaches a peak about a week after the infection or vaccination and then declines steadily, so that after several years it may be barely discernible. However, the appropriate pattern is in some way imposed on the cells of the lymphoid tissue so that on a subsequent re-exposure to the foreign protein, large quantities of antibodies are manufactured immediately without the delay in production seen after the initial exposure (see Fig. 6.7).

A short-lived immunity can be provided sometimes by transfusing blood from an immune individual to another who is not immune. The antibodies in the donor's blood give a *passive immunity* to the recipient until they are destroyed in the course of the normal turnover of the body protein. This practice is useful for protecting an infant who is exposed to a measles infection in his older siblings, or in similar situations.

Allergy. Normally the antigen-antibody reaction takes place in the blood stream, and the debris is removed by the reticulo-endothelial system. If the acquired immunity is only weak, however, a foreign protein may penetrate the interstitial spaces before it comes into contact with an antibody. When the immune reaction occurs in the tissues themselves, the surrounding cells are irritated or even destroyed by the side effects. This can happen when the concentration of antibodies in the plasma is not high enough to react with all of the foreign protein immediately.

Allergies result from weak immunities and the reactions in the tissue give rise to the symptoms of the allergy. In infancy there may be allergic reaction to food proteins because the child has not yet developed the very strong immunity to them possessed by adults. More commonly though, allergic reactions are due to

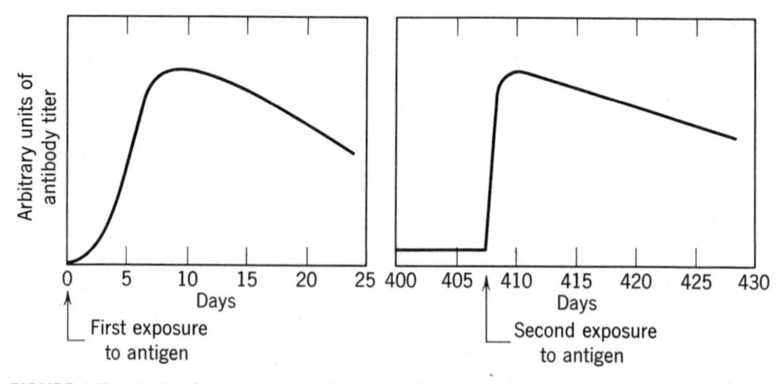

FIGURE 6.7. Antibody production following a first and a second exposure to an antigen.

materials which are sufficiently protein-like to induce some antibody formation but not enough to induce full immunity. These materials are known as *allergens*. There are tremendous numbers of potential allergens, including dust and flower pollens. It is not known why some persons are more prone to allergies than others.

The deletorious effects of the immune reaction in the tissue spaces are of several kinds. The damage to the cells results in the release of *histamine* which is a potent vasodilator and causes redness and swelling in the tissues. This is seen in *urticaria* (hives), and in the irritated membranes of eyes and sinuses in hay-fever. If the histamine release is excessive, a sudden vasodilation can occur throughout the body with the disastrous lowering of blood pressure known as *shock*. This massive response to an allergen is known as *anaphylaxis*. *Anaphylactic shock* is a medical emergency sometimes encountered when a drug or vaccine vehicle is injected into a patient with a weak immunity to that material. In other types of allergy, the immune reaction causes constriction of adjacent smooth muscle fibers, as is seen in *asthma*. The smooth muscle of the *bronchioles* in the respiratory system is constricted to the point where the patient has difficulty in getting enough air in and out of the lungs.

The symptoms of the hay fever type of allergy may be relieved by the antihistaminic drugs and those of asthma by epinephrine. More positive treatment is possible if the allergen itself can be identified by skin tests. If so, small amounts of the allergen can be injected at short intervals in an effort to stimulate antibody formation to the point where immunity is strong enough to prevent the reaction in the tissue.

BLOOD GROUPS

The proteins in the plasma and red cells of each individual are slightly different from those of everyone else. Therefore a transfusion of blood from one person to another can cause an immune-type reaction. From quite soon after birth, an infant's plasma develops antibody-like proteins called *agglutins*. These will react specifically with proteins in the red cells from some other blood and the resulting immune reaction causes clumping (*agglutination*) and *lysis* of the red cells. The red cell proteins involved are the *agglutinogens*. Fortunately, the agglutin-agglutinogen reaction is not universal and it is found that human blood falls into four main types according to their content of agglutins and agglutinogens.

Type A blood contains A-agglutinogens and β-agglutins. Type

B blood contains B-agglutinogens and α-agglutins. If some type B blood cells are introduced by transfusion into type A plasma, then the β-agglutins of the A-type plasma react with the B-agglutinogens causing clumping of the red cells. If the reverse kind of transfusion is made, similar results follow.

Type O blood contains neither A nor B agglutinogens. This means that the transfusion of red cells from this blood type into either A or B types will cause no reaction since there are no agglutinogens to react with the agglutins in the recipients. For this reason, type O individuals are *universal donors*. However, this blood contains both α and β agglutins and the transfusion of either type A or B blood *into* a type O individual will cause agglutination.

The fourth blood type is AB which may be considered as the reverse of type O since it contains both A and B agglutinogens and neither α nor β agglutins. Persons of this blood type can receive a transfusion of any other type since there are no agglutins to react with whatever agglutinogen may be introduced. They are therefore known as *universal recipients*. On the other hand, this type of blood contains both A and B agglutinogens and so will react with any other kind of plasma if given as a transfusion.

The characteristics of these blood types can be set out in a table.

TABLE 6.2 Blood Types

% of Population	Blood Type	Agglutin	Agglutinogen	Remarks
41	A	β	A	
9	B	α	B	
47	O	α and β	—	Universal donor
3	AB	—	A and B	Universal recipient

It is clearly necessary then that blood for transfusion shall match the blood of the recipient. This may be checked by crossmatching the bloods. The serum of one is mixed with red cells of the other and then the procedure is reversed. If no clumping occurs in either procedure, the bloods are of the same type. Alternatively, the blood types of the donor and recipient may be determined in advance, and all blood bank blood is routinely typed before storage. This is done by mixing dilute specimens of blood with two previously prepared sera containing strong α and β agglutins. Such sera are commercially available and are called Anti-A and Anti-B. A small amount of blood from the finger is diluted with an equal volume of saline and two separate drops are placed on a microscope slide.

Then a drop of Anti-A serum is added to one drop and Anti-B is added to the other. Where agglutination occurs it is clearly visible to the naked eye. If the drop receiving Anti-A clumps, the blood is type A and if the other drop clumps, it is type B. If neither drop agglutinates it is type O; if both agglutinate, it is type AB.

TABLE 6.3 Reaction to Testing Sera

Blood type	Anti-A serum (α-agglutin)	Anti-B serum (β-agglutin)
A	+	–
B	–	+
O	–	–
AB	+	+

(+ = agglutination)

If blood of an incompatible type is administered in error, the red cells agglutinate and may lyse immediately. The clumps of cells can block small blood vessels, and the hemoglobin released from the cells can plug the renal tubules, leading to *kidney failure*. These are potentially fatal consequences and must be avoided at all costs.

Rh factor. Besides the four principal blood types already described, there are many more antigen-antibody groups. One of the more important from the clinical standpoint is the Rh factor, so-called because it was first observed in Rhesus monkeys. The Rh factor is an agglutinogen present in the blood of about 85% of the white population. Those with the agglutinogen are said to be *Rh positive*. *Rh negative* individuals do not have a *native* agglutin which will react with the factor in Rh positive blood, but must be first sensitized. A first exposure of an Rh negative individual to the Rh factor may come through a transfusion. The first transfusion causes no visible reaction but the agglutin formation is induced in the same way that antibody formation is induced by infection or vaccination. Subsequent transfusions of Rh positive blood will then cause agglutination and lysis of the red cells.

Another important source of sensitization of an Rh negative person is sometimes seen in pregnancy. The Rh factor is a dominant genetic trait. In a grossly oversimplified fashion it may be said that when one parent is Rh positive and the other is Rh negative, they are more likely to have Rh positive children. When the mother is the Rh negative parent, she can be exposed to the Rh factor by carrying an Rh positive fetus. This seldom has any overt effect during the first and even the second pregnancies, but during these times of exposure the agglutin content is increased in the mother's blood. By the third pregnancy the agglutin content is such that some is able to diffuse across the placental membrane into the fetal blood, and if the fetus is again Rh positive, the agglutins will react with the agglutinogens and cause hemolysis. In consequence, the baby is born in a very anemic condition, and the hemolysis continues so long as the mother's agglutins continue to circulate. This condition is known as *erythroblastosis fetalis.* It is now almost routinely treated by completely replacing the infant's blood by an *exchange transfusion* soon after birth.

Blood Bank

It is almost self-evident that the best treatment for a serious blood loss is to replace it as soon as possible. There are many clinical situations in which a transfusion of blood is positively life-saving. Therefore, blood for transfusion is available at a few minutes' notice in most hospitals. The blood is collected from donors in sterile bottles containing acid citrate and dextrose (glucose). It is commonly known as ACD blood (acid citrate dextrose). The citrate is added to tie up the calcium ions to prevent coagulation. However, the blood is always passed through a coarse filter in the administration apparatus when transfused just in case any clots have formed.

The citrate is in acid form because blood is better preserved in a mildly acid condition. On transfusion, the citrate is rapidly metabolized by the recipient thereby correcting the acidity and freeing the calcium ions. The dextrose is for the nutrition of the red cells since they are alive and continue to respire even in cold storage. Despite this precaution, the red cells deteriorate steadily in storage and blood is not kept for transfusions after about three weeks. In larger centers, the plasma is then separated from outdated blood and preserved by freeze drying (*lyophilization*). Such plasma will keep almost indefinitely without refrigeration and can be reconstituted by merely adding the right amount of sterile water. Although plasma transfusions are not as good as whole blood transfusions in many clinical situations, the convenience of dried plasma makes it ideal for remote situations and the battlefield.

REFERENCES

1. Burnet, Sir Macfarlane, "How antibodies are made," *Sci. American* **191** (5), 74, November 1954.

2. Henderson, Lawrence J., *Blood,* Yale U. P., New Haven, 1928.
3. Kaufman, William, "Asthma," *Sci. American* **187** (2), 28, August 1952.
4. Pappenheimer, A. M. Jr., "The diphtheria toxin," *Sci. American* **187** (4), 32, October 1952.
5. Ponder, Eric, "The red blood cell," *Sci. American* **196** (1), 95, January 1957.
6. Shapiro, Shepard, "The control of blood clots," *Sci. American* **184** (3), 18, March 1951.
7. Singer, S. J., "The specificity of antibodies," *Sci. American* **197** (4), 99, October 1957.
8. Surgenor, Douglas M., "Blood," *Sci. American* **190** (2), 54, February 1954.
9. Wiener, Alexander S., "Parentage and blood groups," *Sci. American* **191** (1), 78, July 1954.
10. Wood, Barry W. Jr., "White blood cells vs. bacteria," *Sci. American* **184** (2) 48, February 1951.

PART 3

ORGANS REGULATING AND DISTRIBUTING THE INTERNAL ENVIRONMENT

CIRCULATION

7

The oxygen and nutrients in the internal environment would be rapidly used up if they were not replenished by the turnover of the interstitial fluid already described. It will be recalled that this is accomplished by the circulation of the blood, which permits the exchange of fluid (and dissolved materials) in the capillaries. The blood pressure provides the driving force for the ultrafiltration of fluid into the interstitial spaces, and the oncotic pressure of the plasma proteins causes a similar return of fluid at the distal ends of the capillaries. For this exchange to maintain the composition of the internal environment effectively, not only must the plasma composition be well regulated by the lungs, kidney, liver, etc., but the plasma must be delivered to the tissues in sufficient *volume* to meet their oxygen and nutrient requirements and at sufficient *pressure* to ensure the all-important ultrafiltration.

The consumption of oxygen and nutrients and the production of waste materials by the tissues increases, for example, in exercise. Therefore, a suitable composition of the internal environment can be maintained only by increasing its rate of turnover, and hence the pressure and rate of delivery of the blood is varied according to the metabolic rate of the tissues. Furthermore, it follows that the regulatory organs cannot correct the composition of the plasma unless it passes through them at a sufficiently rapid rate, which will of course also need to vary with the level of metabolic activity. The whole of the cardiovascular system, in all its complexities of form and function, is devoted to maintaining the fluid and solute exchange in the capillary beds. This should be borne in mind throughout this discussion of the circulation.

It is appropriate to begin with a short description of the general plan of the circulation, shown diagrammatically in Fig. 7.1.

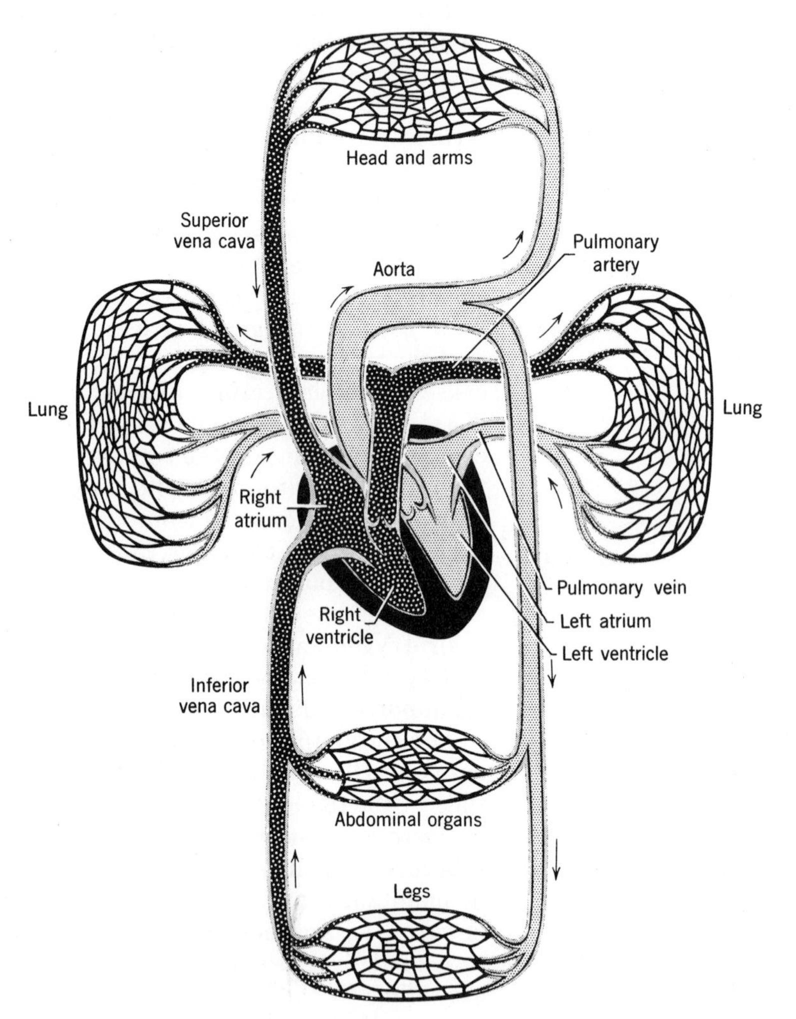

FIGURE 7.1. Diagram of the circulation.

The heart is a muscular organ which can be thought of as two separate pumps in parallel, although they are included in one structure. The *right* heart receives blood from the veins draining the tissues and pumps it via the *pulmonary arteries* to the lungs. Here it passes through capillaries in which it is oxygenated and the carbon dioxide removed (see Chapter 8). The *pulmonary veins* return the blood to the *left* heart which pumps it out again through the *aorta,* from which arise the various *systemic* arteries distributing blood to the other tissues. The *coronary* arteries are small vessels branching from the aorta immediately above the heart. They are important out of all proportion to their relatively small size; they re-enter the heart to supply the *myocardium* (heart muscle) itself with blood.

In all tissues, the arteries subdivide into successively smaller vessels and finally into the capillaries. After passage through the capillaries, the blood is collected in veins of increasing size and eventually reaches the heart again via the two *venae cavae;* one from the anterior (or upper) regions, and the other from posterior (or lower regions). Through these veins, the blood enters the right heart and is recirculated through the lungs. A vessel carrying blood *from* the heart is an *artery* and one carrying blood *to* the heart is a *vein.* The systemic arteries contain oxygenated blood which is bright red in color and often called *arterial* blood; *venous* blood is much darker in color. It is sometimes confusing at first to realize that the pulmonary arteries carry venous blood that has yet to be oxygenated; similarly, the pulmonary veins carry arterial blood. The names of the principal blood vessels are given in the anatomical drawings (Figs. 7.2 and 7.3), and they should be learned.

We see, therefore, that there are two parallel circulations; the *pulmonary* circulation for gas exchange in the lungs, and the *systemic* circulation for maintaining a constant internal environment in the other tissues. The right side of the heart is the pump for the pulmonary circulation and the left for the systemic. Note that *all* the blood goes through *both* systems.

THE HEART

Structure. The structure of the heart is shown in Figs. 7.1 and 7.4. It is conventional to draw the heart from a front view so that the heart's right side is to the viewer's left, as it would be looking into the chest of a man from the front.

Mammalian hearts have four chambers, two on each side. The

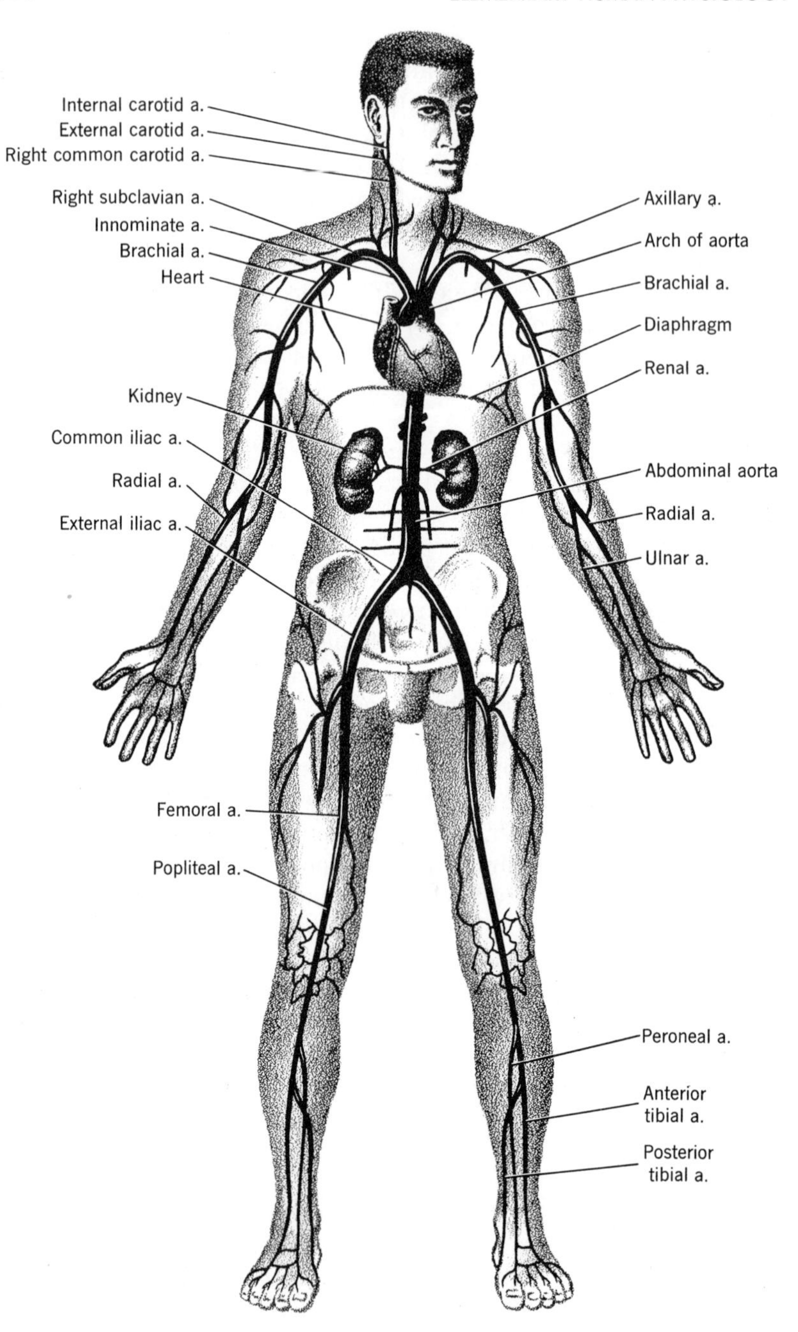

FIGURE 7.2. The principal arteries.

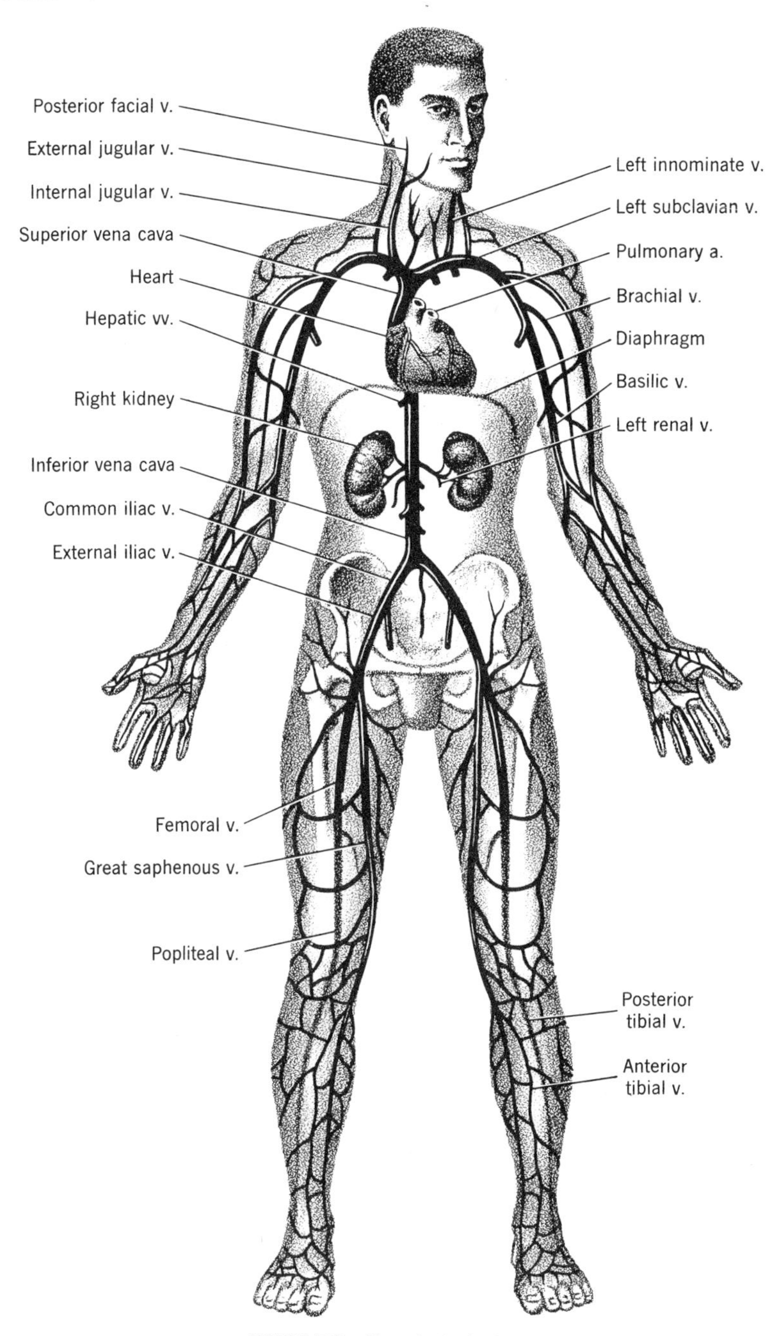

FIGURE 7.3. The principal veins.

smaller, upper chambers are the *atria* (singular, atrium) also referred to inaccurately as *auricles.* They receive the blood from the venae cavae and pulmonary veins respectively. They open into the lower chambers, the larger and more muscular *ventricles,* which provide the propulsive force to circulate the blood. The right atrium connects only with the right ventricle and the left atrium only with the left ventricle; the two sides are completely separated. In frogs and most other cold-blooded vertebrates, there is a single ventricle or an incomplete separation into two ventricles.

In mammals, each side of the heart pumps the same amount of blood, but the left side has to pump it through the systemic vessels in which the total resistance to flow is much greater than in the shorter passage through the lungs. Therefore, the left ventricle has to develop a greater pressure than the right ventricle, and it is a correspondingly larger and more powerful muscle.

The cardiac cycle. The contraction of the heart is known as *systole* and the relaxation as *diastole.* We can consider the events of the heart beat starting from diastole, when the muscle is relaxed and relatively flaccid. Venous blood from the anterior and posterior venae cavae pours into the right atrium, and at the same time, oxygenated blood from the pulmonary veins pours into the left atrium. Meanwhile, the valves between the atria and the ventricles on each side are hanging open so that the blood is able to flow right through the atria into the ventricles. Then, the atria contract and squeeze their blood content into the ventricles, which fills the ventricles more completely than they would be by passive inflow of blood. Very shortly after the atria contract, the ventricles also contract. The rising pressure closes the atrioventricular valves through which the last blood has just been squeezed by atrial contraction, and as the pressure rises still further, it opens the valves on the pulmonary arteries and the aorta, so that the blood is driven out into the circulation. When contraction is completed the ventricles relax. The blood in the aorta is under high pressure and the falling pressure in the ventricles could cause it to reflux, if it were not for the aortic and pulmonic valves which prevent reverse flow. (See Fig. 7.4.)

Even during ventricular contraction, the atria once more relax and venous blood starts to fill them. When the ventricles are completely relaxed, the atrioventricular valves open again, blood pours through to refill the ventricles and the next cycle begins. A common misconception is that the ventricles are empty of blood

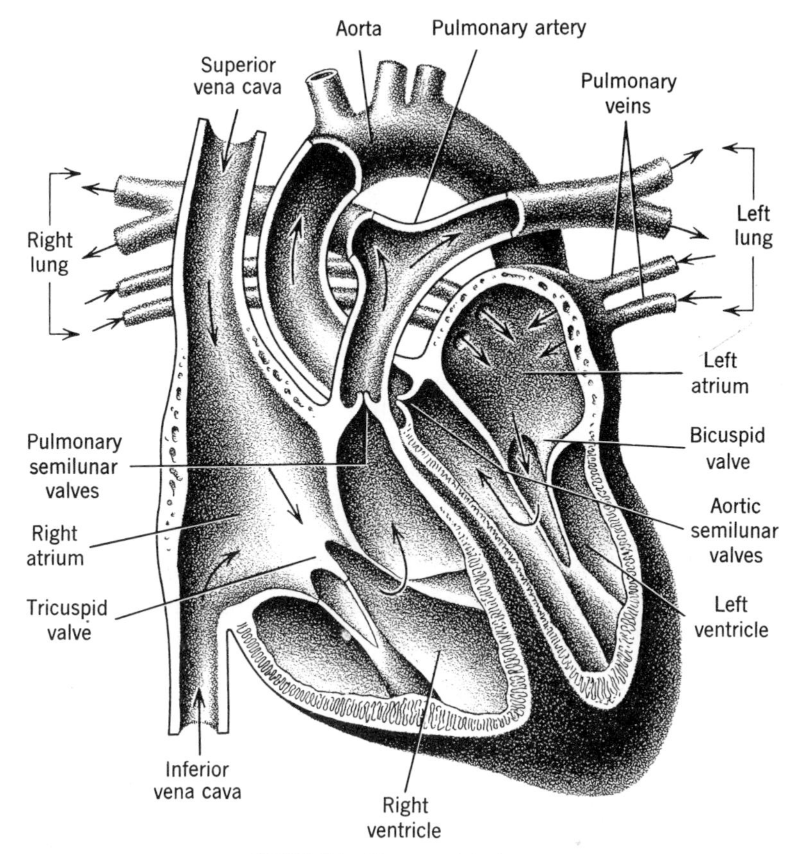

FIGURE 7.4. Diagram of the heart.

until the atria contract and drive blood into them. In fact, of course, the ventricles fill passively at the same time as the atria.

The technical names of the valves are given in the diagram of Fig. 7.4, and these should be remembered. Attention is drawn also to the *chordae tendinae.* These are strings attached to the atrioventricular valves that limit the movement of the valve-leaves so that they cannot be everted by being blown right through into the atria when the ventricles contract.

Atrial contraction is less efficient than ventricular contraction because there are no valves at the entries of the pulmonic veins or the venae cavae. Therefore, at each atrial contraction, a proportion of the blood is regurgitated into the veins rather than forced usefully into the ventricles. The surge of blood into the anterior vena cava can be seen as a *venous pulse* in the neck.

Heart sounds. The noise of the beating heart must have been apparent even to primitive man, and probably the folklore concept of the heart as the seat of the emotions stems from the awareness of changes in heart beat in excitement or fear.

The present-day physician can derive a surprising amount of information by carefully listening to the heart sounds with a *stethoscope.* When this instrument is placed on the chest wall over the heart, a regular "lub-dub," is heard. The second sound follows closely on the first and then there is a rather longer pause before the first is repeated. The first sound comes at the time of ventricular systole and the second at diastole, but the exact cause of these low pitched sounds is subject to some debate. It seems most likely that they result from the vibrations set up in the heart and vessels by the closure of the valves; the first sound being due to the closure of the atrioventricular valves (in systole) and the second to the closure of the semilunar valves (in diastole). Defects in the valves can cause excessive turbulence or regurgitation of the blood, which are audible through the stethoscope as *extra sounds.* The nature of the defects is described in a later section (p. 139).

Rhythmicity of the heart. Unlike skeletal muscle, the heart does not need nervous impulses to start each contraction. It is said to possess *intrinsic rhythmicity,* and this is seen in the continued beating of a heart quickly removed from an animal. As will be described later however, the intrinsic rhythmicity is *modified* by nervous control.

The contraction of the heart is due to a wave of depolarization spreading from a small area of specialized tissue in the right atrium, near the entry of the venae cavae. This is the *sino-atrial node,* the principal *pace maker* of the heart. It depolarizes spontaneously at regular intervals. The fibers of cardiac muscle are branching and the branches unite each fiber with others (see Fig. 7.5). Because of this *syncitial* arrangement, the wave of excitation passes through the atrial muscle to the junction with the ventricles,

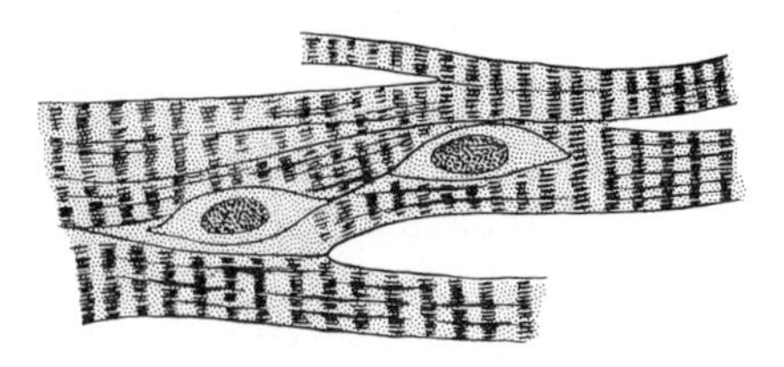

FIGURE 7.5. Cardiac Muscle, showing syncitial arrangement of striated fibers.

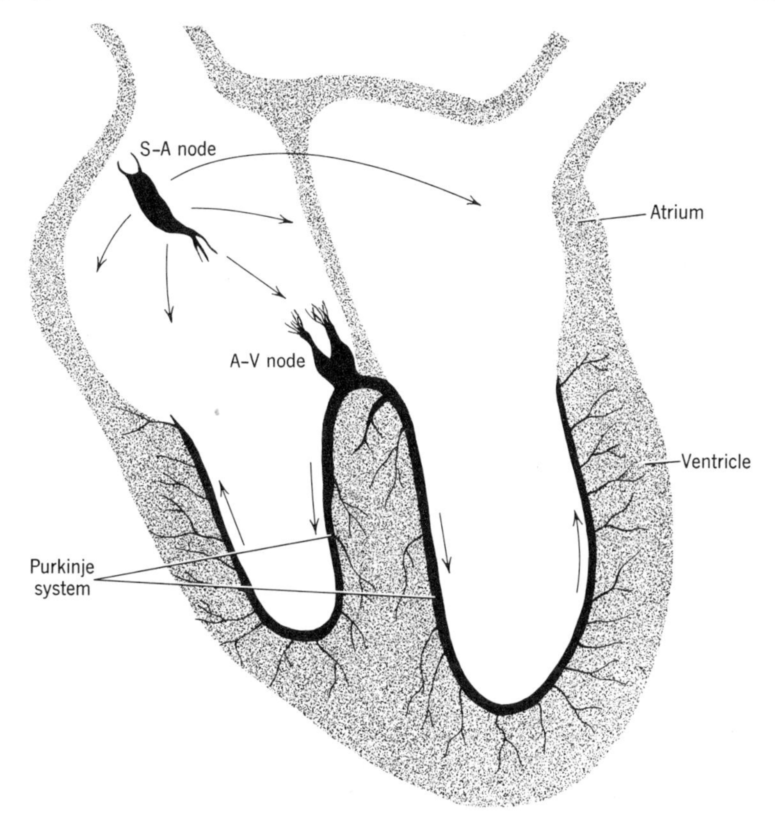

FIGURE 7.6. Diagram of conduction pathways in heart. (From Wiggens, C. J., *Sci. Am.* **196** (5) 74, 1957.)

where the *atrioventricular node,* the secondary pace maker, is located. Here there is a brief pause; then the atrioventricular node fires and the depolarization spreads through the ventricles. There is special conducting tissue (Bundle of His) leading to the inter-ventricular septum, where it divides and spreads through each ventricle (*Purkinje system*) (see Fig. 7.6). The pause at the A-V junction is valuable in that it allows time for the contracting atria to empty their contents into the ventricles. The mechanism of the pause is not understood.

If the S-A node's influence is removed by destruction or other means, the A-V node takes over the initiation of the beat. If so, the depolarization spreads through the ventricles in the normal fashion but reaches the atria by *retrograde conduction.*

Extra systoles. As will be discussed in more detail in chapter 17, a muscle is unresponsive to any further stimulation immediately after contraction. This *refractory period* is especially long in heart muscle and lasts throughout the period of depolarization; about a quarter of a second. This helps to protect the heart from stray impulses which would otherwise cause contraction and disrupt the cardiac rhythm. Occasionally however, a stray impulse does reach the ventricle after the refractory period and before the normal wave of depolarization from the S-A node. This causes a *premature ventricular contraction,* or *extra systole,* which is itself followed by a refractory period. Therefore, the next impulse from the S-A node has no effect and there is a *compensatory pause* until the S-A node fires again, reestablishing the normal rhythm. The sequence of events and the break in the cardiac rhythm is shown in Fig. 7.8. The extra systole expels so little blood from the heart that no pulse is felt for that beat, giving rise to the popular idea that the heart "misses a beat." This is not strictly true, but aptly describes the rather disagreeable sensation. The stray impulses giving rise to extra systoles usually originate within the heart itself at *ectopic foci,* rather than in the pace makers. For no obvious reason, they happen occasionally in normal hearts, but they become more common in electrolyte derangements (which alter membrane potentials) and in inflammation or ischemia of the myocardium. An extra systole does not pump out a useful volume because it follows so closely on the previous contraction that the ventricles have not had enough time to fill with blood and the abnormal excitation leads to a weak beat. This is unimportant except when extra systoles occur at every few beats, especially if the heart is already failing. Therefore, an added hazard in cardiac disease is the occurrence of "runs" of premature ventricular contractions and a still further diminished pumping effectiveness.

Premature *atrial* contractions can arise in a similar fashion but the consequences are relatively minor because the ventricular contraction is normal.

The electrocardiogram. On stimulation of a muscle (or nerve) the cell membranes are depolarized, and on recovery they are *repolarized.* The phenomena are described in more detail in the chapters on muscle and nerve, and at this point, it is sufficient to say that these are *electrical* events. The tissues are electrical conductors, which is to be expected, as they are largely composed of electrolyte solutions. Therefore, the changes in electrical potential associated with muscular activity spread throughout the body and can be detected with suitable instruments on the skin at considerable distances from the sites of origin. In the moving body, the potential changes in the tissues form a bewildering complex resulting from the electrical activities of all the muscles. If a subject lies at rest, however, the potentials associated with the heart can be detected relatively uncluttered by extraneous *muscle artefact.*

In absolute magnitude the heart's potentials are very small when compared with flashlight batteries or other familiar sources of voltage. Therefore, the first obstacle to the investigation of the heart's potentials was the need for a suitably sensitive instrument.

The *string galvanometer* developed by Einthoven in Holland, successfully filled the need for many years. An extremely fine quartz thread is stretched vertically between the poles of a powerful magnet (see Fig. 7.7); connections are arranged so that the currents picked up from the subject's skin flow along the thread (which is gold-plated to make it conductive). This makes the thread act as a minute armature of an electric motor and because it is so light, the tiny magnetic field induced by the flow of a current will deflect it in the main field of the big magnet in a direction depending on the direction of the current. The direction depends on the polarity of the potential (+ or −) being picked up from the skin at that instant and the tension in the thread ensures that it recovers promptly from each deflection as the potential changes. The amplitude of the deflection is of course very small and it has to be magnified by an optical method. To achieve this, a light source is arranged so that the shadow of the quartz thread falls on the sensitive paper in a camera with a slit aperture.

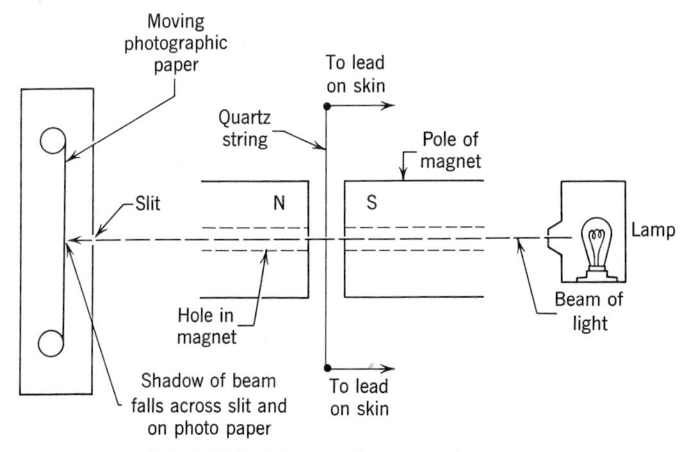

FIGURE 7.7. Diagram of a string galvanometer.

A winding mechanism moves the paper past the slit at a constant speed, and the deflections of the thread's shadow show up as an oscillating line when the emulsion is developed. The resulting picture is called an *electrocardiogram* or ECG. Perhaps in deference to Einthoven, some people refer to it by the initials of the Dutch name, EKG.

Through modern development of vacuum tubes for radio and other purposes, the amplification of small potentials is theoretically almost unlimited and the string-galvanometer has been largely superseded. In vacuum tube instruments, the amplified currents are used to drive pens which write with ink directly on suitable moving paper with obvious advantages in convenience. The amplified potentials may also be applied to the plates of an oscilloscope (see p. 311) and the ECG can be visualized directly on the fluorescent screen. For very exacting work, a photograph of an ECG on an oscilloscope is preferred to a pen-writing machine's record, since the electron beam of the oscilloscope has none of the lag due to inertia and other mechanical disadvantages of the pen system.

Reading the Electrocardiograph

The recording leads are attached to the body by smooth metal plates, moistened with a conductive paste and strapped on. The deflection of the pen is calibrated with a standard source of potential so that a 1 cm deflection corresponds to a 1 mv change in potential. The paper is usually set to travel at 25 mm/sec so that a 1 mm division on the paper represents 0.04 second.

An idealized record is shown in Fig. 7.8. The letters *P, Q, R, S,* and *T* are traditionally used to designate the "waves"; the main

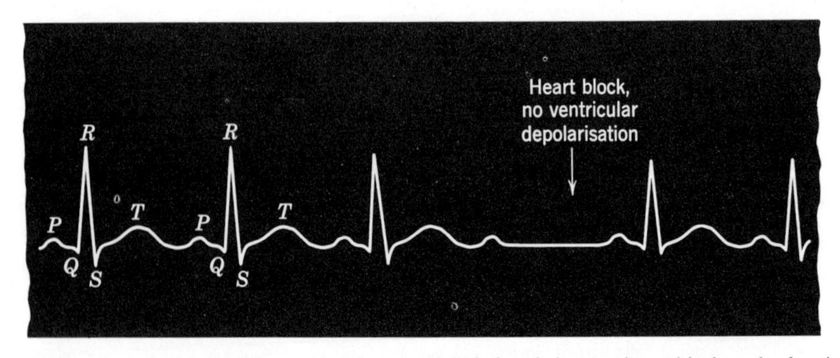

FIGURE 7.8. Idealized ECG record with waves labeled and showing heart block at the fourth atrial beat.

components of the ECG. Another deflection, the *U*-wave, is also seen in the ECG of other animals and in some abnormalities in man. The *P*-wave represents the potential change due to depolarization of the atria (which immediately precedes contraction); the QRS complex is from the depolarization of the ventricles, and the *T*-wave is from the repolarization of the ventricles. The repolarization of the atria is not ordinarily seen because it is masked by the QRS complex.

The amplitude and direction of the waves depend greatly on the position of the recording leads. If the leads are attached in several standard arrangements however, the normal records are surprisingly consistent. The standard leads of the ECG are illustrated in Fig. 7.9, and typical records from three leads are illustrated in Fig. 7.10.

The size and the direction of each deflection vary with the lead and with other less controllable factors, but the *inversion* of one of the waves from its usual direction in a given lead may be clinically significant. The *duration* of each wave, the *intervals* between them and the *rhythm* of their occurrence are of much more clear-cut significance. In oversimplified terms, a longer than normal *P-R* interval would indicate delayed conduction of the wave of depolarization in the ventricles, and a long *Q-T* interval indicates a slow repolarization. These can happen in cardiac lesions or electrolyte defects. The ECG also provides a permanent record of the heart's rhythm which can be heard only fleetingly with a stethoscope. For example, the occurrence of premature ventricular contractions or of periodic blockage of conduction from the atria to ventricles are clearly shown. By considering the various lead records simultaneously, the cardiologist can not only diagnose the presence of damage due to an infarct, for example, but can actually determine the location of a lesion.

No matter how refined the recording methods are, it should be emphasized that the ECG is simply a record of the fluctuations in potential originating in the heart and spreading through the body, which acts as a core conductor. A few theoretical conclusions about the shape of the ECG are justified, but the complexity of the system does not permit a thoroughly quantitative treatment. Some largely empirical rules for interpreting ECGs have been worked out over the years and the intelligent application of this accumulated experience makes the ECG a valuable diagnostic and research tool. We might compare the reading of ECGs to the reading of music. Almost anyone with patience can learn the

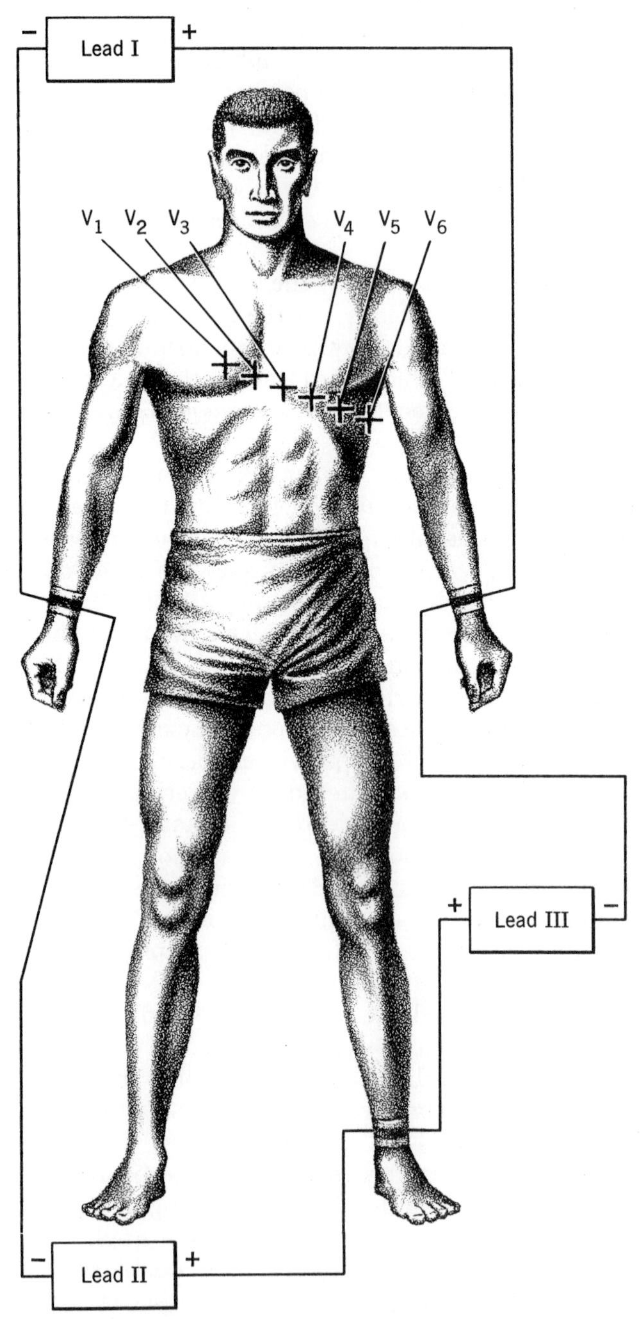

FIGURE 7.9. The standard ECG leads. Lead I, right arm—left arm; Lead II, right arm—left leg; Lead III, left arm—left leg; precordial (or chest) leads are V_1, V_2, etc.

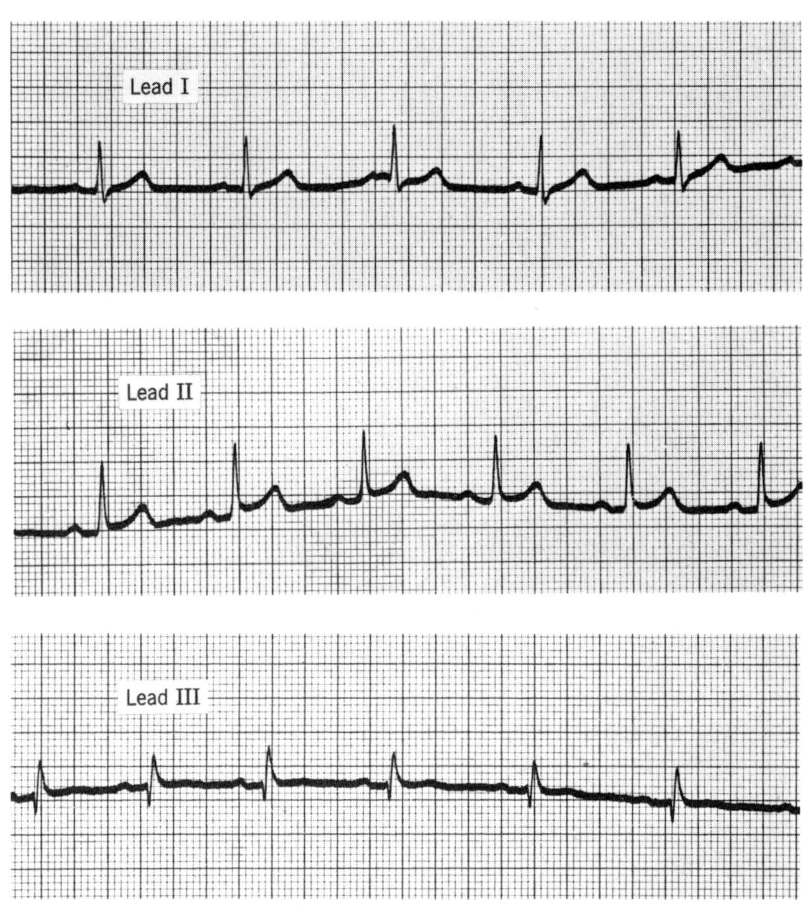

FIGURE 7.10. ECG record.

rules of musical notation, but this is a far cry from being able to read a sheet of music and hear it in the mind. The experienced cardiologist reading an ECG brings far more to bear than the simple rules of thumb.

THE BLOOD VESSELS

The arteries and veins differ in their structure as well as in function. Figure 7.11 shows the differences in the cross-section appearance of the two kinds of vessel. It will be noted that arteries have thicker walls, which contain more elastic tissue and more smooth muscle than do the walls of veins of corresponding capacity. This difference reflects the higher pressure of blood in the arteries.

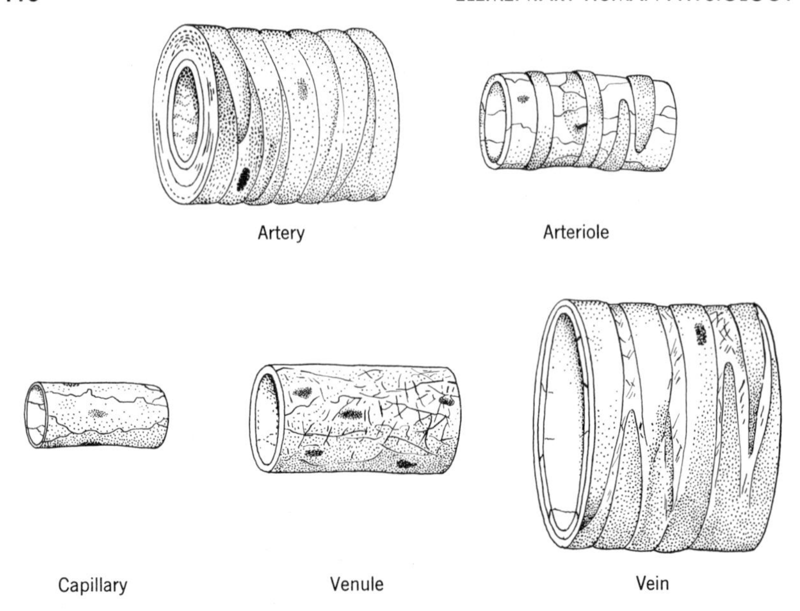

FIGURE 7.11. The structures of blood vessels.

Arteries and arterioles. In the smaller arteries there is proportionately less elastic tissue and more smooth muscle; this is especially true of the smallest arteries, the *arterioles.* As is described later, the degree of contraction in the smooth muscles of the arteriolar walls is controlled by the autonomic nervous system and is one of the most important regulators of blood pressure.

The elasticity of the large arteries is of great importance in maintaining a steadier blood flow than the pulsatile pumping of the heart could accomplish alone. As the blood is forced into the aorta at systole, the great pressure stretches the arteries which temporarily absorb some of the pumping energy. Then, when the pressure falls during diastole, the elastic rebound of the arterial walls compresses the blood and continues to drive it into the tissues, even though the heart is relaxed. The net effect is that the elastic tissue in the arterial walls helps to "smooth out" the fluctuations of flow through the cardiac cycle.

The distension of the arteries during systole extends all the way out to the periphery and can be detected as a *pulse* in several arteries near the surface. One of the most useful is the radial artery at the wrist, but the femoral pulse may also be palpated with ease and is more convenient in infants. The pulse has the same frequency as the heart, but its propagation along an artery is determined by the structure of the vessel; it is quite independent of the velocity of blood flow. The pulse wave travels similarly to the wave resulting when a taut fence wire is plucked. The idea that the pulse results from successive "spurts" of blood is incorrect.

Veins. The unique feature of the veins is the presence of *valves,* made of loose pockets of endothelium, which permit the blood to flow in one direction only—towards the heart (see Fig. 7.12). The veins are of much larger total diameter than the arteries and the blood moves back to the heart relatively sluggishly compared with the surging force in the arteries. The return of venous blood, especially from the feet and legs, is helped by the "milking" action of muscular contractions. When a vein is compressed by movement of a limb, its valves allow the blood to be squeezed only in the right direction.

The capillaries are found between the arteries and veins and they are the smallest vessels of all. It is customary to speak of the capillary beds as if they were composed of uniform vessels, but in fact, the micro-anatomy is very complex. The blood flow through a capillary bed is not only subject to control by the vasoconstriction of the arterioles supplying it (see p. 128), but also by the opening and closing of *precapillary sphincters.* Even under apparently constant conditions, there is a rhythmic constriction and dilation of the small vessels so that the blood flow passes through different vessels from second to second (Fig. 7.13).

BLOOD PRESSURE

Mean arterial pressure. Even with the "damping" effect of the elastic vessel walls, the arterial blood pressure rises and falls with the heart beat; the peak pressure is the *systolic pressure,* the trough is the *diastolic pressure.* In normal young adults these are 120 mm and 80 mm of mercury respectively. The difference between the systolic and diastolic pressures is the *pulse pressure.*

The *mean arterial pressure* is not strictly the arithmetic average of the systolic and diastolic pressures (for normal young adults, $120 + 80 \div 2 = 100$ millimeters), although it is often close to it. It is rather that pressure which, if applied steadily, would have the same *effect* as the fluctuating pressure actually encountered. This depends on the shape of the pressure curves; that is, on the relative duration of the high and low pressure parts of the cycle. In our diagram, the mean pressure is represented by a line whose

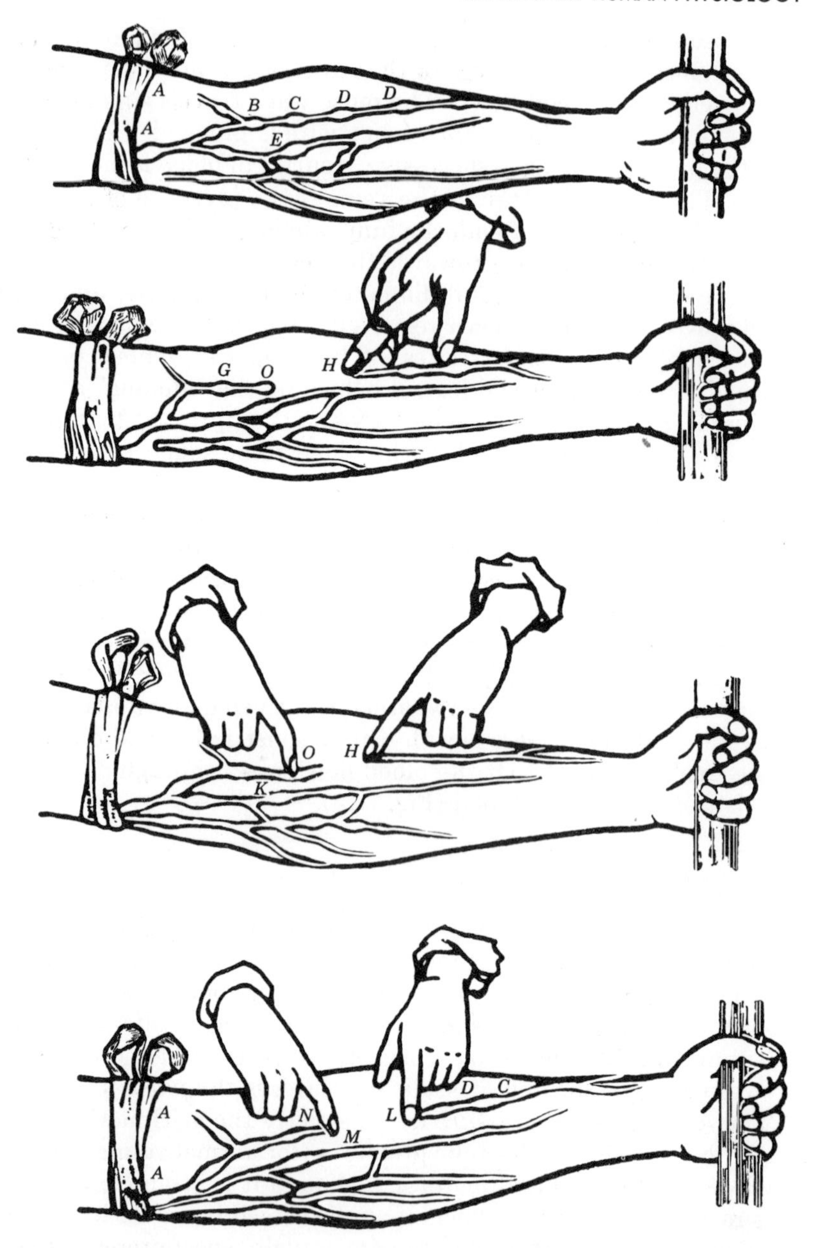

FIGURE 7.12. Harvey's diagram of his experiments demonstrating valves in the veins. Apply a tourniquet above the elbow and apply pressure at *D*, to prevent blood from passing that point. Push the blood upward past C. It will then flow *to*, but not beyond, C, leaving the gap between O and *H*.

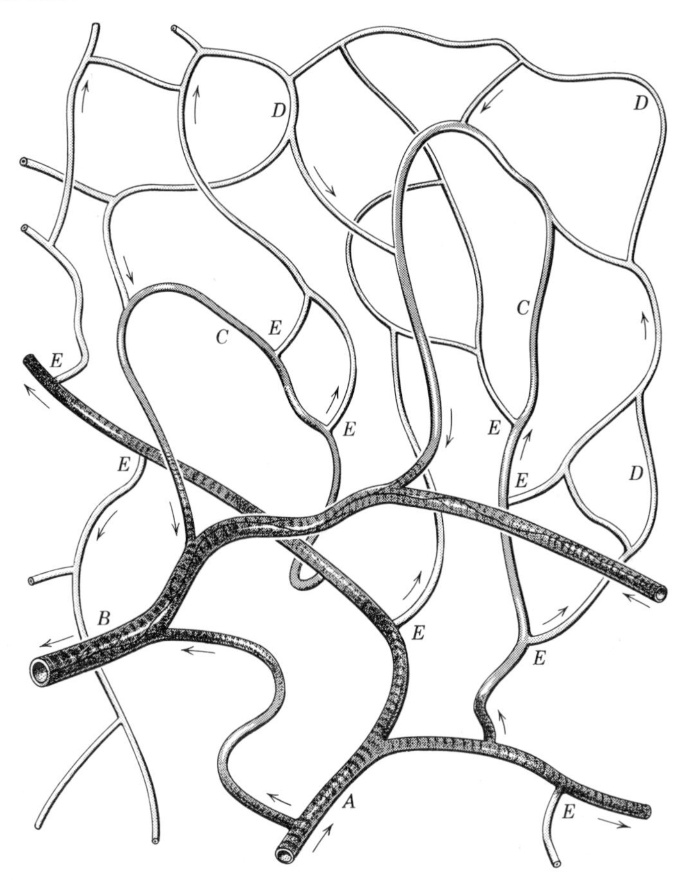

FIGURE 7.13. A capillary bed is depicted in this drawing. The blood flows into the bed through an arteriole (*A*) and out of it through a venule (*B*). Between the arteriole and the venule the blood passes through thoroughfare channels (C). From these channels it passes into the capillaries proper (*D*), which then return it to the channels. The arteriole and venule are wrapped with muscle cells; in the thoroughfare channels the muscle cells thin out. The capillaries proper have no muscle cells at all. The flow of blood from a thoroughfare channel into a capillary is regulated by a ring of muscle called a precapillary sphincter (*E*). The lines on the surface of the arteriole, venule, and thoroughfare channels are nerve fibers leading to muscle cells. At lower left, between the arteriole and venule, is a channel which in many tissues shunts blood directly from the arterial system to the venous when necessary. (From Zweifach, B. W., *Sci. Am.* **200** (1) 54, 1959)

integral is the same as that of the curve of the pressure fluctuations (the area beneath the lines is the same).

The definition of the mean arterial pressure is given in some detail, because with respect to maintaining a suitable exchange of fluid in the capillaries, it is clearly this mean pressure that is of greatest interest. Therefore, in our subsequent discussion of the circulation, the term "arterial pressure" will refer to the mean pressure.

Measurement of blood pressure. A clergyman named Hales, as early as 1733, first measured the great pressure of blood in the arteries. He cannulated an artery in a restrained horse and connected it to a long glass tube held vertically. The blood rose nearly 8 feet in the tube, showing a pressure of approximately 8 feet of water, or 190 millimeters of mercury.

A similarly direct method is still used in following the blood pressure of anesthetized experimental animals in the laboratory. A carotid or femoral artery is cannulated and connected with a *mercury manometer* (see Fig. 7.14). The blood pressure forces the mercury down one side of the *U* tube and up the other so that the vertical distance between the two mercury levels is proportional to the blood pressure. The mercury level bounces up and down with the rise and fall in pressure at systole and diastole. This movement and the absolute height of the mercury column can be recorded on a kymograph by means of a stylus attached to a small plunger which floats on the open surface of the mercury. In practice, the tube between the blood in the carotid cannula and the surface of the mercury is filled with citrate solution. This is to avoid the clotting which would follow if the blood itself filled the space. In research laboratories the mercury manometer is replaced with an electronic pressure transducer, but the principle is the same.

For the routine determination of blood pressure in a human subject, the insertion of an arterial cannula is clearly out of the question. The apparatus used instead is a *sphygmomanometer* (see Fig. 7.15). The cuff (*a*) is placed around the upper arm and inflated with air by squeezing the rubber bulb. The air pressure in the cuff is shown on the manometer (*b*) (some sphygmomanometers have a dial-type pressure gauge instead). When the cuff is sufficiently inflated, the tissues of the arm compress the *brachial artery* against the *humerus* where it runs close to that bone, thereby completely occluding it. A stethoscope is placed over the brachial artery on the inside of the elbow below the cuff. On slowly releasing the pressure in the cuff (by a screw valve on the rubber ball), the escaping blood may be heard coursing in the artery. It follows that when the cuff pressure is diminished until it is exactly equal to the systolic pressure, blood will escape only at the peak of systolic pressure. This is heard in the stethoscope

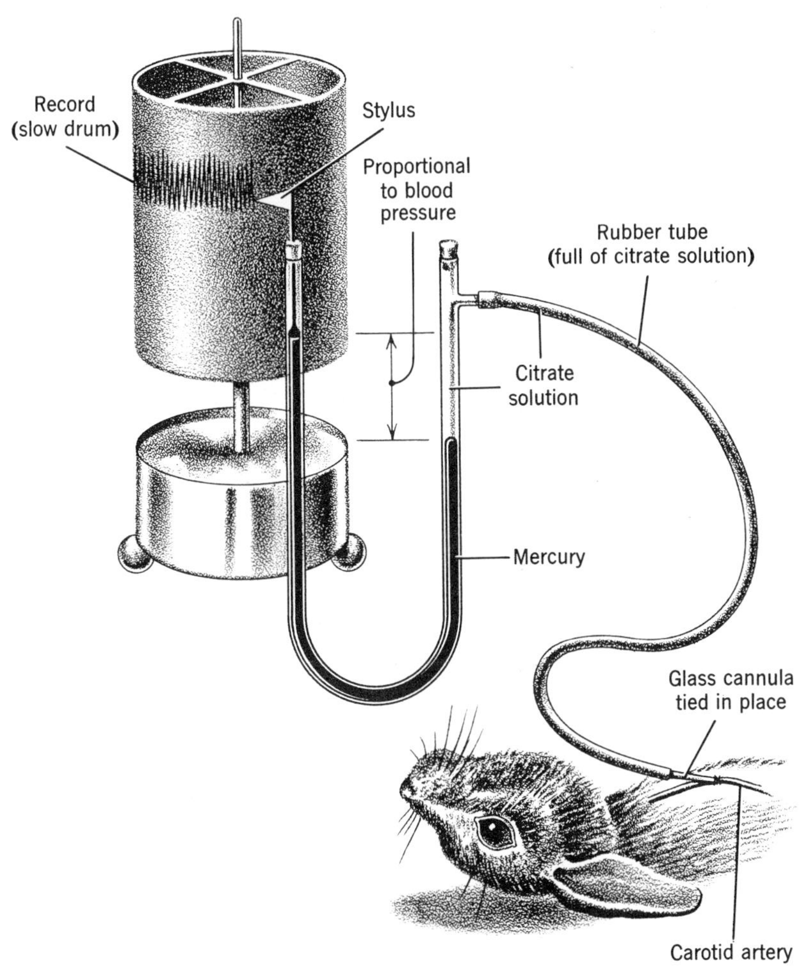

FIGURE 7.14. Scheme showing direct recording of blood pressure.

as a throbbing at the same rate as the heart beat. Further release of the cuff pressure allows blood to pass at lower and lower pressures until it is low enough to allow flow during the diastolic trough. Then the blood flow is continuous and is no longer audible in the stethoscope.

In practice, then, the physician or nurse gradually releases air from the cuff and notes the pressure shown by the manometer at the time the first sounds become audible, and again when the sounds disappear (or when their character changes markedly). These are taken as the systolic and diastolic blood pressures respec-

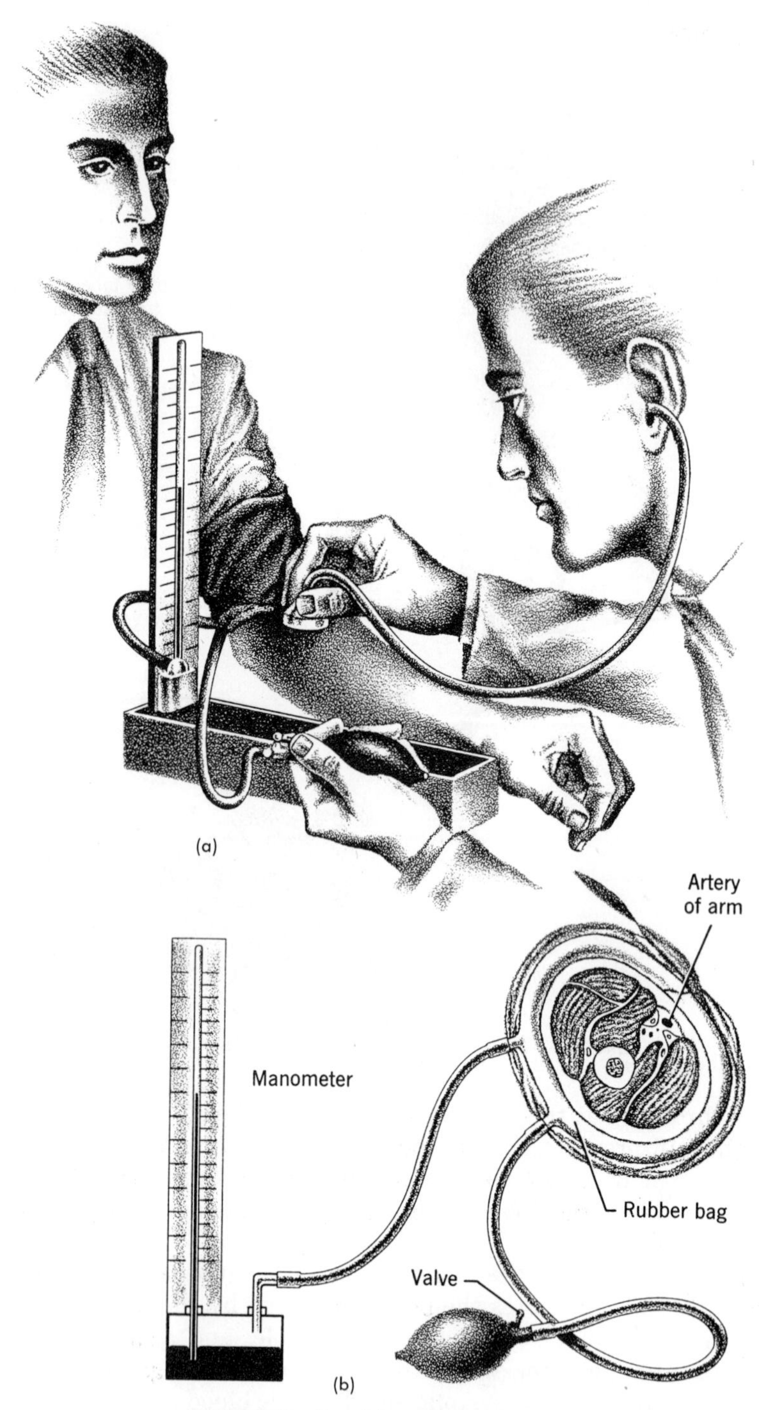

FIGURE 7.15. Use of the sphygmomanometer.

tively. There are many refinements and corrections necessary if the arterial blood pressure is to be determined with absolute accuracy, but in clinical practice it is common to take the blood pressure in a "standard" fashion so that the errors, if any, are always of the same kind.

PRESSURE, FLOW, AND RESISTANCE

For a proper understanding of blood pressure, a few elementary physical principles are essential. The rate at which a liquid passes through a tube is referred to as the flow rate, or simply the *flow*. The flow depends on the pressure in the tube and on the resistance according to the simple equation

$$F = \frac{P}{R}$$

where $F =$ flow, $P =$ pressure, and $R =$ resistance.

The pressure in the tube likewise depends on the flow and resistance, as is seen by rearranging the equation.

$$P = R \times F$$

Resistance depends on the viscosity of the fluid and the diameter of the tube. Tubes of small diameter offer more resistance than larger tubes and so, at a given pressure, the flow in a small tube is less than in a larger one.

In relating these concepts to the circulatory system, it is worth restating the function of the circulation namely, the renewal of the interstitial fluid by ultrafiltration and osmosis in the capillaries. The exchange in the capillaries requires an adequate blood *pressure* for ultrafiltration and adequate *flow* so that enough oxygen and nutrients are delivered to the cells. The two main variable factors which can affect pressure and flow are the amount of blood pumped by the heart (*cardiac output*) and the total peripheral resistance to flow.

The rate of exchange across the capillary membranes can therefore be regulated by integrated control of the many factors affecting cardiac output and peripheral resistance. These are discussed in some detail in the following pages, but as with the discussion of homeostasis, it is probably helpful first to review what these factors are.

Cardiac output. This depends on the rate of the heart beat and the amount of blood pumped at each beat (*stroke volume*). The stroke volume in turn depends on the forcefulness of the heart's contraction and also on sufficient inrush of blood to be pumped.

Total peripheral resistance. The main trunk arteries offer so little resistance to flow that blood reaches the principal regions of the body without much diminution in pressure. The first serious impedance to flow is in the arterioles, and normally they offer more than half the total peripheral resistance; most of the remainder is in the capillary beds. This is shown by a diagram in Fig. 7.16. The total resistance, then, varies chiefly with the number of capillaries open and with the degree of contraction of the smooth muscle of the arteriolar walls. This latter, the *vasomotor tone,* is under autonomic nervous control. The capillary diameters on the other hand, depend largely on local factors; temperature, concentration of metabolites, *and* on the flow of blood through the arterioles supplying that capillary bed.

Finally, it should be noted that although the resistance in the vessels varies chiefly with their diameters, changes in the blood viscosity will also alter the resistance. This is usually fairly constant, however, and only in a few disease states are viscosity changes important.

To achieve a given rate of exchange in the capillaries, all these factors must be under integrated control. While bearing this in mind, it is nevertheless more practical to describe the mechanisms one by one.

Nervous control of the heart. The intrinsic rate of the heart is determined by the pace maker, but it is modified by impulses reaching

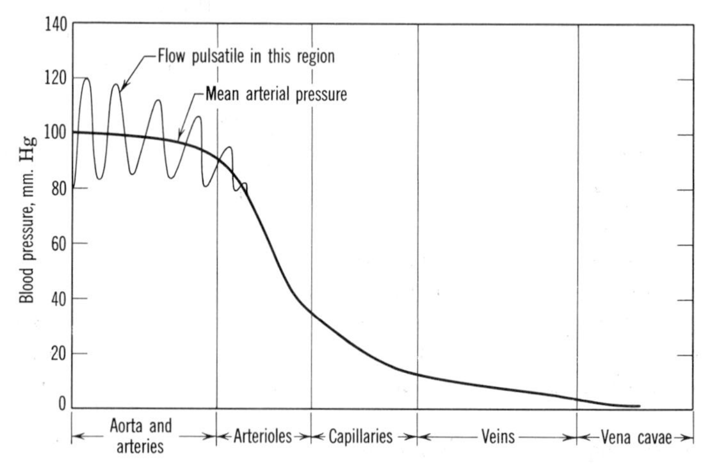

FIGURE 7.16. Diagram showing the fall of blood pressure in the various types of blood vessels.

the nodes over the vagus nerve (containing *parasympathetic* fibers) and over other nerves from the cervical ganglia of the *sympathetic* system. This dual innervation is discussed further in a later chapter on the *autonomic nervous system.* At this stage it is sufficient to state that impulses over the parasympathetic fibers cause the release of *acetylcholine* at their ends (see also Chapter 21). The acetylcholine slows the frequency of impulse formation in the S-A node and also slows the rate of conduction through the myocardium. Such impulses are therefore called *cardio-inhibitory.* The impulses over the sympathetic fibers are cardio-acceleratory in that they cause the release of *nor-epinephrine,* which speeds up both the node rates and the conduction rates.

Because of the opposite effects of the two kinds of nerves, we speak of the *reciprocal innervation* of the heart. An increase in the number of impulses arriving at the nodes over the cardio-inhibitory fibers slows the heart, and a greater number of cardio-acceleratory impulses causes it to speed up. The normal, resting heart rate is the resultant of those two influences, but the vagal inhibition predominates. The resting rate of the heart is therefore slower than the intrinsic rate—a manifestation of *vagal tone.*

The nerves controlling the heart arise from specific regions in the *medulla oblongata* of the brain known as the cardio-inhibitory and cardio-acceleratory centers (Chapter 20). The centers control a feedback mechanism which acts to keep the mean arterial blood pressure constant when the body is at rest.

At the bifurcation of the internal and external carotid arteries in each side of the neck, there is an enlargement of the vessel to form a *carotid sinus* (Fig. 7.17). The walls here contain specialized nerve endings sensitive to changes in pressure, which are therefore called *baroreceptors* or *pressoreceptors.* They send a constant stream of impulses to the cardiac centers in the medulla, and if the arterial pressure rises, the frequency of discharge is increased. In response to this, the cardio-inhibitory center sends impulses to slow the heart, which of course lowers the cardiac output and reduces the arterial pressure. Conversely, a fall in pressure causes fewer impulses from the baroreceptors and a consequent increase in heart rate, which re-establishes the "normal" arterial pressure.

Other arterial baroreceptors are located in the wall of the arch of the aorta and their response is similar to that of the carotid sinus receptors. The reciprocal relationship between blood pressure and heart rate is called Marey's Law, after the man who first described it.

The action of the baroreceptors can be demonstrated by press-

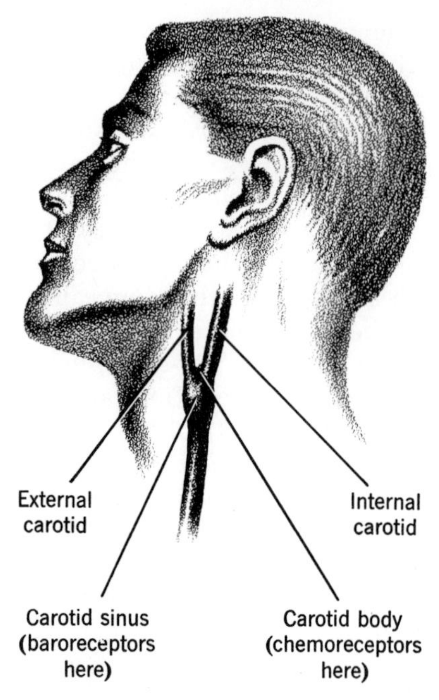

FIGURE 7.17. Location of baroreceptors and chemoreceptors near bifurcation of carotid artery.

ing firmly for a few seconds on each side of the neck just below the jaw. This manual pressure affects the receptors just as if it were increased blood pressure and there is a noticeable slowing of the heart rate. This should not be performed too enthusiastically as cardiac output may be drastically lowered, reducing the blood supply to the brain enough to cause *syncope,* or fainting.

It will be recalled that the cardio-acceleratory fibers act by releasing an epinephrine-like substance at their endings in the myocardium. A release of epinephrine into the blood from the adrenal medulla during excitement or fear has a similar effect (see Chapter 16). We are conscious of increased rate and force of contraction under these circumstances.

There are some rather less effective baroreceptors in the great veins near the heart and in the walls of the right heart itself. These are sensitive to changes of venous pressure, but through them a rise in pressure promotes an *increase* in heart rate, which has the effect of preventing the engorgement of the vessels draining into the heart. This response to increased venous pressure is called the *Bainbridge reflex.*

It should be emphasized that these mechanisms maintain a constant blood pressure *at rest,* and they are overridden by the autonomic changes in exercise.

Response to increased filling pressure. The contractile impulse spreads freely through all of the myocardium and in consequence, either the *whole* heart contracts, or none of it does. Despite this *all-or-none* response, the strength of contraction can still be varied by the degree to which the muscle is stretched immediately before contraction (as well as by humoral effects of acetylcholine and nor-epinephrine). The stretching is due to the inrush of blood during diastole and this varies with the *filling pressure,* or the pressure in the great veins. Over a wide range, the strength of contraction is proportional to the filling pressure, but if diastolic stretching is excessive, the subsequent contractions become weaker (see Fig. 7.18). This relationship was first described and measured by the English physiologist, Starling, and is known as "Starling's Law of the Heart." We see therefore that this mechanism and the Bainbridge reflex work together to increase the cardiac output as venous pressure increases.

It should be noted, however, that Starling's experiments were carried out using a "heart-lung preparation" in which the pulmonary circulation of an experimental animal is left intact, but the output of the left heart is routed through an artificial circulation. This is arranged so that the "peripheral" resistance to flow

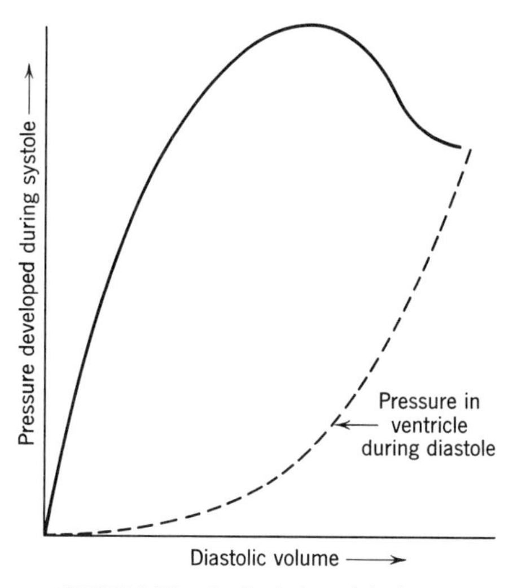

FIGURE 7.18. Starling's law of the heart.

and the filling pressure can be varied by adjusting a clamp and the height of a blood reservoir respectively. Starling's results with the preparation are clear-cut, but it is nevertheless an extremely unphysiological arrangement. Some modern workers have concluded that, in the intact animal, the response described by Starling is relatively unimportant in the minute-to-minute regulation of cardiac output. Output is undoubtedly increased as venous pressure rises, but they suggest that the other mechanisms brought into play are probably enough to account for the response. It must be emphasized then, that whatever the mechanism, an increase in venous return *does* increase cardiac output. Furthermore, the response according to Starling's Law is almost certainly significant in the *permanently* elevated venous pressure which improves the performance of an otherwise failing heart. This point is discussed later in this chapter.

Factors affecting filling pressure. Ordinarily, the venous pressure varies with the "fullness" of the circulation. If the vascular system is full, clearly the great vessels will be full too. This can vary with the absolute blood volume (which depends on the total fluid status) or it can vary with the degree of vasoconstriction in some parts of the circulation. Vasoconstriction in one part of the body (the skin for example) means that the remainder of the circulation contains relatively more blood.

Increases in intrathoracic or abdominal pressure can reduce venous return by simply squeezing the great veins. This is apparent in the slight fluctuations of cardiac output in the ordinary breathing cycle as the pressures within the chest change. The *Valsalva maneuver* is the forceful expiration against a closed glottis which helps defecation and which is also frequently performed in a reflex manner when lifting a heavy weight. Under these conditions, the intrathoracic pressure may be raised sufficiently to prevent venous return altogether for a brief period, and the consequent fall in cardiac output can be enough to cause syncope.

Vasomotor control of arterial pressure. The most "controllable" aspect of the peripheral resistance is the tone of the smooth muscles in the arteriolar walls.

Impulses causing *vasoconstriction* are carried over sympathetic nerve fibers which arise from the same center in the medulla as the accelerator nerves supplying the heart. The center is responsive to the pattern of afferent impulses from the baroreceptors in the aorta and carotid sinus, and sends vasoconstrictor impulses appropriately. Thus, a fall in arterial pressure causes a reflex vasoconstriction which, together with the cardio-accelerator reflex already described, helps to restore pressure to normal. Conversely, a rise in blood pressure causes peripheral vasodilation which, along with the reflex slowing of the heart, helps to restore normal pressure from the other direction. It is supposed that reflex vasodilation is accomplished largely through inhibition of the *vasoconstrictor* center, rather than through the activities of the much sparser *vasodilator* fibers.

Control of the circulation in exercise. Hitherto we have discussed the feedback mechanisms controlling circulation in terms of maintaining a *steady state.* Clearly, the tissues' requirements for oxygen and nutrients are not always the same; for example, they are greatly increased during muscular exercise. (The oxygen consumption of a muscle increases tenfold during activity.) Under these conditions, the control mechanisms must promote an increased flow of blood at the same, or even higher, pressure.

Carbon dioxide. Probably the most important stimulus for the circulatory changes in exercise is the increased concentration of carbon dioxide in the body fluids resulting from the stepped-up metabolism in the muscles. The cardio-accelerator center in the medulla is extremely sensitive to the carbon dioxide content of the plasma, and it promptly responds by sending impulses to increase the heart rate and induce vasoconstriction, thereby raising the blood pressure. In exercise, the vasoconstrictor impulses are sent out in a discriminatory fashion so that the constriction is in the *cutaneous* and *splanchnic* portions of the circulation (skin and viscera). The vessels in the muscles are not constricted. At the same time, the increased carbon dioxide production in the working muscles has a *local* effect in actually dilating the arterioles and capillaries.

The vasoconstriction in the non-exercising tissues diminishes the blood content of those parts of the circulation, helping to increase the filling pressure of the heart and so increase the cardiac output. We can see therefore that the "shunting" of blood flow to the regions where it is needed the most is an economical and effective device.

Effect of higher centers. There is evidence that impulses arising from the higher parts of the brain directing the muscular activity (see Chapter 20) are relayed to the cardiac centers and reinforce the cardio-acceleration. It is frequently overlooked that the cardiac and respiratory centers are parts of the *brain* and are as subject to influence from other areas of the brain as they are from peripheral receptors.

Epinephrine. In violent exercise, especially if there is any element of fear or excitement, the epinephrine released from the adrenal medulla further augments cardio-acceleration and vasoconstriction. The peripheral vasoconstriction is clearly seen in the paleness associated with fear. In the coronary circulation, epinephrine causes vasodilation, improving the blood flow to the myocardium.

Splenic Reservoir. The *spleen* normally contains a reservoir of two or three hundred milliliters of blood of very high hematocrit. In a general sympathetic discharge, the muscular capsule of the spleen contracts and ejects the stored blood into the circulation. This addition to the blood volume increases the filling pressure and so enhances the cardiac output.

Pulmonary flow. It will be recalled that the pulmonary blood flow is the same as the systemic. Therefore, the increased cardiac output in exercise presents more blood to the lungs for oxygenation and removal of carbon dioxide. This helps to meet the special demands of exercise, especially since the greater pulmonary perfusion is accompanied by a reflex acceleration of breathing, also induced by carbon dioxide in a fashion similar to the circulatory reflex (see Chapter 8). The *net effect* of these various mechanisms is to increase enormously the blood flow through the muscles and lungs, partly at the expense of blood flow through the viscera and skin. Because of the great vasodilation in the muscles, the total peripheral resistance is not greatly increased, and the greater flow can be accomplished by a smaller increase of blood pressure than would be necessary if the vasoconstriction were indiscriminate.

Training. Regular exercise tends to enlarge the heart and certainly increases its capacity for more forceful contraction. In general, athletes respond to an exercise load by a greater increase in stroke volume and a lesser increase in rate than do untrained individuals. Furthermore, after a period of exertion, the heart rate of an athlete returns to normal more quickly than that of a person in poor condition. This is the basis of a simple test used in the Armed Forces and college-entrance physical examinations. The subject's heart rate is counted at rest, again after a half minute's hopping or stepping on and off a box, and again at three half-minute intervals during recovery.

The circulatory adaptation to exercise is not unlimited. As the heart rate increases, diastole becomes shorter and shorter, so that at very rapid rates, the time for filling the heart is too brief to be effective. Even with increased venous return, the brief filling time becomes a limiting factor, and at heart rates over about 180 per minute, there is incomplete filling of the ventricles. Accordingly, the cardiac output actually falls off with further increases of rate.

Control of circulation for temperature control. The *hypothalamus* is part of the brain which may be regarded as the chief autonomic control center (see Chapter 21). It contains cells that are extremely sensitive to the temperature of the blood. When the body temperature rises, as in exercise or a warm environment, the hypothalamus sends impulses to the vasomotor center causing dilation of the blood vessels of the skin. This has the effect of shunting more blood to the surface of the body where its heat may be dissipated more readily. The cutaneous vasodilation accounts for the flushed appearance of the skin after a few minutes' exercise. Conversely, there is a vasoconstriction in the skin and a conservation of blood heat when the body temperature would otherwise fall (see Chapter 13). It should be noted that the *initial* response in exercise is for vasoconstriction in the skin, and that vasodilation occurs later as the temperature tends to rise.

During intense vasoconstriction of the skin vessels, as in severe cold for example, the blood supply is diminished enough to make the tissue somewhat anoxic, and there is an accumulation of carbon dioxide, lactic acid, and other end-products of metabolism. These have a vasodilatory effect which is overridden by the central control so long as the cold conditions continue, but when the subject enters a warm room, the vasoconstrictor impulses from the brain are diminished and the local effects become apparent as a *reactive hyperemia,* or flushing. Another common experience is that hot or cold compresses cause local vasodilation and constriction respectively.

The relationship of the circulation to interstitial fluid. Each of the control mechanisms discussed so far in this chapter is prompt in its action and helps to regulate the circulation from second to second. There are other, equally important, but slower-acting mechanisms that maintain an optimal *volume* of blood in the circulation. It is apparent that the venous return, and therefore the cardiac output, depends largely on the "fullness" of the vascular system. The blood volume in turn depends on the total fluid content of the body, and particularly on the equilibrium between the plasma and the interstitial fluid. These relationships are conveniently illustrated by describing the changes resulting from blood loss, or *hemorrhage.*

The circulation is a *closed system* of pumps and pipes, and thus it follows that the removal of some of the fluid contents must lead to a decrease in pressure. The falling arterial pressure in hemorrhage first of all evokes the immediate neurogenic responses already described—vasoconstriction and cardio-acceleration. In consequence, the arterial pressure is restored to near normal and the available blood is shunted to the more vital organs. The contraction of the spleen is particularly valuable in hemorrhage.

Fluid Shift

If a hemorrhage is severe, say more than 500 milliliters in a man, the total effect of these "emergency" responses is still not enough to maintain arterial pressure. In this event, and as a *result of* the lowered pressure, a *fluid-shift* response is elicited.

References have been made to the exchange of fluid between the capillary plasma and the interstitial fluid. The steady state of the exchange normally represents a balance between the ultrafiltration pressure at the proximal ends of the capillaries and the oncotic pressure of the plasma proteins. Even in a severe hemorrhage, the oncotic pressure remains the same, since the blood loss does not immediately affect the protein *concentration.* A lowered arterial pressure, however, causes a reduction in ultrafiltration, and so the return of fluid due to the oncotic pressure exceeds the squeezing out of fluid by ultrafiltration. In other words, more fluid enters the vascular system than leaves it. The blood volume is thus augmented by interstitial fluid (which has essentially the same composition as plasma except that it contains no protein). As the blood volume increases, the arterial pressure rises until the filtration pressure again balances the oncotic pressure. It is clear that this occurs at an arterial pressure somewhat below normal, as the addition of protein-free interstitial fluid to the blood lowers the oncotic pressure.

Within half an hour of a fairly severe hemorrhage therefore, the arterial pressure and blood volume may be almost normal, although specific gravity determinations show a diminished plasma protein concentration. The splenic reserve of blood contains such a large proportion of red cells that even with the hemodilution by interstitial fluid the hematocrit is restored to an approximately normal value.

In this state, then, the circulation of the blood is maintained, but nevertheless there is a total body deficit of water, salts, and plasma-protein and hemoglobin. The plasma-protein and hemoglobin are normally manufactured by the liver and the bone marrow within the next few days, although this depends on adequate nutrition.

The "volume" deficit is made up by the renal retention of ingested water and salts, presumably through the same poorly understood mechanisms that control the elimination of an isotonic salt-water load (see Chapter 9).

The *renal control of circulation* lies in its regulation of total fluid volume, which largely determines the blood volume and the filling pressure. It is accomplished by varying the extent of sodium reabsorption in the kidney. If there is a net retention of sodium, water accompanies it by obligate reabsorption and we may think of sodium being retained as an isotonic solution of sodium chloride. Hence the increase in fluid volume.

The reabsorption of sodium in the kidney is controlled by the mineralo-corticoids secreted by the adrenal cortex, *aldosterone* in particular. An increased secretion of aldosterone after a hemorrhage has been demonstrated, but the means by which the adrenal cortex is "informed" of the volume deficit is still an open question.

Shock

If a hemorrhage is so severe that none of the compensatory mechanisms can maintain cardiac output, the oxygen supply to the tissues is endangered. One of the most sensitive tissues in this respect is the brain, and the patient usually loses consciousness. The oxygen shortage in the vasomotor centers themselves induces a massive vasoconstriction throughout the body as if in a "last-ditch" effort to restore arterial pressure, and is evidenced by an extreme pallor and cold skin. As the situation continues, the oxygen shortage affects the vasomotor muscles so that they cannot remain contracted, the arterioles begin to dilate and the blood pressure is lowered further. The heart muscle is similarly affected, and when it fails as well, the patient is not far from death. When shock has progressed to this point, the hypoxia seems to cause the capillary walls to deteriorate so that they become extremely permeable; then, even massive blood transfusions do not help and the patient is said to be in *irreversible shock.*

Before this extreme situation, the best treatment for hemorrhagic shock is clearly the replacement of the lost blood volume by transfusion of whole blood or plasma. The infusion of isotonic saline is better than nothing, but since it contains no protein, it is immediately distributed throughout the whole of the extracellular space and only one fifth of it remains in the vascular system to increase the blood volume. Because of the difficulties of blood and plasma transfusions on the battlefield, during World War II there was intensive research into the possibilities of *artificial plasma expanders* —solutions with large colloidal molecules that can be retained in the vascular system. Some of these like "dextran," are certainly more effective than saline but none is completely satisfactory.

In concluding this discussion, it should be pointed out that loss of blood is only one of several possible causes of shock. Any condition leading to inadequate cardiac output and subsequent ischemia of the tissues leads to shock; as for example in the acute failure of the heart due to a coronary thrombosis. Similarly, a massive allergic response causes extreme vasodilation and frequently heart block as well. This condition is called *anaphylactic* shock.

Hypertension

The mean resting arterial pressure is about 75 millimeters of mercury in infants, 100 millimeters in young adults and rises to 120 or even 130 millimeters in old age. The rising pressure in old age can be partly attributed to the thickened walls and loss of elasticity in the arteries associated with the aging processes; the blood has to be pumped at a higher pressure to maintain flow in arteries of reduced diameters.

Increases of blood pressure of various degrees are also seen in younger people, and when the systolic pressure exceeds 140 millimeters, we rather arbitrarily term the condition *hypertension.* This condition can develop in several fairly well-understood disease states, but the great bulk of hypertensive patients suffer from *essential hypertension* for which there is as yet no satisfactory explanation. It seems as if the vasomotor feed-back control becomes adapted to a resting state at a higher pressure.

Arterial pressure is notably increased in emotional crises, and in some patients, hypertension may be reasonably attributed to chronic emotional stress. There is evidence that hypertension of fairly recent origin can be sometimes alleviated by psychotherapy or by the patient's merely changing to a more congenial occupation. Investigation has shown that a predisposition to hypertension is strongly hereditary. A psychogenic origin is at least consistent with this fact, but of course the emotional make-up of an individual is only one of many hereditary factors. Considerations of this nature offer many tempting avenues of speculation about the effect of the emotions on the physical state of the body, and vice versa. An essential hypertension that develops and worsens markedly over a period of only a few months (as opposed to years) is termed *malignant hypertension.*

One of the less common forms of hypertension results from renal diseases that leave too little normal kidney tissue for effective filtration of the plasma (see Chapter 9). In such cases, the arterial pressure is increased and the greater renal blood flow partly compensates for the defect. A widely accepted theory proposes that the inadequately perfused kidney secretes a substance, *renin,* which acts through complex pathways on the vasomotor muscles to induce the compensatory hypertension. *Experimental hypertension* can be induced in dogs by a partial clamping of the renal arteries.

Hypertension can also follow a vascular defect or tumor in the medulla, which reduces the blood flow through the vasomotor center. The ischemic vasomotor center "blindly" responds by inducing a general vasoconstriction just as if the ischemia were due to a falling cardiac output (as in shock).

Finally, disorders of the adrenal cortex involving oversecretion of aldosterone can also cause hypertension through the excessive retention of salt. The consequent fluid retention, which is desirable after a hemorrhage, thus becomes pathological. This topic is discussed more fully in the section on cardiac failure which follows. Hypertension is alleviated to some degree by a low-salt diet, presumably because of lessened salt retention, even where there is no gross malfunction of the adrenal cortex. A typical American or Western European diet contains 8 or 10 grams of salt per day, whereas the strictly nutritional requirements are met by 1.5 grams or less. The "taste" for salt food is progressive and seems to increase with age, so that in middle age (when hypertension is most likely to be a problem) the salt intake may be very high and any subtle defect in its excretion can lead to fluid retention. Common sense would indicate a conscientious avoidance of a more and more salty diet in middle life, without necessarily resorting to a truly low-salt diet which is undoubtedly miserably insipid.

Nicotine is a potent vasoconstrictor agent which contributes to hypertension. Those predisposed to the condition should therefore avoid excessive smoking. In gross obesity, the circulation of blood through worthless fat deposits imposes an extra load on the heart; for this and other reasons to be discussed, obesity is also conducive to hypertension.

Consequences of Hypertension

In order to maintain adequate blood flow through the arterioles with greater vasoconstrictor tone, the hypertensive subject's heart must work harder and so gets bigger, just like any other muscle that is worked hard. This *hypertrophy* is actually restricted to the left ventricle, as the resistance in the pulmonary circulation increases only in more advanced vascular degeneration.

The work capacity of the normal heart ranges from that required for an adequate cardiac output at rest all the way to the maximum needed in severe exercise. The hypertensive's heart is already working part way up this range even at rest. Therefore the *cardiac reserve* is diminished and the capacity to undertake exertion is limited. If, as is often so, hypertension accompanies some form of cardiac insufficiency, the heart is unable to increase output to meet even moderate extra demands.

Chronic hypertension subjects the arteries to a strain that is only briefly encountered in spells of exercise in the normal state; this strain induces degenerative changes in the walls, especially of the arterioles. The small arteries on the retina of the eye, for example, develop characteristic hemorrhages and surrounding edema readily apparent to examination. Most insidious of all is that at an early age, the arteriolar walls show the same kind of thickening and the loss of elasticity characteristic of old age. Furthermore, as hypertension continues, the thickening progresses not only by the increase of normal tissue, but also by the deposition of fatty material and later of calcareous plaques. These further reduce the diameter of the arteries and the elasticity of their walls, thereby exacerbating the hypertension.

Arteriosclerosis and atherosclerosis. The thickening and degeneration of the arteries from any cause is called *arteriosclerosis.* The commonest form is that involving fatty deposits, which is known as *atherosclerosis.* The two terms are frequently (and erroneously) used interchangeably, but in fact atherosclerosis is a special case of arteriosclerosis.

As indicated, arteriosclerosis is normally associated with old age but can occur in younger people. Hypertensives are particularly susceptible to the atherosclerotic form. It is probable that the high blood pressure damages the arterial walls and that the microscopic lesions are the sites for subsequent fatty deposits. This is hard to prove experimentally, however, and all that can be said with certainty is that animals made hypertensive by surgical manipulation develop atherosclerosis more readily than do normal animals.

There are two main dangerous consequences of arteriosclerosis. Very high blood pressure is needed to maintain adequate flow through the partially occluded arteries and the lessened elasticity also prevents their "giving" with the increased cardiac output of exertion or excitement. In consequence, the pressure may shoot up to very high levels in a sudden effort or an outburst of anger, leading to the rupture of an artery. If the rupture occurs in the brain, the hemorrhage and subsequent ischemia of the brain tissue served by the artery have profound neurological consequences. This is one form of *cerebrovascular accident* (C.V.A.), or *stroke.* Equally serious and more frequent is the blockage of an artery by a calcareous plaque that has broken away from the arterial wall. It may lodge in a cerebral artery with similar consequences to a hemorrhage, or it may occlude a coronary artery and cause ischemia and the death of the heart muscle supplied by the artery. This is called a *myocardial infarcation.*

Sometimes the blockage may be severe and cause a dramatic and fatal C.V.A. or "heart attack." On the other hand, there may be a succession of minor infarcts over a period of months or years which gradually destroy the heart muscle until it can no longer maintain an adequate output and goes into *failure.* Similarly, in very old people, the muddle headedness and tendency to become lost or confused are probably attributable to the deterioration of the brain following a series of minor strokes, which individually pass almost unnoticed.

The roughness of calcareous plaques on arterial linings presents a further hazard. The mechanical damage to passing platelets may be severe enough to set off the clotting reactions (see Chapter 6). The intravascular clots or *thrombi* are as dangerous as plaques which break away.

Arteriosclerosis and fat intake. Arteriosclerosis is the underlying cause of the cardiovascular diseases which account for more deaths than cancer, tuberculosis, and the other "killer diseases" put together. Consequently, it was with great interest that clinicians noted that the deaths from cardiovascular disease decreased in Britain and Germany during the war years, despite the severe stresses of mass air raids, long working hours, separation of loved ones, and personal bereavement. The decrease in these diseases seemed to be correlated with the low fat content of the wartime diet in those countries. Furthermore, it is also apparent that in poorer countries with a relatively low-fat diet, there is a smaller incidence of cardiovascular disease than in the United States, which has the largest per capita fat consumption in the world. (The greater life expectancy in this country is mainly due to better survival in infancy. For example, an American baby has a greater chance of living to the age of 50 than its Italian cousin, but an American of 50 has more chance of developing a fatal cardiovascular disease than an Italian of 50).

These correlations have provoked intensive research into the fundamental relationship of dietary fat and atherosclerosis. Although the interpretations are not yet universally accepted, much research suggests that the *cholesterol* intake is critical. Cholesterol is a lipid found in the fats of eggs, milk, and meat, which the body uses in modest amounts as the precursor of the *steroid* hormones (see Chapter 16). Certain patterns of the size of lipoprotein particles in the blood are correlated with atherosclerosis and these patterns tend to occur if the dietary cholesterol intake is high. On reducing the cholesterol intake, the blood-lipid pattern changes to one not correlated with atherosclerosis. The effect of cholesterol intake is exacerbated by a caloric intake in excess of requirements and conversely, a more frugal regime with the same cholesterol content is as good as reducing the total cholesterol intake. This has been demonstrated in feeding an identical diet (high calorie, high cholesterol) to two groups of young men. Those who carried out their normal sedentary activities showed the sinister changes in blood lipids, whereas the other group on an intensive work and exercise regime did not.

The final conclusions about the relationships between cholesterol and cardiovascular disease must await the results of careful statistical studies, now in progress, of the populations of whole cities. The evidence at present supports the traditional medical observation that obesity and a sedentary life predispose to "heart trouble" and common sense suggests an avoidance of too much dairy produce in middle age.

Heart Failure

The term, "heart failure," means that the heart is unable to achieve an adequate cardiac output to maintain the oxygen and nutrient supplies to the tissues. Almost by definition, this must have fatal consequences if uncorrected, but ordinarily a failing heart can achieve enough output for resting requirements if the various mechanisms for increasing the output during exercise are brought into play. In addition, the slow-acting renal mechanisms increase the extracellular fluid volume and elevate the filling pressure. When a near-normal output is achieved by these means the heart is said to be in "compensated failure." It will be noted that the compensation is adequate only for rest or minimal activity, because the normal reserve mechanisms are already in use and are not available for meeting the requirements of exercise. The failure itself can be due to many causes, some of which are listed.

i. *Valvular defects* which cause inefficient pumping and regurgitation. These may be congenital in nature but more frequently they are a legacy from rheumatic fever. The course of this serious disease is marked by inflammation and ulceration of endothelial tissues such as those lining the joint cavities, the blood vessels, and the valves of the heart. After the acute phase of the illness, the valve tissue is often so scarred or destroyed that the valves no longer shut tightly when pressure closes them; this is called valvular *incompetence.* Alternatively, the leaves may be stuck together by scar tissue so that the valve does not open properly; this is valvular *stenosis.*

Mitral incompetence for example, causes some of the blood to reflux into the left atrium instead of being forced into the aorta during ventricular systole. In consequence, the left ventricle has to do more pumping than would otherwise be necessary to achieve a given cardiac output. In *mitral stenosis* the ventricle fills too slowly for an effective stroke volume to be achieved. Similar defects in the other cardiac valves lead to comparable losses of efficiency.

ii. *Muscular defects* of the heart typically follow an infarction because the death of part of the myocardium leaves too little muscle for adequate pumping. Even without gross blockage, severe arteriosclerosis of the coronary circulation can cause ischemia and general weakness of the myocardium. Weakness of the heart muscle can also accompany severe electrolyte defects (see Chapter 5).

iii. *Conduction defects* are usually associated with the muscle damage in an infarction. The conductive tissue in the ventricles may be destroyed by the ischemia so that the ventricles contract independently, or at every second or third beat of the atria. This makes for marked loss of efficiency.

An extreme conduction defect is ventricular *fibrillation.* Impulses arise in an apparently random fashion so that various parts of the ventricles contract rapidly and independently of adjacent parts. The result is an aimless quivering of the ventricles with no pumping value whatsoever. Defective hearts sometimes go into short spells of fibrillation which correct themselves, but the condition is fatal if extended beyond a very few minutes. The fatal result of a severe electric shock is often the fibrillation which follows the disruption of normal conductivity.

iv. *Overdilation* by a long-standing venous engorgement can also weaken the myocardium.

Acute failure. The cardiovascular system is a virtually closed circuit and ordinarily the heart can only pump out as much as it receives and it can only receive as much as it pumps out. Over a short period though, cardiac failure can be marked by a venous engorgement due to the inability of the heart to "get rid of" the venous return. If the failure is chiefly in the left heart, with the right heart relatively normal, the blood piles up in the pulmonary circulation, leading to edema and congestion of the lungs. This impairs the gas exchange in the alveoli and the consequent anoxia and build-up of carbon dioxide in the blood make the patient feel he is "fighting for breath" (see Chapter 8). The situation is worse when the patient is lying down, and *paroxysmal nocturnal dyspnea* (P.N.D.), a characteristic symptom of left heart failure can arise. This is an unpleasant and terrifying sensation.

Chronic failure. *Chronic failure* is better described as *compensated.* This is also marked by venous engorgement, but not for the reasons described previously. In the long run, the reduced output of the failing heart triggers the mechanism controlling the secretion of aldosterone and the retention of salt, just as if the falling output were due to a hemorrhage. This "blind" response of the homeostatic control is effective in that it increases blood volume and cardiac output, but at the expense of edema due to the massive increase in interstitial fluid.

REFERENCES

1. Bing, Richard J., "Heart metabolism," *Sci. American* **196** (2), 50, February 1957.
2. Colp, Ralph Jr., "Ernest Starling," *Sci. American* **185** (4), 56, October 1951.
3. Fine, Jacob, "Traumatic shock," *Sci. American* **187** (6), 62, December 1952.
4. Kilgour, Frederick G., "William Harvey," *Sci. American* **186** (6), 56, June 1952.
5. McKusick, Victor A., "Heart sounds," *Sci. American* **194** (5), 120, May 1956.
6. Page, Irvine H., "Angiotensin," *Sci. American* **200** (3), 54, March 1959.
7. Rosenthal, Sanford, "Wound shock," *Sci. American* **199** (6), 115, December 1958.
8. White, Paul D., "Coronary thrombosis," *Sci. American* **182** (6), 44, June 1950.
9. Wiggers, Carl J., "The heart," *Sci. American* **196** (5), 74, May 1957.
10. Zweifach, Benjamin W., "The microcirculation of the blood," *Sci. American* **200** (1), 54, January 1959.

RESPIRATION

8

We have seen that the cells require a close control of the oxygen and carbon dioxide in the internal environment. This important aspect of homeostasis is accomplished by allowing the blood to equilibrate with "fresh" air in the lungs, where the oxygen content is replenished and the carbon dioxide diminished. In the broadest sense, respiration refers to all of the oxidative processes concerned with the release of energy from nutrients; in the narrowest sense it is sometimes used merely as a synonym for *breathing*. The oxidative processes within the cells are briefly discussed in a later chapter, and so the present discussion will cover the delivery of oxygen to the cells and the simultaneous removal of carbon dioxide.

All of the gas exchanges in the lungs and tissues take place in *solution* and it is therefore appropriate to review some properties of gases in solution.

PRESSURE AND PARTIAL PRESSURE

The earth's atmosphere at sea level exerts a pressure of 760 millimeters of mercury. That is, the weight of the air pushing down on a surface of mercury will support a column of mercury 760 millimeters high. Because water is thirteen times lighter than mercury, atmospheric pressure is equivalent to more than *thirty feet* of water. Oxygen comprises about 20% of the air, and so it exerts 20% of the total pressure, or 156 millimeters of mercury. This is the *partial pressure* of oxygen in the air. At a high altitude where the atmospheric pressure might be 600 millimeters of mercury instead of 760 millimeters, the partial pressure of oxygen would be only 120 millimeters. Furthermore, if a chamber is filled with pure oxygen, the partial pressure of oxygen is the same as the total atmospheric pressure in the chamber.

Henry's Law. The amount of a gas dissolved in a liquid is proportional to its partial pressure in the gas mixture in equilibrium with the liquid. This is shown in the diagram in Fig. 8.1. Obviously, it may take considerable time for equilibrium to be reached if the volume of liquid is large and there is no agitation. Different gases have different solubilities in water, therefore the actual *amount* dissolved may bear no relation to the proportion of gases in the equilibrium mixture. Each gas in solution, nevertheless, is in *equilibrium* with that in the mixture above the surface. Therefore, it is convenient to speak of the partial pressure, or *tension,* of a gas in solution rather than its *concentration.* Thus, the oxygen tension

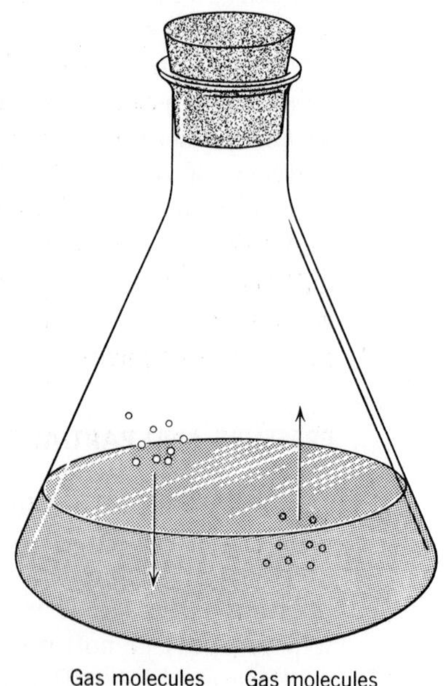

FIGURE 8.1. Henry's law.

of water equilibrated with air at atmospheric pressure is about 150 mm.

Diffusion

A gas dissolved in water behaves very much like a crystalloid, because it will diffuse from a region of high concentration (tension) to a lower. The simple unicellular animal in the ocean or lake is surrounded by water with a practically constant oxygen tension. As oxygen is used up inside the cell a tension gradient is produced, causing oxygen to diffuse into the cell from the environment. Similarly, the build-up of carbon dioxide within the cell makes a gradient in the other direction so that it diffuses outwards. In the *steady state,* the oxygen tension within the cell is slightly lower than that of the environment and the carbon dioxide tension is slightly higher.

Once again, the relatively small volume of the internal environment of the higher animals would be rapidly depleted of its oxygen content if it were not constantly renewed. The renewal is accomplished in the lungs where the blood equilibrates with "fresh" air. The blood flows through capillaries in the walls of millions of tiny air sacs, or *alveoli,* which provide a diffusion surface area equivalent to that of a tennis court. The diminished oxygen tension in the venous blood arriving at the lungs causes diffusion of oxygen into the blood from the alveolar air; conversely, its increased carbon dioxide tension leads to diffusion in the opposite direction (normal air contains only 0.003% carbon dioxide). As is true of other regulatory processes, the equilibration between the blood and the interstitial fluid is so rapid that the regulation of the blood alone is enough to control the composition of all of the internal environment. In the lungs, the oxygen tension is kept at a high level by breathing. The "used" air is expired and repeatedly replaced by inspiring fresh air; at the same time, this process keeps the alveolar carbon dioxide tension low. The renewal of air in the lungs is called *ventilation.*

THE UPPER RESPIRATORY TRACT

The anatomy of the nose and mouth is extremely complicated although the diagram (Fig. 8.2) presents a simplified version. Air drawn in through the nose passes over a complex series of surfaces made up of the *nasal septum* and the *nasal turbinates.* These surfaces not only warm and moisten the air but also act as baffles to

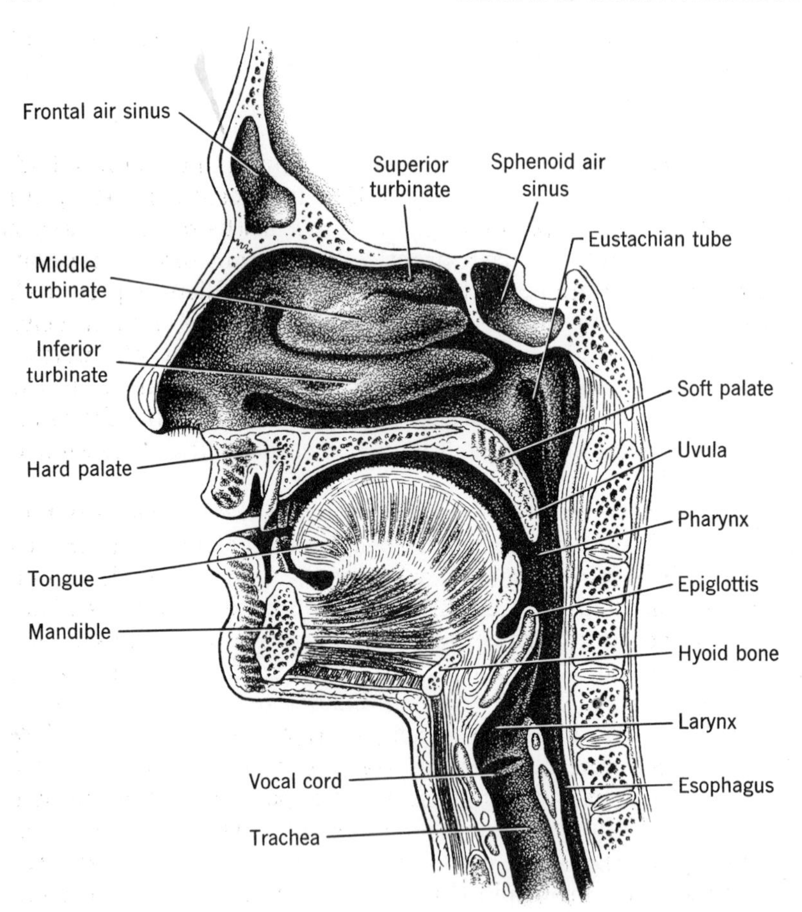

FIGURE 8.2. Diagrammatic section of the upper respiratory tract.

its flow, causing dust particles to settle out on the mucus films lining the nasal passages. Large parts of the surfaces are made up of ciliated epithelia (see Chapter 3), and the beating of the cilia moves the dust-bearing mucus backwards to the pharynx where it is swallowed (or expectorated). Breathing through the mouth results in relatively cold and unfiltered air reaching the lungs.

The nasal and buccal cavities both open into the *pharynx,* and the *esophagus* and the *larynx* descend from the lower part of the pharynx. The *trachea* (wind pipe) in turn descends from the larynx. The *epiglottis* is a flap of tissue which reflexly closes over the larynx when food or drink is being swallowed.

ANATOMY OF THE LUNGS

Figure 8.3 shows the structure of the lungs and their location in the thorax. The trachea descends from the larynx and just below the base of the neck, divides into the two *bronchi* (singular, bronchus). These tubes divide again and enter the various *lobes* of the lung where they repeatedly subdivide into *bronchioles,* which ultimately connect with the alveoli. See Fig. 8.4. These respiratory passages are also lined with ciliated epithelium and the ciliary movement conveys dust particles upward to the pharynx in a slow current of mucus. The thorax is divided from top to bottom by a double wall, the *mediastinum.* The heart, great vessels, esophagus, and many nerves lie within the mediastinum, and the lungs are in the *pleural cavities* on each side.

As described in the chapter on circulation, the pulmonary arteries arise from the right ventricle of the heart. They subdivide throughout the lungs and fill the capillaries in the alveoli, where the blood equilibrates with the inspired air. The oxygenated blood returns to the left atrium by the pulmonary veins and is then pumped out to supply oxygen to the tissues.

Breathing. The lung tissue is elastic but is unable to expand or contract of itself. The mechanism by which the lungs are filled is best illustrated by a model like that in Fig. 8.5. The balloons are connected to the outside air by the Y-shaped tube which passes through the stopper of the jar; the space *a* between the outside of the balloons and the inside of the jar contains air, but is quite closed off from the outside. The floor of the jar is made of rubber, and when this is pulled down, *b,* the same amount of air in the jar fills a greater space and therefore exerts *less pressure* on the outside of the balloons. The inside of each balloon is still subject to atmospheric pressure, and as this is now relatively higher, air enters the balloons and inflates them until the pressure in the space *a* and the elastic tension of the balloons is again equal to atmospheric pressure. When the rubber floor is allowed to return to its original position, the rising pressure in the space *a* squeezes air out of the balloons.

The filling of the lungs is similar but more complex. The rib cage, or *thorax,* is quite air tight and is analogous to the jar; the diaphragm closes off the floor of the thorax as does the rubber in the model and of course, the bronchi and trachea connect the lungs with the atmosphere. In the body, the downward movement of

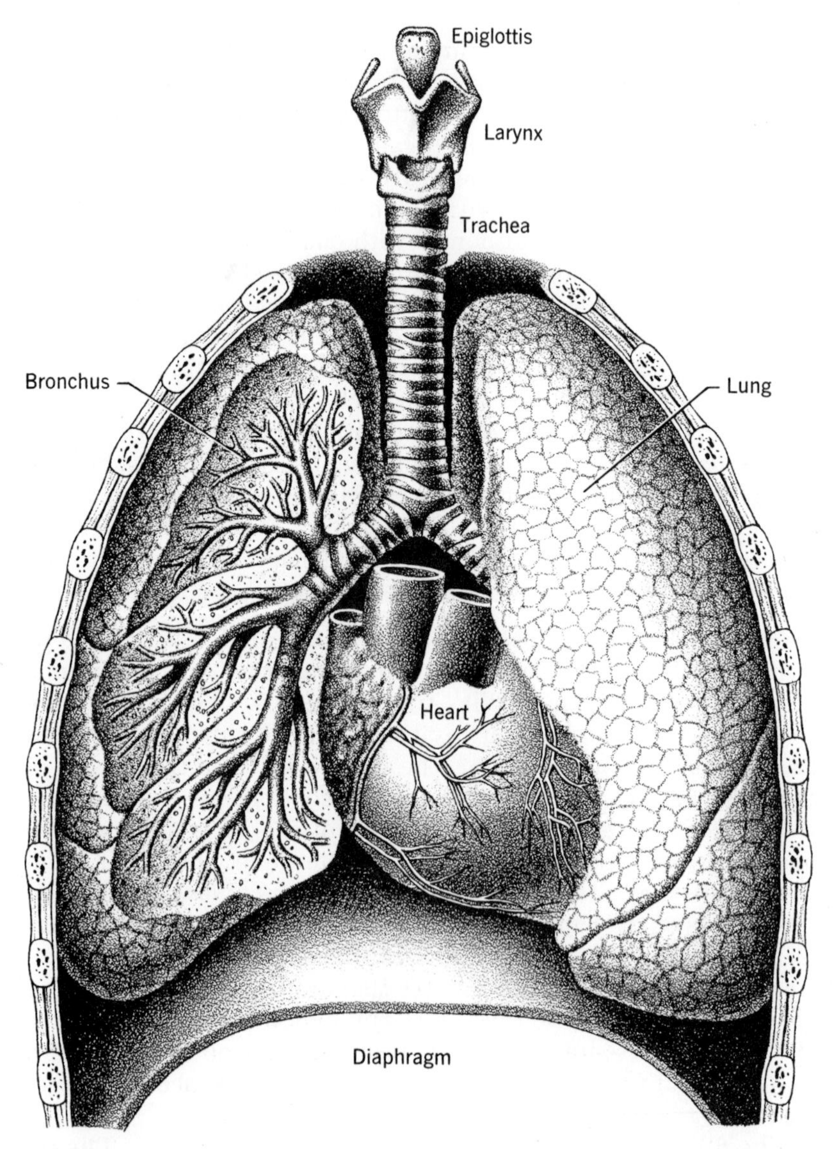

FIGURE 8.3. The respiratory organs *in situ*. At the left side of the drawing the great vessels of the heart and part of one lung have been cut away to show how the trachea branches into a bronchus, which branches in turn throughout the lung into bronchioles.

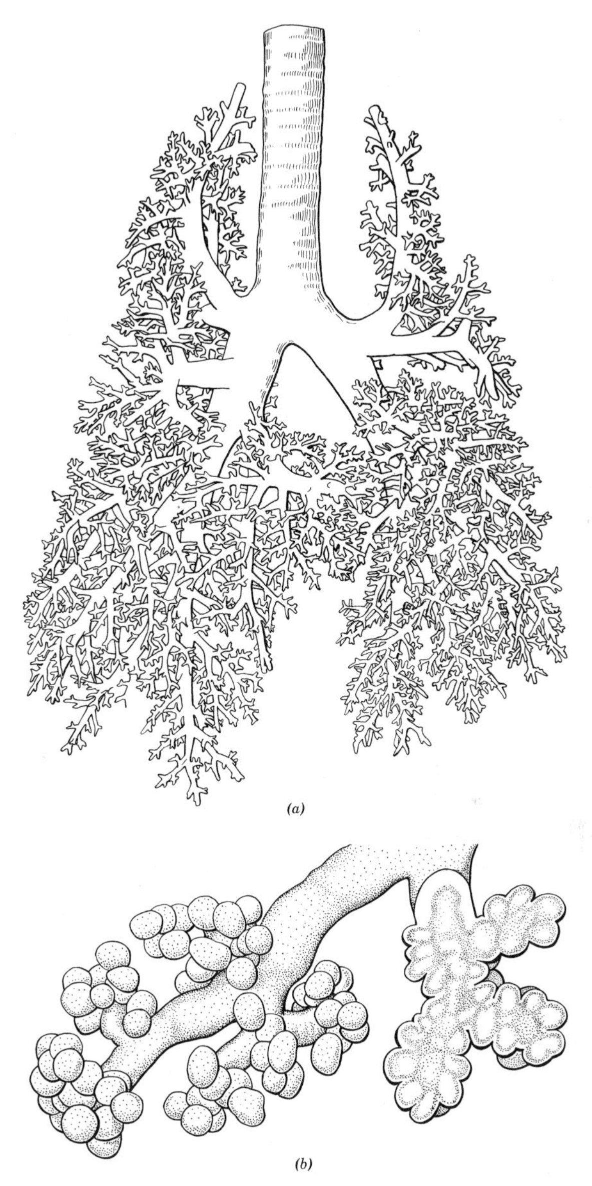

FIGURE 8.4. (*a*) The branched pattern of the airway as seen in a plastic cast. (*b*) The arrangement of the alveoli at the ends of the bronchioles. (From Fenn, W. O., *Sci. Am.* **202** (1) 138, 1960.)

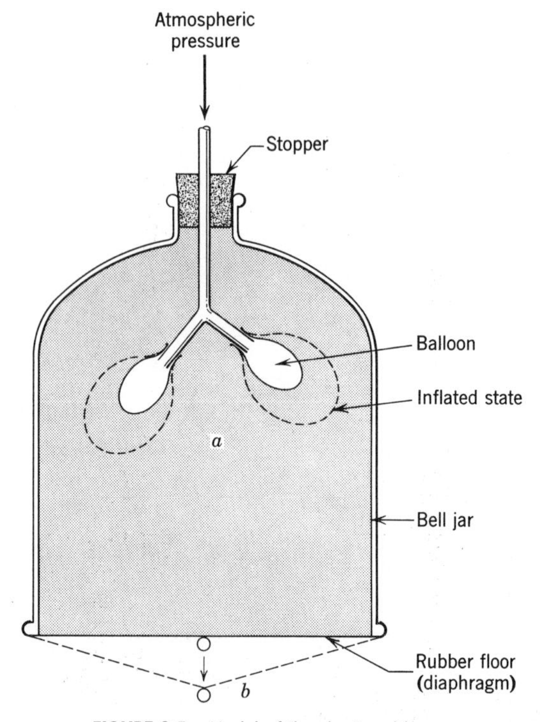

FIGURE 8.5. Model of the chest and lungs.

the diaphragm causes inspiration, but since the rib cage is not fixed, it too can be lifted and expanded to decrease the pressure outside the lung walls. An important difference from the model lies in the fact that the lungs always completely fill the thorax so that the *pleural space* (analogous to *a*), is filled only by a thin film of fluid between the lungs and the chest wall. In spite of its minute volume, it is the pressure changes in this space that cause inspiration and expiration.

Respiratory muscles. The diaphragm is the most important single respiratory muscle. At rest, it forms a dome over the liver and stomach. When it contracts it makes the floor of the thorax slightly flatter, thereby decreasing the intrapleural pressure. Contrary to popular belief, the ribs are not rigidly fixed to the spinal column, they are hinged in a way that allows them to be lifted and rotated by the contraction of the intercostal muscles (and some of the other thoracic muscles). This movement is especially important in deeper breathing (see Figs. 8.6 and 8.7).

Expiration

If the lungs are removed from the body, their elasticity causes them to collapse to a smaller volume than they ever occupy in life. This tendency to collapse and pull away from the chest wall causes the intrapleural pressure to be rather less than atmospheric pressure at rest. Furthermore, this elasticity, or lung *compliance,* means that expiration can occur passively and that a simple relaxation of the inspiratory muscles is all that is necessary. In faster expiration, as in blowing or sneezing for example, compliance is aided by muscular contractions which actively reduce the volume of the thorax and increase the intrapleural pressure.

Pneumothorax. If there is a hole in the chest wall, the intrapleural space is open to the atmosphere and the lowering of the diaphragm does not decrease the intrapleural pressure. Consequently the lung on that side of the thorax does not inflate. In cases of multi-

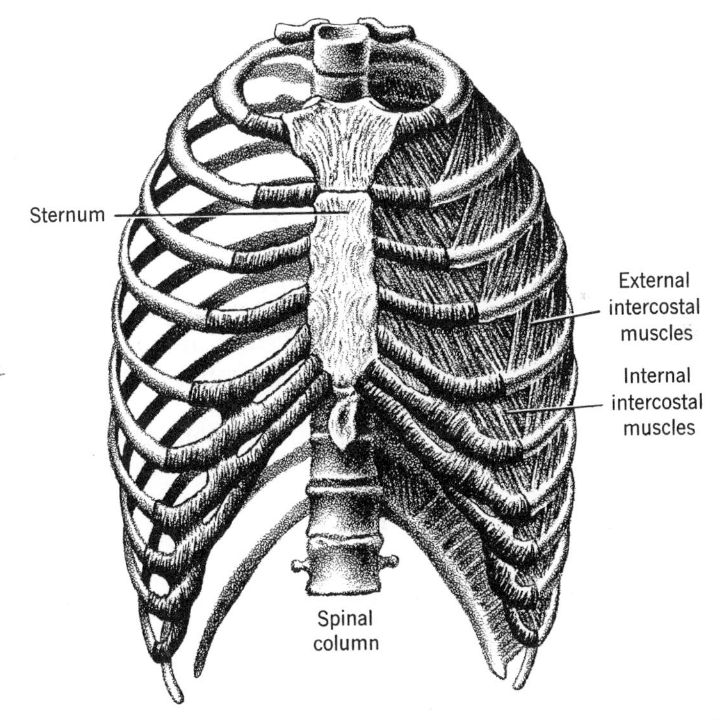

FIGURE 8.6. The rib cage is hinged to the spinal column and to the sternum. Adjacent ribs are connected by two sets of muscles; the external intercostal muscles lift the ribs, the internal intercostal muscles lower them.

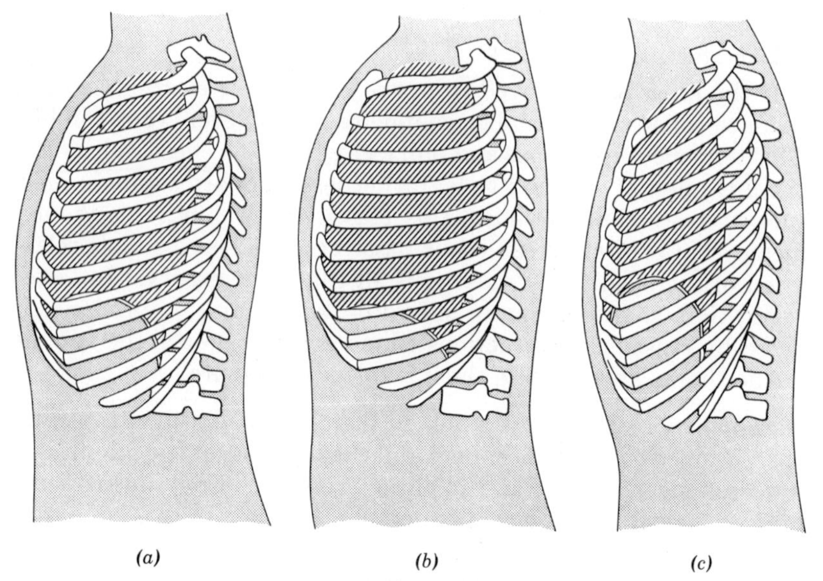

FIGURE 8.7. Chest and diaphragm positions in breathing. (*a*) The end of a normal expiration. (*b*) Maximum inspiration. (*c*) Maximum expiration.

ple gunshot wounds of the thorax, the patient can die of anoxia unless the holes are covered with tight dressings. Before the development of drugs effective against tuberculosis, this disease was commonly treated by a surgical *pneumothorax*. Some sterile air (or other gas) was injected into the pleural space on one side causing that lung to collapse. This enforced "rest" for the diseased tissue was considered of value. In any surgical operation involving the opening of the chest, the patient cannot breathe for himself and must be artificially ventilated.

The volume of air moved at each inspiration, or the *depth* of breathing, is determined by the extent of the excursions of the intercostal muscles and the diaphragm. The total ventilation of the lungs depends on the *depth and rate* of breathing, and these are controlled by the *respiratory centers* in the brain. Their function is discussed later. In ordinary quiet breathing, the rate and depth are automatically adjusted to give adequate ventilation with the least muscular effort.

LUNG VOLUMES

Much can be learned of lung function by measuring the volume of air moved at each breath. This is conveniently carried out using a *spirometer,* (see Fig. 8.8). The bell *a* is partially full of air and is connected by a tube *b* to the subject's mouth. At inspiration, the removal of air causes the bell to sink; expiration causes the bell to rise. The movement can be observed on the scale *c,* which can be calibrated for volume of air moved. An added refinement is the use of a pen writing on a rotating drum (kymograph), which gives a continuous record of volume changes within the bell.

An idealized record is shown in Fig. 8.9. The relatively small oscillations at the left are due to normal quiet breathing; the volume they represent is the *tidal volume.* A maximal inspiration followed by a maximal expiration give the next part of the record. The greatest volume that can be inspired from the tidal end-expiratory position is the *inspiratory capacity.* The greatest volume that can be expired from the tidal end-expiratory position is the *expiratory reserve.* The total of inspiratory capacity and expiratory reserve is the *vital capacity.*

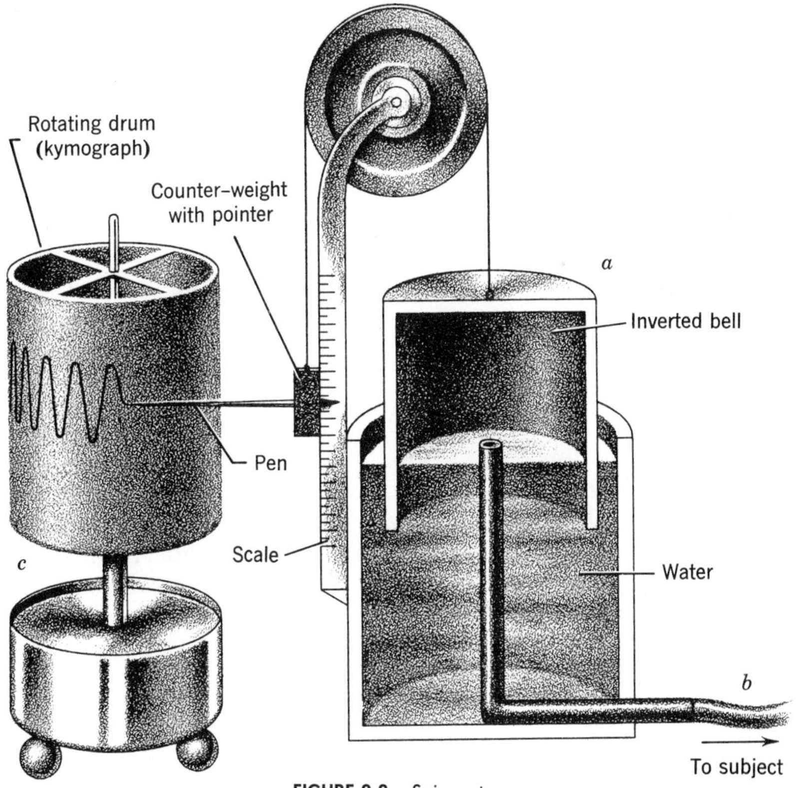

FIGURE 8.8. Spirometer.

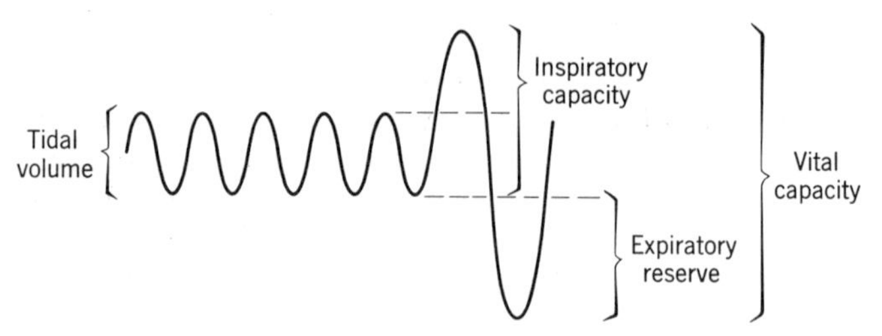

FIGURE 8.9. Idealized spirometer record.

On the whole, trained athletes tend to have larger than normal vital capacities. Many pulmonary diseases are characterized by small vital capacities. There is great variation among normal individuals however, and only very small vital capacities are necessarily sinister. *Changes* in the vital capacity of a subject are frequently useful in following the progress of pulmonary (or cardiac) disease, because the increase or decrease of pulmonary edema is reflected in the vital capacity.

In addition to the lung volumes described, the *residual air* is that still contained in the lungs after maximal expiration. It is sometimes of clinical interest to measure this. One method for this determination is an application of the dilution method described for measuring body fluids (see p. 68). The subject rebreathes from a spirometer a known volume of a gas mixture containing helium or some other inert gas. The original concentration of helium is *diluted* by the air in the subject's lungs which of course originally contains no helium. The extent of the dilution is proportional to the subject's total lung volume, which can therefore be calculated.

Dead space. At inspiration, not all of the fresh air gets down to the lungs where gas exchange with the blood can take place; some of it remains in the bronchi and trachea. Conversely, at expiration, these tubes remain full of "used" air, which is taken back into the lungs at the next inspiration. Consequently the air actually in the alveoli always has a lower oxygen content than room air because each breath includes some air which has already been in the lungs. The volume of air in the bronchi, trachea, etc., is called the *dead-space* volume. Some typical values for these various lung volumes are given in Table 8.1.

Ventilation and perfusion. The amount of oxygen entering the blood in any period will depend on, among other factors, the steepness of the diffusion gradient in the lungs and the amount of blood passing through the lung capillaries. The oxygen tension in the lungs

TABLE 8.1 Some Typical Pulmonary Data for a 70 Kilogram Man

Tidal volume	500 milliliters
Inspiratory capacity	3500 milliliters
Expiratory reserve	1200 milliliters
Vital capacity	4700 milliliters
Dead space	150 milliliters
Residual volume	1200 milliliters

can be raised by increasing the ventilation rate, which has the reciprocal effect of lowering the carbon dioxide tension. As a result, the oxygen tension in the blood leaving the lungs is increased and the carbon dioxide tension is decreased. This is exactly what happens in exercise; the increased breathing (*hyperventilation*) helps to meet the tissues' increased oxygen demands and to get rid of their extra carbon dioxide production. Furthermore, the increased cardiac output in exercise means that more blood is re-oxygenated per unit time, and so more oxygen is delivered to the tissues. As is discussed later, the hyperventilation in exercise is chiefly in response to the increased carbon dioxide tension of the blood rather than to the lowering of the oxygen tension.

Artificial respiration is often necessary in accident victims, paralytic diseases or open chest surgery. Prompt application can save the life of a nearly-drowned or electrocuted person for example. A small degree of ventilation can be achieved by the classic *Schafer* method. The subject is laid face down and pressure is applied periodically by pushing with the hands in a downward and forward direction on the lower ribs. Slightly better is the *Holger-Nielsen* method in which expiration is achieved in a similar way, but inspiration is aided by lifting the subject's arms from the elbow (see Fig. 8.10).

Studies on the tidal volumes achieved by these methods in curarised volunteer subjects show that they are disappointingly small, and often barely exceed the dead space. The method of choice is now the mouth-to-mouth method. Its principle is that the subject's nose is closed by the fingers or a clamp and the operator breathes forcefully through his own mouth directly into the subject's mouth. In this way, a substantial volume of air is passed by positive pressure. It has long been realized that this is the best method, but esthetic and possibly deeper psychological objections have prevented its being widely advocated. In many hospitals, hard rubber tubes are available which can be inserted into the sub-

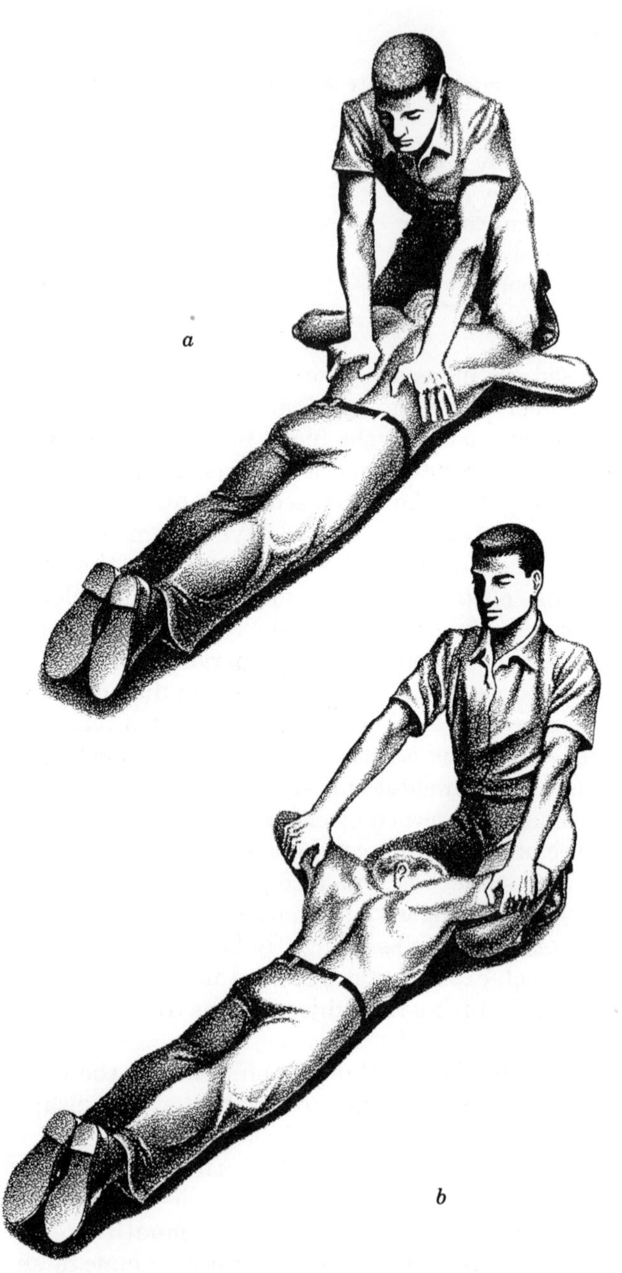

FIGURE 8.10. The two basic positions of the Holger Nielsen method of resuscitation. (*a*) Expiration. Rock forward, keeping your elbows straight. Press with the weight of the upper part of your body, slowly and evenly downward on the victim's back. This empties the lungs. (*b*) Inspiration. Draw the victim's arms upward and toward you until you feel a slight resistance at his shoulders. Do not bend your elbows. Lifting the arms expands the chest so that air can get into the lungs. Drop the arms.

ject's mouth to hold down the tongue as well as providing an airway through which the operator can forcefully expire. This method is extremely effective but takes a certain amount of skill and practice.

During surgical procedures in which the patient is curarised, or where the chest is open, artificial respiration is provided through an *endotracheal* tube—a tube passed under the epiglottis into the trachea. The tube is attached to a rubber bag which the anesthetist can squeeze, or to a simple mechanical pump. For prolonged artificial respiration, as needed in paralytic diseases, the patient is placed in a tank with only his head projecting. By means of an electric pump and valves, the pressure in the tank is periodically lowered so that atmospheric pressure inflates the lungs, just as when the excursions of the respiratory muscles lower the intrapleural pressure. In this case, however, the pump also applies positive pressure to aid expiration. The apparatus is popularly known as an "iron lung."

RESPIRATORY FUNCTION OF THE BLOOD

The oxygen *dissolved* in the blood is almost completely equilibrated with the alveolar air in the lungs, but the *amount* of oxygen carried in solution would be quite inadequate to meet the enormous demands of the tissues. The oxygen-carrying capacity of the blood is increased sixty-fold by the loose combination of oxygen with hemoglobin. As oxygen dissolves in the plasma, it diffuses into the red cells and combines with the hemoglobin to a degree which, up to a saturation point, depends on the oxygen tension. In the tissues, the dissolved oxygen diffuses into the interstitial fluid and then into the cells, causing the hemoglobin to "unload" its oxygen as the tension falls. Hemoglobin also combines with carbon dioxide, and therefore in passage through the capillaries, the red cells unload oxygen and take up the carbon dioxide diffusing out of the cells.

The relationship between oxygen tension and the uptake of oxygen by hemoglobin is best illustrated by the *oxygen-hemoglobin dissociation curve* in Fig. 8.11. The shape of this curve is of a profound importance which should be clearly understood. In this graph we see that the hemoglobin is 95% saturated at an oxygen tension of 90 millimeters of mercury, and that further increases in the gas tension do not increase the saturation very much. On the other hand, when the pressure falls below 60 millimeters of mercury, the saturation decreases sharply. That is to say, oxygen-

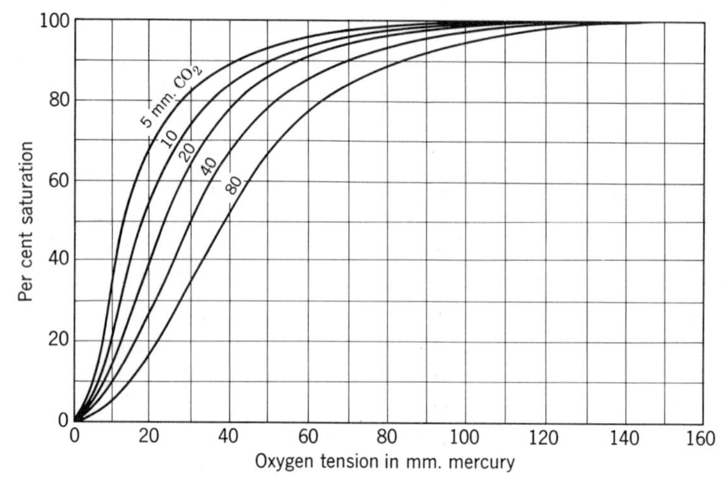

FIGURE 8.11. Oxygen dissociation curves of a dog's blood at different CO_2 tensions. (Krogh)

saturated hemoglobin unloads its oxygen very rapidly at partial pressures below 60 millimeters of mercury.

These properties are valuable in delivering oxygen to the cells. The hemoglobin can be oxygenated in the lungs by a partial pressure of only 90 millimeters almost as well as it can if the alveolar air has as much as 700 millimeters (which is the theoretical maximum when breathing pure oxygen). When the blood reaches the capillaries, the much lower oxygen tension (due to oxygen consumption by the cells) leads to the immediate release of a great deal of oxygen. This unloading effect is slightly enhanced by the simultaneous uptake of carbon dioxide, because the hemoglobin-carbon dioxide combination has a smaller oxygen-carrying capacity. Conversely, in the lungs, the uptake of oxygen and its combination with the hemoglobin greatly reduces the carbon dioxide capacity, thereby facilitating the unloading of carbon dioxide. The effect of oxygen uptake on carbon dioxide dissociation is greater than the reverse effect in the tissues. The effect of carbon dioxide on the hemoglobin's capacity to combine with oxygen is known as the *Bohr* effect. Conversely, the interference of oxygen with the carbon dioxide combination is the *Haldane* effect.

In some primitive animals, the oxygen dissolved in the internal environment is sufficient for the demands of the tissues. Others have hemoglobin (or similar respiratory pigments) in solution in their plasma. Dissolved hemoglobin has a lesser affinity for oxygen than hemoglobin contained in discrete red cells, as is found in all the vertebrates.

Hypoxia. There are many situations in which too little oxygen is delivered to the tissues, and a brief discussion of some of them will help to point up the normal functions. Tissue hypoxia can result from one or more of five kinds of defects:

i. Inadequate ventilation leading to a low partial pressure of oxygen in the alveoli.
ii. Obstructed gas diffusion or reduced lung volume for gas exchange.
iii. Inadequate perfusion in which too little blood is brought into equilibrium with alveolar air.
iv. Inadequate transport.
v. Inability of tissues to use the oxygen even if delivered.

Inadequate ventilation occurs at its grossest when a patient is not breathing. In a few minutes, the oxygen in the alveolar air is depleted so that its partial pressure is similar to that in venous blood and there is no diffusion gradient. This situation is encountered to a lesser degree in an asthmatic attack, in which even though the respiratory muscles are making desperate efforts, the constriction of the bronchioles prevents the flow of air. The anoxia resulting from high altitudes may be considered a special case of reduced ventilation, although breathing is unobstructed. With the low atmospheric pressure at altitudes over about 15,000 feet, the partial pressure of oxygen in the alveoli is too low to insure adequate loading of the hemoglobin. This will be discussed in more detail.

Gas diffusion in the lungs can be obstructed by fluid accumulation, as in *pneumonia* and some phases of heart failure, or by a fibrous thickening of the alveolar walls as in *emphysema.* In both conditions, the lung may be adequately perfused with blood, but because this blood does not come into suitably close contact with the alveolar air, there is inefficient gas exchange. The slower diffusion of oxygen means that the hemoglobin does not become saturated during its rapid passage through the lungs, and tissue hypoxia results. Patients with impaired diffusion derive great benefit from oxygen therapy. By raising the oxygen content of their inspired air, the oxygen pressure gradient in the lungs is increased, thereby giving greater "drive" to the diffusion of oxygen into the blood.

When a part or whole of one lung has been removed, the total alveolar surface available for gas diffusion is reduced but the whole of the pulmonary blood flow passes through the intact tissues. This extra perfusion helps compensate for the reduced diffusion surface. Individuals with only one lung are able to maintain arterial hemoglobin saturation under most conditions.

Inadequate perfusion results from circulatory disorders. A reduced cardiac output causes less blood to be oxygenated per unit time and so the tissues become hypoxic. Some of the most extreme perfusion inadequacies are seen in certain congenital heart defects, as for example, a hole between the right and left atria. In this condition, a proportion of the venous blood passes over to the left side of the heart and is pumped out into the systemic circulation without first going through the lungs. A similar result is seen in the persistence of the *ductus arteriosus* (a vessel normally found only in the fetus), which shunts blood from the pulmonary artery directly to the aorta again bypassing the lungs. Perfusion defects lead to anoxia whether the inefficient circulation is in the lungs or the tissues. A limb with impaired circulation (as from a tourniquet) will become anoxic no matter how well oxygenated is the blood delivered to other parts of the body. Perfusion may be impaired with dramatic suddenness when a clot, or *thrombus,* blocks a vessel.

Inadequate transport is most commonly encountered in anemia. If there is too little hemoglobin, the blood must be circulated faster to carry the same amount of oxygen per unit time. Ventilation rate is also increased to raise the oxygen tension and so improve the loading of the hemoglobin available. The same results can follow if the hemoglobin is "poisoned" by combination with other materials. In particular, hemoglobin has a high affinity for carbon monoxide, a constituent of coal gas and automobile exhausts. This gas is *not* unloaded in favor of oxygen and therefore, that proportion of the hemoglobin combined with carbon monoxide is unavailable for oxygen transport. A prolonged exposure to carbon monoxide results in tying up progressively more and more hemoglobin until there is too little for oxygen transport. Death then follows.

Hemoglobin can be oxidized to *methemoglobin* by many industrial chemicals, notably phenylhydrazine and nitrobenzene. In this form it again has no affinity for oxygen and this kind of poisoning is similar to carbon monoxide poisoning. A small degree of methemoglobin formation is accepted as a side effect in the use of several otherwise useful drugs.

Various drugs and poisons affect the ability of the cells to utilize oxygen. These usually act by disrupting the functions of the *cytochrome* systems responsible for electron transport. Cyanide is the most spectacular poison in this respect.

NERVOUS CONTROL OF RESPIRATION

As is illustrated by the model in Fig. 8.5, the movement of air in and out of the lungs depends on a rhythmic contraction and relaxation of the diaphragm, intercostals, and other respiratory muscles. Because muscles contract in response to volleys of impulses over the nerves serving them, it is clear that breathing originates with some "on-off" nervous control. This control is quite automatic and unconscious; indeed, if we pay attention to our breathing it is difficult to breathe "naturally." We still do not understand all the nervous factors in respiration, but the following concepts are widely accepted and offer a partial explanation. For simplicity, it is convenient to consider the rhythmicity of breathing from two aspects, although it cannot be too strongly emphasized that they function smoothly together. We will consider first the mechanism of the basic on-off rhythm and then the factors that control the rate and amplitude of this rhythm.

Rhythmicity of breathing. Within the medulla of the brain there are some paired structures which are together known as the *respiratory centers* (see Fig. 8.12). On each side there is an *inspiratory* center and an *expiratory* center. The inspiratory center sends out nerve impulses causing the contraction of the respiratory muscles and, if left to itself, would keep the chest in a constant state of inspiration. Within the lung tissues themselves, however, are *stretch receptors*—specialized nerve endings which are stimulated by stretching with the expanding lung tissue and which send impulses back to the expiratory center. These impulses from the stretch receptors stimulate the expiratory center, which in turn sends signals to *inhibit* the inspiratory center. When inhibited, the inspiratory center no longer sends out impulses keeping the chest muscles contracted; consequently, the lungs deflate passively as described previously. As the lungs empty, the stretch receptors are no longer stimulated, and the inhibitory effect of the expiratory center dies away rapidly. A few seconds later, the impulses from the inspiratory center are renewed and the cycle begins again.

Through the stretch receptors, the act of inspiration is automatically stopped after it has gone so far. This is known as the *Hering-Breuer reflex*. In addition to this reflex, there is another automatic mechanism which also periodically shuts off the *tonic* excitation of the inspiratory center. Higher in the brain, in the *pons,* the *pneumotaxic* center is located (see Fig. 8.12). Some impulses from

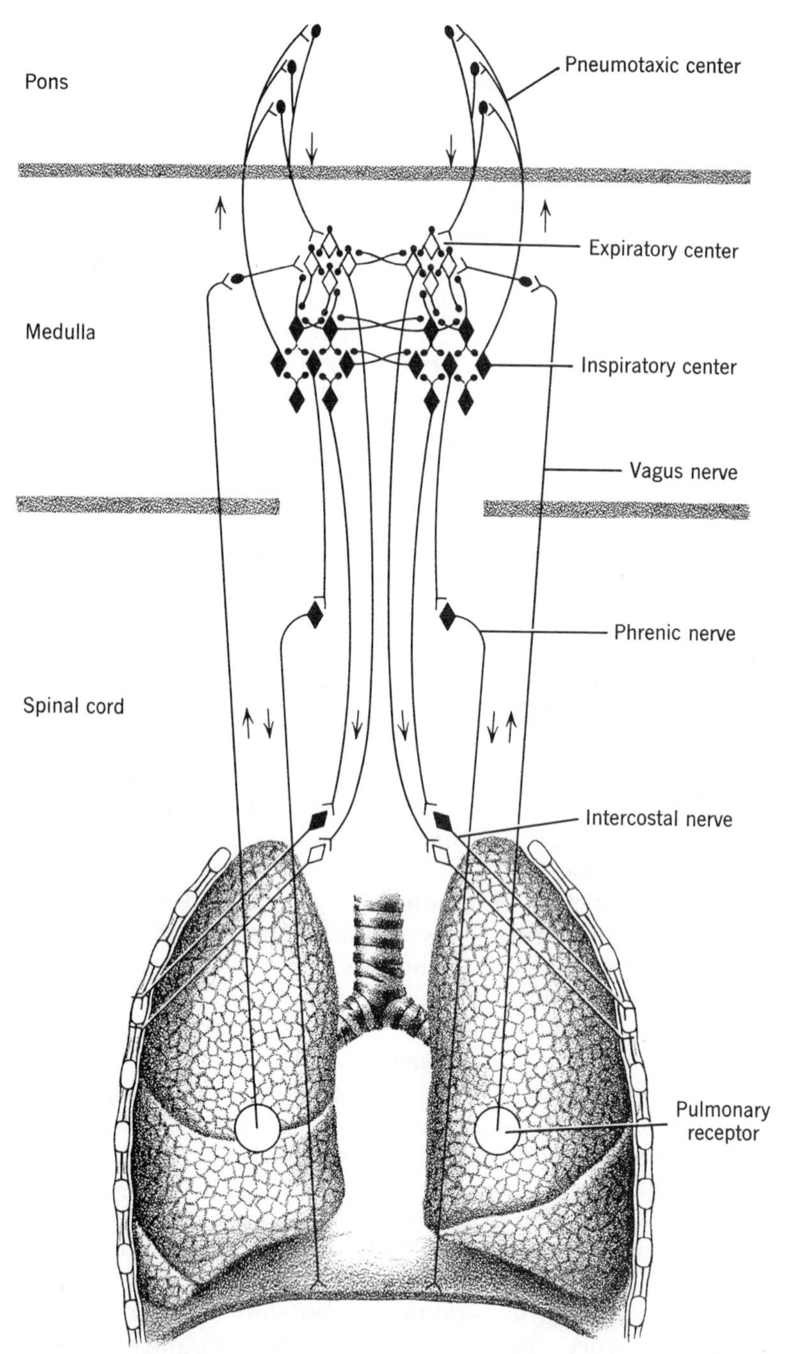

FIGURE 8.12. A schematic outline of the nervous mechanism of breathing. At the bottom are the trachea, lungs, rib cage, and diaphragm. The arrows indicate the direction in which nerve impulses travel. (From Kaufman, W., *Sci. Am.* **187** (2) 28, 1952.)

the inspiratory center *ascend* the brain stem and progressively excite the pneumotaxic center to a degree at which it discharges signals to the expiratory center, in much the same way as do the stretch receptors in the lungs. The excited expiratory center then inhibits the inspiratory center. Once this has been completed, the pneumotaxic center requires another series of impulses from the inspiratory center to excite it again.

Experiments have shown that breathing continues after the vagus nerve (containing part of the Hering-Breuer reflex arc) is cut, but the respiration rate is slowed considerably. After this, though, the destruction of the pneumotaxic center stops breathing entirely. There are therefore, at least two neural mechanisms for maintaining the rhythm of breathing.

Ventilation Rate

The oxygen and carbon dioxide tensions in the internal environment depend on the partial pressures of those gases in the alveolar air with which the arterial blood is equilibrated. The gas exchange in the lungs continually diminishes the oxygen content of the alveolar air and adds to the carbon dioxide. This effect is offset by the influx of fresh air with each breath. The average composition of the alveolar air depends, therefore, on the rates of oxygen consumption and carbon dioxide production on the one hand, and on the rate and depth of breathing on the other. The oxygen consumption and carbon dioxide production depend on the level of metabolic activity in the tissues; for example, they both increase in muscular exercise. In order to maintain near constancy of the respiratory gas tensions in the blood, the alveolar air composition must be maintained constant. Therefore, at each level of exercise, there is a ventilation rate that will "match" the metabolic rate.

The regulation of the ventilation rate is largely accomplished through the sensitivity of the respiratory center to the arterial carbon dioxide tension. When this tension begins to increase, as in exercise, the center is excited and the ventilation rate is stepped-up. The increased ventilation permits the blood to lose carbon dioxide more readily, and consequently, the arterial carbon dioxide is kept at its resting level, although it may be *produced* at several times the resting rate. Similarly, the inspiration of air which already contains a small percentage of carbon dioxide induces a faster rate of breathing, or *hyperventilation.*

It is now known that the increased carbon dioxide in exercise is not the only stimulus for the hyperventilation. The respiratory center is also sensitive to the increased acidity of the plasma under these conditions, resulting not only from carbonic acid (formed by the reaction of $CO_2 + H_2O$) but also from the lactic acid produced by exercising muscles. There is also good evidence that the respiratory center becomes generally more excitable (by carbon dioxide for example) in exercise. It is supposed that this excitability results from afferent nerve impulses arising from the exercising muscles or impulses from the motor cortex.

Surprisingly enough, the lowering of the blood-oxygen tension by increased consumption in exercise is not an important factor inducing hyperventilation. Reference to the oxygen-hemoglobin dissociation curve (Fig. 8.11) shows that minor decreases in the partial pressure of oxygen in the alveoli have little effect on the oxygen uptake by the blood. With a normal alveolar oxygen pressure of 90–100 millimeters of mercury, hemoglobin is about 95% saturated, but the partial pressure can fall as low as 60 millimeters of mercury and the saturation is still 90%. Only at oxygen pressures below 60 millimeters of mercury is the oxygen-carrying capacity of the blood seriously affected. The respiratory center is so sensitive to increased carbon dioxide, that in exercise, the hyperventilation induced by the extra CO_2 production is more than enough to maintain oxygen tensions in the broad optimal range.

There is however, a relatively insensitive mechanism which invokes hyperventilation if the arterial oxygen tension falls seriously. Within the wall of the arch of the aorta and in the bifurcation of the common carotid arteries (see Fig. 7.17), there are *chemoreceptors,* contained in the *aortic* and *carotid* bodies respectively. They are sensitive to the oxygen (and carbon dioxide) levels in the blood, and if the arterial oxygen falls too low, they are stimulated to send impulses to the medulla which speed up ventilation. As indicated previously, the response of the respiratory center to increased carbon dioxide seldom permits the less sensitive chemoreceptors to respond in exercise. At high altitudes, however, the blood is exposed to a low alveolar oxygen tension with *no corresponding increase in carbon dioxide.* The chemoreceptors are then stimulated and their impulses cause hyperventilation, which helps to raise the arterial oxygen tension. This hyperventilation has the effect of simultaneously lowering the alveolar carbon dioxide tension, thereby causing the carbon dioxide to fall *below* optimum in the internal environment. As is discussed in a subsequent section on acid-base balance, this fall below optimum is as deleterious as *excess* carbon dioxide and leads to hyperexecitability of nerve and muscle tissue. It is changes of this nature that cause the type of altitude sickness encountered at about 10,000 feet.

At still higher altitudes, the oxygen tension in the atmosphere is so low that no amount of hyperventilation can raise the alveolar oxygen enough to insure adequate loading of the hemoglobin. At these heights, as in the Himalayas, for example, altitude sickness includes tissue hypoxia as well as hypocapnia (excessive loss of CO_2). These effects can be avoided by breathing pure oxygen, or in practice, air enriched with oxygen. By this means, the alveolar oxygen tension is raised to suitable levels so that there is no hypoxia or hyperventilation. Natives of the high Andes and the Tibetan Plateau show some adaption to their low atmospheric oxygen pressure. Their blood has an increased red cell (and hemoglobin content) and their vital capacities are significantly larger than normal. Lowlanders develop an increased red cell content after residing at high altitudes for several weeks, but their physical performance never matches that of life-long mountain dwellers.

At still greater altitudes of over 30,000 feet, the total atmospheric pressure is so low that even with an all-oxygen atmosphere, the oxygen pressure is too low for adequate saturation of the hemoglobin. In aircraft and space vehicles this is overcome by using air-tight cabins artificially maintained at the atmospheric pressure of a much lower altitude.

REFERENCES

1. Comroe, Julius M. *et al., The Lung,* Yearbook, Chicago, 1955
2. Fenn, Wallace O., "The mechanism of breathing," *Sci. American* **202** (1), 138, January 1960.
3. Gray, George W., "Life at high altitudes," *Sci. American* **193** (6), 58, December 1955.
4. Ingalls, Theodore M., "The strange case of the blind babies." *Sci. American* **193** (6), 40, December 1955.

KIDNEY

9

In the brief review of homeostatis on p. 73, we have seen that the kidney is the prime regulatory organ. By excreting unwanted materials and by retaining others, it controls the composition of the internal environment, particularly with respect to: 1. the osmotic pressure, 2. the concentration of toxic materials, 3. the suitable ratio of electrolyte concentrations, and 4. the acid-base balance.

URINE

The excretion from the kidney is by way of the urine, which contains the surplus and toxic materials. Since these are excreted in *solution,* the formation of the urine requires water. Normally we drink more water than we need to replace evaporative losses; the surplus is excreted by the kidney and is the solvent in which the solid materials are dissolved and excreted. Even when the water intake is marginal, however, some precious water is used for urine formation, otherwise the waste products would accumulate in the extracellular fluid to reach toxic proportions, with results more serious than those from a mild dehydration.

ANATOMY OF THE URINARY SYSTEM

The regulatory mechanisms of the kidney are best considered in terms of the anatomical arrangements. The details and terminology outlined below are the minimum for a proper understanding of the physiology of the regulation of the blood plasma by the kidney. The anatomical names should be learned.

Superficial gross anatomy (see Fig. 9.1). The kidneys are a pair of bean-shaped organs attached to the dorsal (posterior) wall of the abdomen. Each is supplied with a large *renal artery,* which arises from the abdominal aorta at about the same level as the kidney itself, and it enters the kidney at the concavity on the medial surface (the *hilum*). Under resting conditions, about ¼ of the blood pumped out by the heart passes through the kidneys for regulation. The blood leaves by way of the *renal veins* which return to the *inferior vena cava* alongside the renal arteries.

From each kidney, close to the point of entry and exit of the

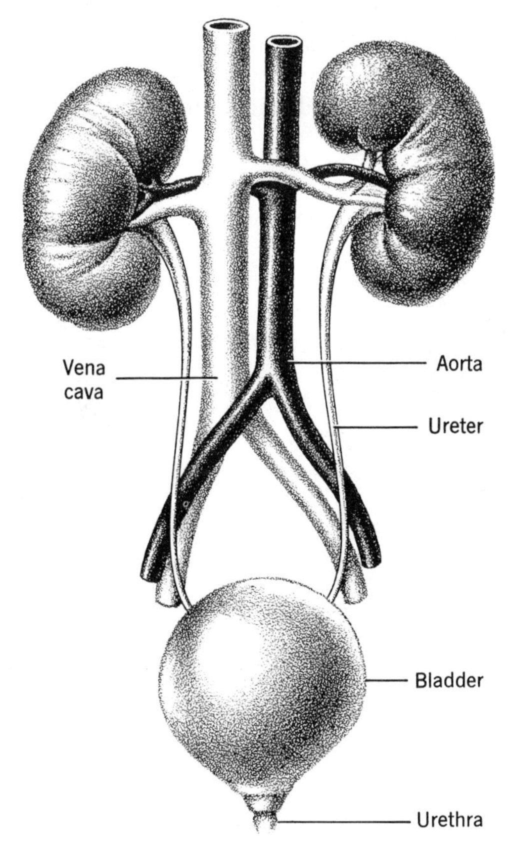

FIGURE 9.1. Kidneys.

blood vessels, arises the *ureter*. By peristaltic contractions, this tube conveys the formed urine from the kidney to the *bladder,* which lies lower in the abdomen, actually within the bowl of the *pelvis.* Urine production is almost continuous and it accumulates in the bladder until the increased pressure invokes the nervous reflex that causes the bladder to empty. When the muscles of the bladder contract, the urine leaves by way of a single tube, the *urethra,* which discharges at the *urethral orifice* just anterior to the vagina in the female and through the penis in the male. The discharge of the urine is known as *urination* or *micturition.*

As mentioned previously, micturition is a reflex act, but in the adult human the reflex is normally inhibited by the will, until a socially opportune occasion presents itself. In small children and other animals, the voluntary inhibition of the reflex is less reliable or absent altogether.

Detailed gross anatomy of the kidney (Fig. 9.2). In section, the "meaty" part of the kidney, known as the *parenchyma,* is a darker brown at the outer edges than towards the center. The darker part is called the *cortex* and the inner part the *medulla.* The

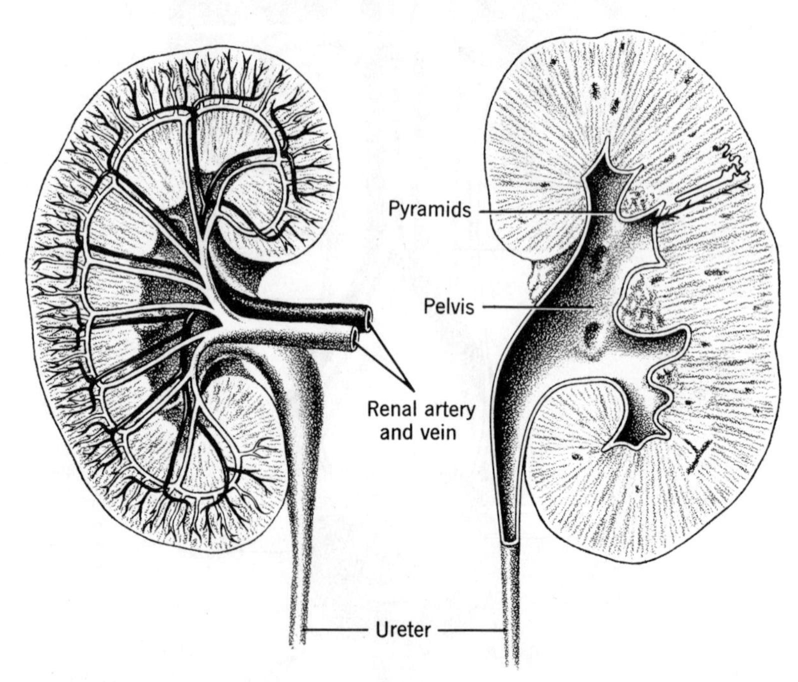

FIGURE 9.2. Kidney sections showing internal blood vessels and arrangement of nephrons.

inner surface of the medulla is convoluted into projections, the *pyramids,* which project into the collecting space called the *pelvis* of the kidney. Numerous *collecting tubules* bring the urine from its sites of formation in the parenchyma to orifices at the apices of the pyramids (*papillae*), where it is discharged into the pelvis. The ureter is the extension of the funnel-shaped pelvis.

Microscopic anatomy. The functional unit of the kidney is the *nephron,* a minute tubule about 12 mm long. The parenchyma of the kidney is made up of the closely-packed nephrons with their associated blood vessels, lymphatics, nerves and collecting tubules. There are about one million nephrons in each kidney and each one can perform part of the work of excretion. Therefore, a study of the function of a single nephron is the essence of renal physiology.

Nephron; Epithelial Portion

The structure of the nephron is shown in Fig. 9.3. Essentially, it is a tube about 12 mm long with thin walls made up of epithelial cells. It is open at one end, where it connects with a collecting tubule which also serves other nephrons. At the closed end, the wall forms a concavity similar to that made by pushing in a finger tip of an empty rubber glove. This *invagination* is known as *Bowman's capsule.* The next part of the nephron is the *proximal convoluted tubule,* which is followed by a straighter portion, the loop of Henle. In general, the capsules of Bowman lie in the renal cortex and the loops of Henle dip down into medulla. The section after the loop of Henle is the *distal convoluted tubule,* which joins with similar tubules to form *collecting tubes.* The walls of the various parts of the nephron have characteristic cellular structures related to their functions described below. The structural details are omitted in the interests of simplicity, but they may be found in any histology textbook.

Nephron; Blood Supply

It is obvious that the blood supply to a nephron is for more than the mere nourishment of those cells, since the function of the nephron is to adjust the composition of the blood delivered to it. The renal artery subdivides within the kidney, and a small vessel (*the afferent arteriole*), enters each Bowman's capsule where it forms a tuft of capillaries (the *glomerulus*), which entirely fills the concavity of the capsule. The blood leaves the glomerulus by way of the *efferent arteriole,* which vessel again subdivides into a mass

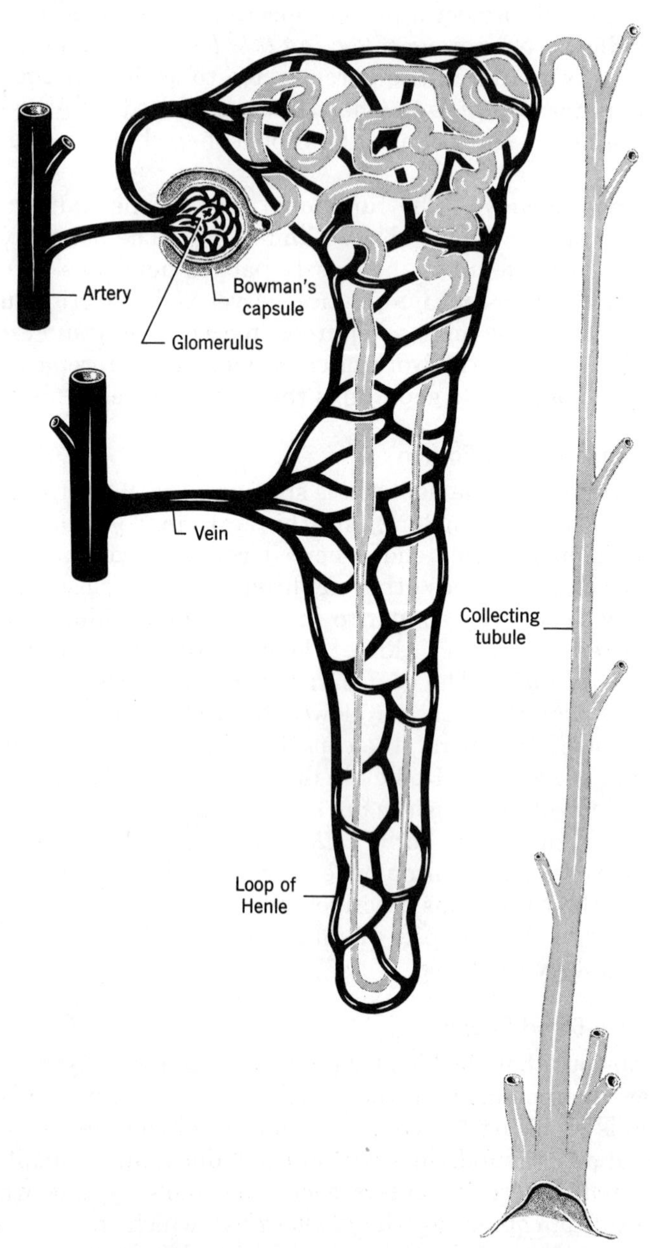

FIGURE 9.3. Diagram of a nephron. (After Homer Smith)

of capillaries that closely encompasses the proximal and distal tubules. The capillaries reunite as veins, which eventually join the renal vein and this returns the blood to the general circulation.

FUNCTIONS OF THE NEPHRON

It is customary to discuss the function of the nephron in terms of the formation of urine, but it cannot be too strongly emphasized that the real function of the nephron is the adjustment of the *plasma* composition. A comparable example is the apple grader in a fruit packing plant. He is primarily engaged in producing boxes of graded apples; the production of boxes of culls is only incidental.

In this section, the formation of the urine will be discussed in the following order:

A. A description of the processes in the nephron.
B. The factors that affect each of the processes, and the way in which they can be controlled so that the kidney excretes or retains material in the maintenance of homeostasis.
C. Renal regulation of acid-base balance.
D. Urine characteristics.
E. Urine characteristics in disease and the consequences of renal failure.
F. Kidney tests.
G. Diuretics.

The processes in the nephron. Figure 9.3 should be referred to in the following discussion:

Filtration. The first process in the formation of urine is *glomerular filtration.* The walls of the glomerulus are similar to the other tissues of the body in that they are permeable to water and other small molecules. Since the blood entering the glomerulus is under pressure (relative to the *lumen* of Bowman's capsule), some of the nonprotein constituents pass through by *ultrafiltration,* which in this connection is usually called glomerular filtration.

The blood cells, of course, cannot pass across the membranes, and they pass on in the capillaries. The glomerular filtrate has then a composition similar to that of the plasma except that it contains no proteins or other large molecules. It contains the waste products which it is the kidney's job to remove from the plasma and in this respect one part of the job is already accomplished by the simple process of filtration.

The newly produced glomerular filtrate amounts to about 180 liters a day, however, and obviously this could not be excreted without depleting the body of water and other plasma components at a fantastic rate. The function of the nephron distal to the capsule of Bowman is to reabsorb the useful materials and to leave the waste products in the tubule for ultimate excretion. As the filtrates moves down the tubules, the reabsorption is accomplished by a variety of processes carried on simultaneously at different levels. The processes must be described singly, but it is important to bear in mind that the overall changes are carried out continuously as the tubular fluid flows past, and not by a succession of batch operations.

Reabsorption of water in the proximal tubule. The blood in the capillaries surrounding the proximal tubules is that which was just previously subjected to ultrafiltration in the glomerulus. Its protein content is intact, but the water and salt content is diminished by the fraction that was filtered out in the glomerulus. The total osmotic pressure of the plasma in the capillaries is therefore increased, and furthermore, the hydrostatic pressure is lower than in the glomerulus. This state is highly comparable to that in the distal ends of the capillaries in the other tissues, described in the section on the Starling principle in a previous chapter (see p. 87). In the nephron, the cells separating the blood from the newly formed glomerular filtrate are permeable to water, and since the osmotic pressure of the filtrate is lower than the now-concentrated plasma, there is a diffusion of water back into the plasma. This is known as the *obligatory reabsorption* of water; the term is self-explanatory.

Active reabsorption in the proximal tubule. At the same time, specialized cells in the walls of the proximal tubules are engaged in the *active transport* of electrolytes and other valuable materials back into the plasma. Active transport refers to the movement of a solute against a concentration gradient, (see p. 66), and this means that the salts, sugar, etc. are moved from the tubular lumen into the plasma, even though the concentration in the tubule is the same as that in the plasma. The active transport processes are not completely understood, but they are known to require energy and can only proceed if the tubular cells are metabolically intact.

The substances reabsorbed in this fashion include the important electrolytes sodium, potassium, calcium, etc., and also glucose, amino acids, and other nutrients. Curiously enough, the list also includes urea, which is one of the waste products, but its reabsorption is only partial. This would appear to be an error in design, but it is logical from considerations of the evolutionary development of the mammalian kidney.

The initial obligatory reabsorption of water renders the glomerular filtrate isosmolar with the plasma. That is, the osmotic pressure of the *tubular urine* is the same as that of the plasma in the capillary. The active reabsorption of salts, sugar, etc. also tends to lower the osmotic pressure of the tubular urine, so that even more water diffuses back into the plasma at a rate that keeps the tubular urine isosmolar. This is also referred to as obligatory water reabsorption. It is thought that the simple diffusion of water occurs particularly easily in the thin-walled section of the loop of Henle.

For simplicity, we have discussed the relationships between the plasma and tubular urine as if they were separated by a single layer of cells. In fact, of course, the materials actively or passively leaving the tubules pass through the tubular cells into the interstitial fluid of the kidney, and from there they must enter the plasma by diffusion through the walls of the capillaries.

Tubular secretion. At the same time that reabsorption is proceeding in the tubules, other compounds are actively *secreted* by the tubular cells into the tubular urine. These are active transport processes in the same way as is the reabsorption of glucose, but they operate in the other direction. Among the substances actively secreted are potassium and creatinine. Both of these materials are also filtered in the glomerulus, and in the case of potassium, most of the amount filtered *and* secreted must be reabsorbed. This apparently useless arrangement is another legacy from the evolutionary development of the kidney. Some "unnatural" compounds such as Diodrast are actively secreted by the tubular cells and the efficiency of their excretion of an injected load is used as an index of the integrity of renal function. (See the section on renal function tests later in this chapter).

Water reabsorption in distal tubules. After the obligatory reabsorption of water in the proximal tubules, the urine is still approximately isosmolar to plasma and in considerable volume. Active transport processes in the distal tubules (and collecting ducts) remove more water, thereby rendering the urine hypertonic to plasma and reducing the volume. The extent of this active reabsorption of water is varied according to the overall fluid status of the body, (see Factors affecting nephron functions).

To summarize the processes in the nephron; the fate of any individual plasma constituent may be:

i. Filtration without reabsorption or tubular secretion, for example, inulin.

ii. Filtration with subsequent reabsorption in part or *in toto,* for example, urea or glucose.

iii. Filtration with tubular secretion, with or without simultaneous reabsorption, for example, potassium, creatinine.

Factors affecting nephron functions. If the kidney is to regulate the plasma composition, the preceding processes described must be sufficiently flexible to compensate for the changes in the internal environment which the vagaries of life may present. The control mechanisms and other factors affecting the urine production are the following.

Filtration rate. Since filtration is the first process in the kidney, the filtration rate determines the amount of plasma treated. In a normal man, the blood entering the glomeruli amounts to 1200–1300 milliliters a minute. This contains 600–700 milliliters of plasma, and of this, about 125 milliliters a minute are filtered. The most obvious factor likely to affect the glomerular filtration rate is the number of functioning glomeruli, which may be reduced by disease. Otherwise, the blood pressure and flow in the glomeruli are the principle determinants of the filtration rate.

The blood pressure is the driving force for the ultrafiltration. Therefore, with a higher blood pressure, the fraction of the plasma filtered at the glomerulus is increased. The rate of blood flow through the glomerulus determines how much plasma will be filtered at any particular pressure.

However, the kidneys maintain a constant filtration rate over a surprisingly wide range of systemic blood pressure, because both the afferent and efferent arterioles are provided with smooth muscle. By appropriate vasoconstriction and vasodilation, the blood flow through the kidney may be adjusted to maintain a constant filtration rate, despite fluctuations of pressure. Nevertheless, the blood flow is still subject to the general nervous control of the circulation. It is normally reduced in exercise, for example, by the shunting of blood away from the viscera to the active muscles. In shock, the fall in blood pressure is greater than can be compensated for by the vasodilation of the afferent arterioles and glomerular filtration actually ceases.

Tubular reabsorption. The active transport in the tubules depends on the integrity of the cells around the lumen. If the metabolism of the cells is inhibited by disease, drugs, or poisons, the active transport of the compounds normally reabsorbed will stop, and the diffusion of the water that normally accompanies these compounds in order to maintain isosmolarity in the tubular urine will also stop. The result is a more copious urine containing wasteful amounts of salts, etc., which should have been retained. This situation can result from anoxia or other general upset of the tubular cells, but the inhibition of active transport can also be more specific. Good evidence exists that there is a separate enzyme system for the transport of each compound (or set of related compounds) that is reabsorbed. These enzymes may be selectively poisoned or may even be missing entirely in an individual, as the result of an hereditary defect.

The actual capacity of each transport system is limited and can be overloaded if the amount of material presented is too great. For example, if the glucose concentration in the plasma is normal (at 100 milligrams percent), the proximal tubules are presented with 125 milligrams of glucose per minute. The transport system can handle this so effectively that no glucose appears in the urine at all. If the plasma glucose concentration is greatly increased by disease or artificial means, the transport system is able to deal with the extra glucose up to about 300 mg per minute (or about 240 milligrams percent in the plasma). In excess of this, some glucose escapes reabsorption and the amount appearing in the urine is that presented in excess of the limit of reabsorption. This limit is known as the *tubular mass* for reabsorption, or the Tm_G, where the subscript "G" refers to glucose. It will be noted that the units of the tubular mass are "mg/min" and do not refer to a plasma concentration. The plasma glucose *concentration* at which the tubular mass is reached depends on the glomerular filtration rate. The concentration at which this normally occurs in an individual is known as the *threshold* for glucose.

The presence of excess glucose in the tubular urine means that less water may diffuse back into the plasma if isosmolarity is to be retained. Thus, the volume of urine presented to the distal tubules is increased. If the increase exceeds the reserve capacity for the reabsorption of water in the distal tubule, a more copious urine will be excreted. An increase in urine volume is called a *diuresis,* and the type just described is a *solute diuresis,* or an *osmotic diuresis.* It is caused by the retention of water in the proximal tubules in order to keep the tubular urine isosmolar with plasma, despite its increased solute content. The resulting volume presented to the distal tubule is greater than the concentrating mechanisms can handle and the volume of urine is consequently increased.

These considerations are important in *diabetes mellitus.* Because of a defect in carbohydrate metabolism in this disease, glucose is continually released into the plasma without being used. The concentration rises until the threshold is exceeded and there is an osmotic diuresis. Uncontrolled diabetes is characterized, therefore, not only by sugar in the urine, but also by a continual diuresis, or *polyuria.* The loss of water in the urine causes great thirst, which is another of the signs of this disease.

The filtration of any of the normally reabsorbed compounds in excess of their tubular mass results in a diuresis; urea is another example and is sometimes injected to induce a diuresis.

Reabsorption in the distal tubule. With the exception of the vasomotor control of glomerular filtration, the factors so far described as affecting the filtration and proximal tubular functions cannot be strictly described as *control* mechanisms. The volume and composition of the tubular urine delivered to the distal tubule is not subject to fine controls which can adjust surpluses or deficits in the internal environment. These adjustments are made in the distal tubule.

The control of the water excretion may be considered first. If water intake is in excess of the requirements, the internal environment becomes slightly diluted, and in this respect strays from the optimum. The dilution of the extracellular fluid lowers its osmotic pressure relative to the intracellular fluid, and water diffuses into the cells in consequence. This causes all the cells to swell, as can be seen in experiments with red blood cells (see Chapter 4, p. 55). In the *hypothalamus* of the brain there are cells which are particularly sensitive to changes in the osmolarity of the extracellular fluid and these *osmoreceptors* are in neural connection with the *neurohypophysis,* or *posterior pituitary gland.*

Under "normal" circumstances of optimal hydration, the neurohypophysis releases a hormone known as the *antidiuretic hormone* or ADH. This travels by way of the blood to the distal tubules and, in a manner that is a complete mystery at present, promotes the *active* reabsorption of water in the distal tubules and in the collecting ducts. This active transport of water is called *facultative* reabsorption as opposed to the *obligatory* reabsorption previously discussed. In overhydration, the osmoreceptors are stimulated by the relative dilution of the internal environment and send nervous impulses to the neurohypophysis which *inhibit* the release of ADH.

The lower concentration of ADH in the blood causes a less efficient reabsorption of water in the distal tubules and results in an increased urine volume. This means that more water is removed from the plasma passing through the kidneys; the relative dilution of the internal environment is corrected, and it returns to its optimal concentration. Conversely, if the water loss exceeds the intake, the extracellular fluid tends to become more concentrated. This hypertonicity has the reverse effect on the osmoreceptors and their inhibition of ADH release is removed. Then, as the concentration of ADH in the blood increases, the reabsorption of water in the distal tubules increases and water is conserved. The control is so fine that the increased reabsorption of water is initiated when the concentration of the extracellular fluid is ideal, so that the conservation of water is already started when dehydration begins. As is discussed in a later chapter on the hormones, ADH is actually synthesized in the hypothalamus and only stored in the neurohypophysis.

The effects of this control mechanism are clearly seen in a *water diuresis.* After drinking a large volume of water, the output of urine rapidly increases and then returns to normal after the excess fluid has been excreted. It is important to distinguish a water diuresis from a solute diuresis described previously. In a water diuresis, the extra urine formation results from reduced active transport of water in the distal tubules; in a solute diuresis it results from the retention of water in the proximal tubule in excess of the subsequent concentrating ability of the distal tubule.

The differences can be shown experimentally. If one subject drinks 1 liter of water and another subject drinks 1 liter of 1.8% urea solution, both will develop a diuresis. If both are then injected with an extract of the neurohypophysis (containing ADH), the diuresis will promptly diminish in the subject who drank the water. The other subject will be relatively unaffected and the diuresis will continue (Fig. 9.4).

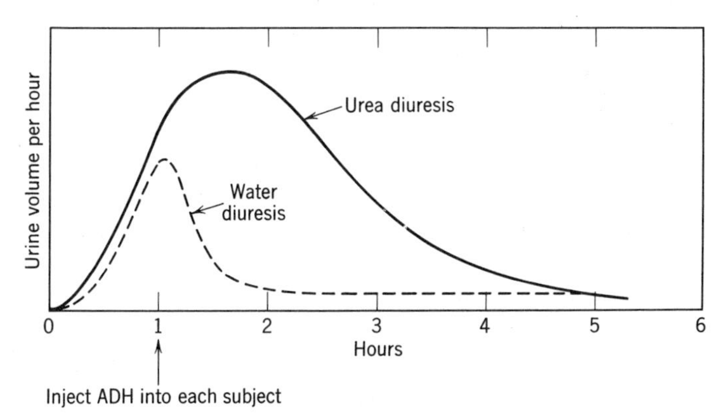

FIGURE 9.4. Water diuresis and osmotic diuresis. An ADH injection will diminish a water diuresis but has no effect on an osmotic diuresis.

It sometimes happens that the neurohypophysis or the hypothalamus is damaged by an injury or tumor, so that the normal ADH secretion is reduced or absent altogether. Without the potentiating effect of the hormone, the reabsorption of water in the distal tubule is always at its minimal level and the unfortunate patient produces a urine volume as high as 20 liters per day. This condition is known as *diabetes insipidus.* The name is a legacy from the medicine of earlier times when the superficial resemblance of the polyuria of diabetes mellitus caused the two diseases to be associated. The names mean "sweet" and "tasteless" respectively, referring of course to the presence or absence of glucose in the urine. The similarity of the names is also unfortunate in that the nature of the polyuria is different in each case. In diabetes mellitus the patient suffers a solute diuresis and in diabetes insipidus the kidney responds to the absence of ADH as if there were a water diuresis.

The upper limit of facultative water reabsorption in the distal tubule is determined by the enzymes of the active transport systems in the same way as is the reabsorption of glucose in the proximal tubules. Even in the most acute water shortage, urine production continues, although it is highly concentrated. The capacity of the active transport system for water is one of the limiting factors for survival in dehydration. Some mammals have adapted to desert life by their ability to excrete a highly concentrated urine, thus minimizing this route of water loss. Insects and some reptiles are particularly successful in this respect, although they have an added advantage in excreting their nitrogen waste in a different form. The response of the distal tubules to minute fluctuations in the osmolarity of the internal environment is an elegant example of a "feed-back mechanism." Most physiological regulations are of this "push-pull" nature and they are becoming widely used in industry with the growth of automatation.

Electrolyte Regulation

Sodium is the electrolyte in the greatest concentration in the tubular urine. It is actively reabsorbed in both the proximal and distal tubules, and the proximal reabsorption carries with it much of the obligatory water absorption.

The completeness of the reabsorption of sodium is affected by several of the steroid hormones secreted by the adrenal cortex. These include desoxycorticosterone and cortisone, but aldosterone has by far the greatest effect. An increase in the secretion of this hormone from the adrenal cortex causes an increased reabsorption of sodium, partly at the expense of a less complete reabsorption of potassium. The mechanism is brought into play following a low sodium intake or excessive loss of sodium in sweat. If a subject is placed on a sodium-free diet, the excretion of sodium is stopped almost immediately. The conservation of potassium is not so prompt and it takes several days on a potassium-free diet before the potassium excretion reaches a minimum.

The fine control of the sodium concentration in the internal environment is effected by the influence of the adrenal cortical secretions on the tubular cells. Since the sodium ion is one of the principal contributors to the osmotic pressure of the extracellular fluid, this mechanism shares with the osmoregulatory center the control of osmotic pressure. Except that potassium regulation is to some extent reciprocal to that of sodium, very little is known about the control of the absorption and excretion of the other electrolytes.

After a hemorrhage, the body fluid volume is reduced but it is still of normal osmolarity. It is presumed that this situation stimulates the secretion of aldosterone, because it has been noted that the reabsorption of sodium is increased after a hemorrhage. The extra reabsorption of salt also increases the obligatory water resorbtion so that the body fluid volume is restored. It will be recalled that the cardiac output can be increased by raising the filling pressure of the heart, which can be accomplished by increasing the total extracellular fluid volume. Therefore, in heart failure we find a retention of sodium and increased extracellular fluid volume even to the point of gross edema. It seems that the adrenal cortex blindly responds to the lower output of the failing heart as if the decreased output were due to hemorrhage. The edema of heart failure is uncomfortable for the patient and can lead to pulmonary congestion and death. It can sometimes be relieved by administering diuretics.

Sodium retention by the kidney tubules and the consequent edema is also seen in cirrhotic liver disease and presumably is also due to increased aldosterone output. Curiously enough, the same response is seen in the immediately premenstrual period in many normal women. They notice swelling of the hands and ankles.

Renal regulation of acid-base balance. The whole topic of acid-base regulation is taken up in the next chapter, in which it is related to the separate contributions by the body's buffer systems, the respiratory processes, and the renal mechanisms. At this stage it may be said that the kidney is the final arbiter of body pH because it alone can *excrete* excess acid or base derived from the food. The other mechanisms can merely minimize the effects of temporary excesses until the kidney is able to dispose of them.

Urine characteristics. The urine formed by the processes described above is a hypertonic solution of inorganic salts, nitrogenous waste products, and traces of other materials. A good index of the degree of concentration is in the specific gravity. This is measured with a hydrometer (see Fig. 9.5), which is similar in principle to the device used for testing wet batteries. The normal range of specific gravity is between 1.016 and 1.020 (water, of course, is 1.000). Higher and lower values are encountered quite normally, however, depending on the fluid intake. Indeed, if the specific gravity does *not* change with different fluid intakes, kidney damage is indicated. In general, good kidney function results in a urine with a high specific gravity when the fluid intake is low.

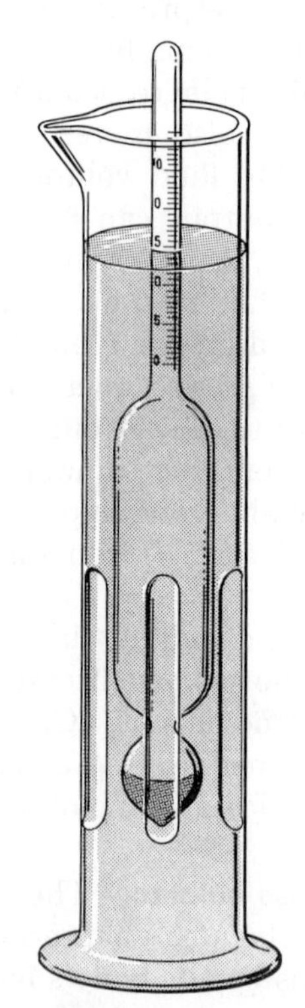

FIGURE 9.5. Urinometer.

An adult usually excretes about 1,500 milliliters of urine per day, but this is again affected by the fluid intake as well as by renal pathology. The typical composition of urine is shown in Table 9.1, but it should be emphasized that this is subject to great variation depending on the nature of the diet. For example, the large excretion of sodium chloride is merely because the diets of our Western civilizations contain many times the minimal requirements of this salt and the excess must be excreted by the kidneys.

TABLE 9.1 Typical Composition of One Day's Urine Output

Constituent		Grams
Water		1400.0
Total Solids		70.0
(including)	Urea	45.0
	Creatinine	1.4
	Ammonia (combined)	0.8
	Phosphate	1.3
	Sulfur	1.0
	Chloride (as NaCl)	14.0
	Sodium	4.5
	Potassium	2.5
	Calcium	0.3
	Magnesium	0.2

Abnormal constituents of urine. Glucose is normally completely reabsorbed in the proximal tubules by the active transport mechanisms described in the preceding section. It will appear in the urine (glycosuria) whenever the T_{max} for glucose is exceeded. This can result from:

i. Renal damage which interferes with the enzyme systems transporting glucose from the lumen of the tubules to the plasma. In this case, the threshold for glucose is actually lowered.

ii. A very high intake of sugar can temporarily raise the blood sugar above the threshold, causing a transient excretion of glucose in the urine.

iii. A severe emotional upset or fright causes an epinephrine discharge (see Chapter 16), which results in elevated blood sugar levels with results similar to those described in ii above.

iv. In diabetes mellitus, carbohydrate metabolism is deranged so that glucose accumulates to concentrations above the threshold (see Chapter 12).

Ketone bodies are also found in the urine in uncontrolled diabetes mellitus. One of them is acetone, and so their presence in the urine is called acetonuria as well as ketonuria. The defective glucose metabolism causes incomplete combustion of fats, with an accumulation of the fatty acid residues. These react together to yield keto-acids and acetone. The keto-acids render the plasma (and the urine) more acid and the acetone gives a characteristic fruity odor to the urine, and even to the breath of the patient.

Albuminuria, or the presence of albumin in the urine, is abnormal but not necessarily ominous. In active young people, a small amount of plasma albumin is sometimes filtered, but only when the subject is in the upright position. If albumin also appears in the nighttime urine, some damage to the glomeruli or nephrons is indicated.

Hematuria and pyuria. Blood and pus appear in the urine during kidney or bladder infections. They are sufficiently important signs to warrant further tests and examinations.

Consequences of Renal Failure

The functions of the kidney are to remove waste products from the plasma and to maintain optimal concentrations of electrolytes and hydrogen ion. Any interference with kidney function will lead to a toxic accumulation of waste products, ionic imbalance, and disruption of acid-base equilibrium. Not only do urea and other organic materials accumulate, but also the retention of fluid makes the patient grossly edematous. In a normal diet, an upset in acid base regulation results in an accumulation of excess acid, which is *acidosis.* The cells of the brain and heart are particularly sensitive to changes in the pH of the internal environment, and the acidosis of renal shutdown may be regarded as the main cause of death in this condition.

A discussion of the large number of diseases which can impair renal function is beyond the scope of this text, but a few principles may be stated. The lesion may affect the glomeruli, tubules, or both. It may be due to an infection, specific poison, ischemia from renalvascular occlusions, or from shock. If the damage is extensive, *renal shutdown* ensues and there is no excretion at all. Less severe damage may lead to only diminished efficiency and partial retention of waste products. Many of the tests to be described are designed to evaluate the efficiency of the kidney in various respects.

Minor damage to the kidneys sometimes has no visible effects, because their regulatory capacity is much greater than is normally required. In fact, after the removal or complete destruction of one kidney, the other is able to carry the load. This is made possible by an increased blood flow through the intact kidney. Similarly, partial destruction of each kidney can be compensated for by increased blood flow to the intact portions. The increased blood flow is partly accomplished by increased arterial blood pressure, and so kidney disease can be one cause of *hypertension.* The partial occlusion of renal arteries by arteriosclerotic plaques will also cause hypertension of renal origin. Conversely, hypertension from other causes can lead to the rupture of small blood vessels in the kidney and the death of the kidney tissue served by them. This, of course, reduces the number of functioning nephrons and the blood pressure has to be raised still further in compensation. Thus hypertension can be self-exacerbating.

The possible destruction of kidney tissue by infection is too obvious to merit separate discussion.

In shock, the tubular cells become anoxic, and if the blood pressure remains low for very long, the cells die. Minor fluctuations in blood pressure have little effect on nephron function, because the vasoconstriction and dilation of the afferent and efferent arterioles can maintain glomerular filtration at a constant level over a wide range of systemic blood pressure.

Renal Shutdown from Hemolysis

Following a transfusion reaction (see p. 95) or a severe crushing injury, there is widespread hemolysis of the red cells. Some of the hemoglobin released into the plasma is filtered in the glomeruli and appears in the tubular urine. The reabsorption of water in the tubules concentrates the hemoglobin solution to the point where it precipitates, plugging the tubules and preventing the further passage of tubular urine. If this happens throughout both kidneys, complete shutdown follows. This is a common fatal sequel to injuries from automobile accidents and the battlefield.

Protein Loss

If the glomeruli are damaged by infection or poisons, the membranes become more permeable to protein. If severe, the loss of plasma proteins lowers the oncotic pressure of the plasma and edema results. Even when the loss is minor, the presence of protein in the nighttime urine is a useful indication of some glomerular abnormality.

Kidney tests. Kidney tests fall into two classes with respect to their practicability. Some can be performed simply and rapidly; they are therefore useful as routine diagnostic procedures. Others are probably more informative but are sufficiently complex that their use is restricted to research. In the category of the clinical tests, the following are commonly used.

Phenolsulfonphthalein Test (PSP)

If the dye of this impressive name is injected, it is rapidly removed from the plasma by both glomerular filtration and tubular secretion. Normally in the first hour, about 50% of the injected dye is excreted, which can be determined by the colorimetric analysis of the urine collected for that period. If the amount recovered in this time is significantly lower than 50%, it is an indication of glomerular or tubular malfunction.

Urea Clearance Test

In this connection, clearance means the number of milliliters of plasma from which all the urea is removed in one minute. In actual fact, of course, the removal of urea from the plasma is far from complete. The urea clearance value is a useful number which represents the volume of plasma containing the amount of urea that is removed by the kidneys in one minute. In practice, the patient is instructed to drink enough water to start a moderate diuresis, and then the urine is collected through about half an hour. At the same time, a sample of blood is drawn and the urea content of both the plasma and urine is determined. The urea excretion per minute is calculated by dividing the total amount by the number of minutes in the collecting period. The plasma concentration is divided by the amount excreted per minute and the result is the urea clearance. It will be noted that the dimension of clearance is in milliliters per minute. The normal value lies in the neighborhood of 70 milliliters per minute and in renal damage, this volume decreases.

Blood Urea Nitrogen Concentration (BUN)

This value can be determined by chemical analysis of the plasma from a small blood sample. It is a useful guide to renal function since the kidneys are mainly concerned with the excretion of urea. If the kidneys are not functioning adequately, the urea nitrogen will rise from its normal value of 15 milligrams per cent. This test is interpreted with caution, however, because the urea level is subject to great variation with diet.

Nonprotein Nitrogen Cencentration (NPN)

This test is used in the same manner as BUN determinations. As the name implies, the chemical method determines other nitrogenous materials as well as urea, so that the normal value is about 25 milligrams per cent. Again, this is subject to great variation, and only increases of over 75 milligrams per cent are necessarily indicative of failing renal function.

Creatinine levels in blood are informative to the same degree and are actually subject to less day to day variation.

Pyelograms are X-ray photographs taken of the kidneys when some radio-opaque material is being excreted (see Figure 9.6). Diodorast and other iodine-containing compounds are opaque to X-rays and are rapidly excreted by filtration and tubular secretion. As they are excreted, they fill the renal pelvis so that this region is clearly outlined and any anatomical anomalies or tumors can be seen. The degree of opacity in the pelvis of each kidney can be compared to determine whether the kidneys are carrying an equal load.

Diuretics. A *diuretic* is anything that will cause an increased flow of urine, or a *diuresis.* Obviously, the simplest diuretic is a large drink of water. Extra water intake lowers the osmotic pressure of the internal environment and through the osmoregulatory mechanism causes less reabsorption of water in the distal tubules until the excess water has been excreted. Other diuretics include injected mannitol or inulin, which are filtered but not reabsorbed and thus cause a solute diuresis by virtue of their high concentration in the proximal tubular urine.

In treating edema and some renal diseases, it is frequently desirable to start a diuresis. The injection of inulin, etc., is sometimes resorted to, but specific drugs are more frequently used. The mercuric group of diuretic drugs inhibit the reabsorption of sodium in the proximal tubule, thereby reducing the obligatory water reabsorption. The effect may be described as a solute diuresis caused by lowering the T_{max} for sodium. Another important group of diuretics inhibit the enzyme, *carbonic anhydrase,* which is involved in the normal exchange of hydrogen ion for sodium in the tubules. Especially good results are sometimes obtained from a combination of these two groups of drugs. Still another group of

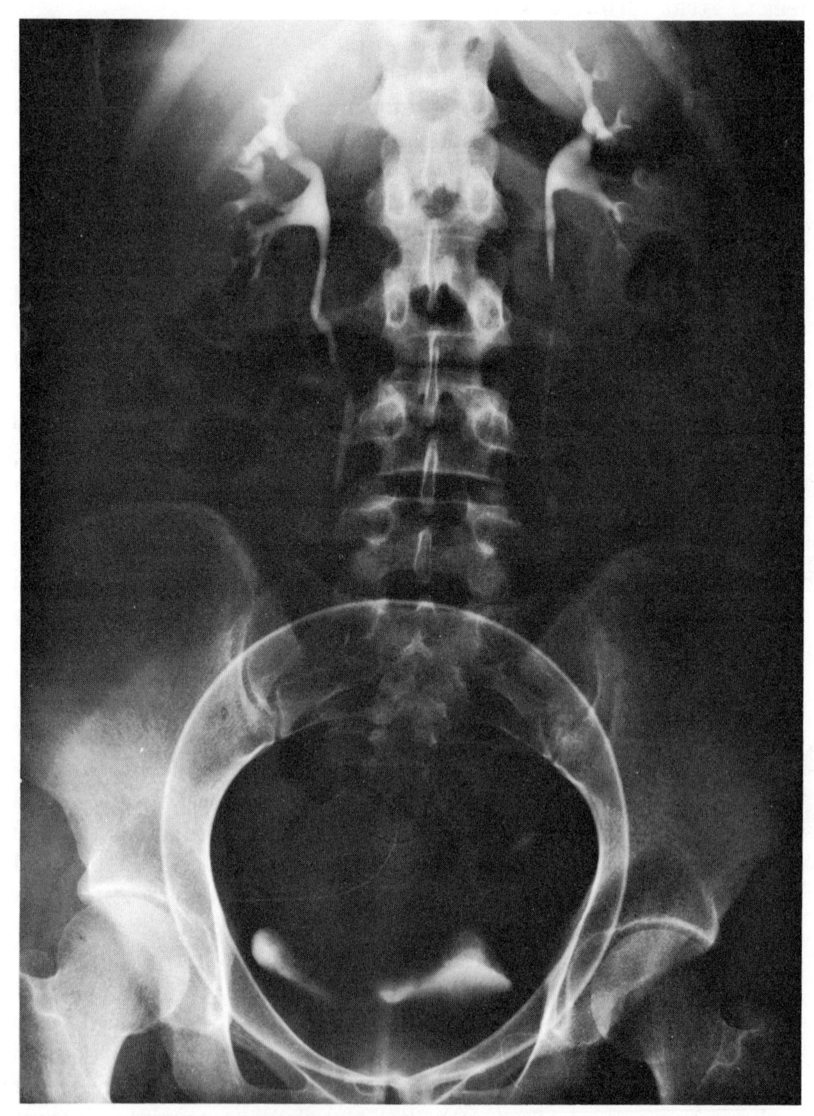

FIGURE 9.6. Pyelogram. The pelvis of each kidney is clearly visible in the upper portion of the picture. The large sphere lower down is a balloon inflated within the bladder to prevent too rapid a loss of the radio-opaque material into the bladder. (Photo by courtesy of Radiology Department, Stanford University School of Medicine.)

drugs increases the urine volume by increasing the glomrular filtration rate to a level at which the tubular reabsorptive processes are overloaded. It is presumed that these drugs act by dilating the afferent arterioles of the glomeruli.

After a crushing injury or a transfusion reaction, the precipitation of freed hemoglobin in the renal tubules can be prevented if a sufficiently massive diuresis can be maintained. A large volume of dilute urine is sometimes enough to keep all the hemoglobin in solution.

REFERENCES

1. Robinson, James R., *Reflections on Renal Function,* Blackwell, Oxford, 1954.
2. Salisbury, Peter F., "Artificial internal organs," *Sci. American* **191** (2), 24, August 1954.
3. Schmidt-Nielsen, Knut and Bodil, "The desert rat," *Sci. American* **189** (1), 73, July 1953.
4. Smith, Homer W., "The kidney," *Sci. American* **188** (1), 40, January 1953.
5. Wolf, A. V., "Thirst," *Sci. American* **194** (1), 70, January 1956.

ACID-BASE REGULATION

10

In an earlier chapter outlining some principles of elementary chemistry, we referred to acids as molecules that can dissociate to yield hydrogen ions in solution. The "acidity" of a solution is determined by the hydrogen ion concentration, conveniently expressed as the pH, which is the negative logarithm of the hydrogen ion concentration in mols per liter. Furthermore, in our discussion of the principles of homeostasis, the pH of the internal environment was listed as one of the factors which is closely regulated for the optimal metabolic conditions of the cells.

As with the nutrient and salt concentrations of the internal environment, the prime determinant of the pH is the nature of the diet. The metabolism of some foods leads to the production of acids which would lower the pH (increase the acidity) of the internal environment if they were not controlled by the proper homeostatic mechanisms. Most proteins include some of the sulfur-containing amino acids, and their catabolism leads to the production of *sulfuric acid.* The weaker *phosphoric acid* is also produced in protein catabolism. Curiously enough, the foods popularly regarded as "acid" do not yield acid products when metabolized. The acids in citrus fruits and vegetables are all of an organic nature (like citric acid), and since they are metabolites, they are completely combusted to carbon dioxide and water in the body. As they exist in the fruits, these organic acids are chiefly combined with potassium, and so their combustion leaves a surplus of potassium—a *basic* cation. Therefore, the superficially acid foods like fruit are actually conducive to alkalinity in the overall economy of the body, and it is a meat diet which leads to the greatest production of acid.

The maintenance of a constant pH, therefore, depends on a suitable renal excretion of the hydrogen ion derived from the diet. As with the nutrient supply in the internal environment, the irregularity of dietary intake would lead to wild fluctuations in the pH unless there were mechanisms to deal with it on a temporary basis. The principal agents for the short term regulation of pH are the *buffer systems* of the body.

BUFFERS

In our earlier discussion of acids, it was mentioned that strong acids are, by definition, completely dissociated in solution and weak acids are relatively weakly dissociated. Carbon dioxide dissolved in water forms carbonic acid, which was given as an example of a weak acid. The formation and dissociation of carbonic acid can be represented by the following equations

$$CO_2 + H_2O \rightleftharpoons H_2CO_3 \rightleftharpoons H^+ + HCO_3^-$$

The actual concentration of carbonic acid in the solution depends on the partial pressure of carbon dioxide in equilibrium with the solution and on the solubility of carbon dioxide in water at that particular temperature.

Physiologically, the most important salt of carbonic acid is sodium bicarbonate. Unlike the acid, the salt dissociates completely in solution, according to the following equation:

$$NaHCO_3 \rightarrow Na^+ + HCO_3^-$$

If a strong acid, hydrochloric acid for example, is now added to a solution of sodium bicarbonate and carbonic acid, part of the reaction is as follows:

$$(Na^+ + HCO_3^-) + (H^+ + Cl^-) \rightarrow (Na^+ + Cl^-) + H_2CO_3$$

The ionized sodium chloride thus formed is neutral and the carbonic acid formed by combination of the hydrogen ions with bicarbonate ions is less strongly dissociated (and therefore less acid), than was the hydrochloric acid originally added. The essential point is that the free hydrogen ion of the HCl is "sopped up" by the bicarbonate ions to form weakly dissociated carbonic acid. This is the *buffering* of the added hydrochloric acid.

Many buffer systems of the body are of this kind, a mixture of a weak acid and its salts. The addition of a strong acid leads only to the formation of more of the weak acid, so that the increase in hydrogen ion concentration is relatively small. If a base, that is, a molecule yielding free hydroxyl ion, is added to a solution containing sodium bicarbonate and carbonic acid, the reverse buffering follows. The added base diminishes the free hydrogen ion concentration to a relatively small degree by reacting with some of the carbonic acid, according to the following equation:

$$H_2\,CO_3 + Na\,OH \rightarrow Na\,HCO_3 + H_2O$$

The concentration of carbonic acid is reduced and that of the sodium bicarbonate is increased.

The sodium bicarbonate-carbonic acid buffer system is important in the extracellular fluids of the body. If we consider once more the addition of hydrochloric acid to such a system, a further reaction must be taken into account. The hydrochloric acid combines with the bicarbonate to yield sodium chloride and carbonic-acid. At a given carbon dioxide tension, however, the concentration of carbonic acid in solution depends on the solubility of carbon dioxide. Therefore, the production of extra carbonic acid leads to the evolution of carbon dioxide, which in the intact animal is expired through the lungs. Therefore, the net effect is due to both the buffering of the added hydrochloric acid and to the outright neutralization of some part of the hydrochloric acid at the expense of the bicarbonate reserve of the extracellular fluid.

A buffer system has its greatest flexibility when the weak acid is half-dissociated because then it can take up either hydrogen ions or hydroxyl ions with the least change of pH. In the *titration curve* of a buffer solution, the pH is plotted against the ratio of acid to salt. This is shown for carbonic acid in Fig. 10.1, in which the pH is plotted against the per cent of the total carbon dioxide of the system which is present as bicarbonate. Note the small increment of pH between bicarbonate CO_2 contents of 20% and 80%.

The phosphates in the body fluids are also important buffers.

Phosphoric acid not only is another weak acid, but it also has *three* hydrogens which can dissociate, according to the equations:

$$H_3PO_4 \rightleftharpoons H^+ + H_2PO_4^- \rightleftharpoons H^+ + H^+ + HPO_4^{--} \rightleftharpoons H^+ + H^+ + H^+ + PO_4^{---}$$

Proteins as buffers. The plasma proteins and the hemoglobin of the red cells are weak acids and have dissociation characteristics similar to those of carbonic acid (see Fig. 10.2). Therefore, they also act as buffers and contribute to homeostasis by smoothing out temporary variations in the hydrogen ion concentration of the extracellular fluid. In addition to these extracellular buffer systems, the proteins of the *intracellular* fluid also participate in buffering. Excess hydrogen ion in the internal environment diffuses into the cells, where it is buffered by the anion groups of the cell proteins, usually with a concomitant displacement of intracellular potassium.

It will be appreciated that the buffer reserves of the body can be "used up" much as the absorbing capacity of a sponge can be used up. The maintenance of a constant pH in the internal environment depends therefore on the ultimate *excretion* of the

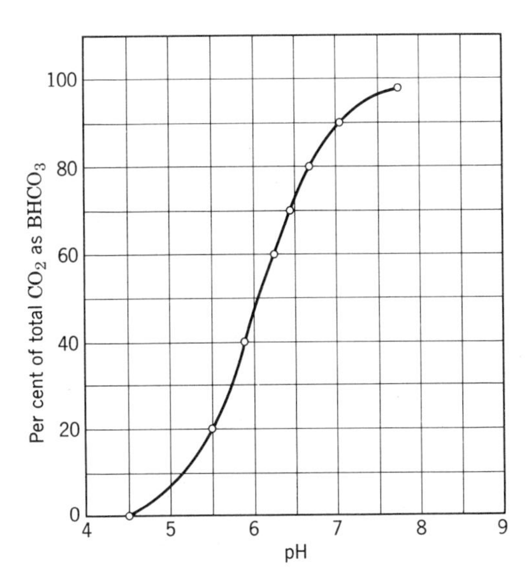

FIGURE 10.1. Action of $NaHCO_3$:H_2CO_3 buffer, showing maximum buffer effect at middle of curve when $NaHCO_3$:H_2CO_3 ratio = 1.

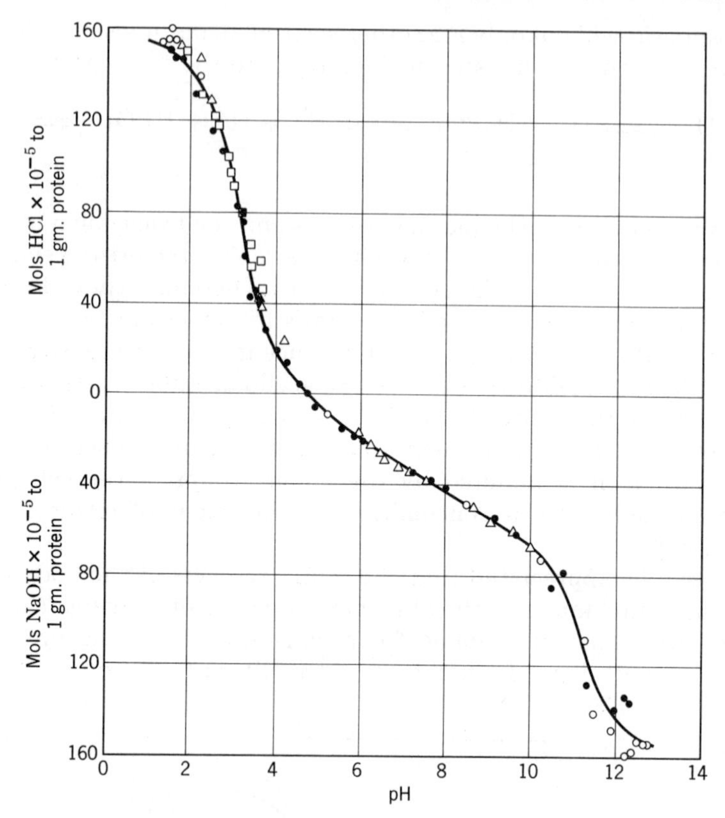

FIGURE 10.2. Titration curve of serum albumin. Reproduced by permission from L. J. Henderson, *Blood,* Yale University Press, 1926.

excess hydrogen ion. The buffer systems provide only temporary control of pH.

RESPIRATORY MECHANISMS IN ACID-BASE CONTROL

There is an important respiratory mechanism assisting the temporary control of pH. We have indicated that the carbonic acid concentration in the blood depends on the carbon dioxide dissolved in the fluids, and that this in turn depends on the carbon dioxide tension in equilibrium with it. The carbon dioxide content of the body fluids, however, is determined by its rate of production in the tissues and its rate of dissipation in the lungs. The latter can be increased by increasing the ventilation rate; if the alveolar air is exchanged more rapidly with fresh air from outside, less carbon dioxide accumulates in the alveoli and so the tension of carbon dioxide in the body fluids falls. Conversely, hypoventilation will permit the accumulation of carbon dioxide in the alveoli, which in turn raises the carbon dioxide tension in the body fluids.

The addition of an acid load to the internal environment leads to the production of carbon dioxide through the reaction of the acid with sodium bicarbonate and the carbon dioxide is expired through the lungs. Also, the lower pH of the plasma stimulates the respiratory centers in the medulla to increase the ventilation rate. The resulting decrease in carbon dioxide tension lowers the carbonic acid concentration in the blood, thereby helping to bring the pH back towards normal. Even without an acid load, voluntary hyperventilation can lead to an undue excretion of carbon dioxide, a lowering of the carbonic acid concentration, and a consequent lowering of the hydrogen ion concentration which, in effect, is an alkalosis.

THE EXCRETION OF ACID

Lungs. The carbon dioxide continually produced by the oxidative metabolism of the cells can be regarded as an acid inasmuch as it reacts with the body water to give carbonic acid. Therefore the excretion of carbon dioxide through the lungs is part of the acid-excretory mechanism of the body. The variation of ventilation rate in the temporary adjustment to pH changes described in the previous section is superimposed on the normal and continual excretion of carbon dioxide. The carbon dioxide tension in the body fluids is in a *steady state,* depending on its rate of production (oxidative metabolic rate) and the ventilation rate.

Kidneys. The other main route of acid excretion is, of course, through the kidneys, which excrete what are called the *fixed acids* (and *fixed bases*). The presence of excess acid in the body fluids can be thought of in terms of the presence of extra anions of these acids; for example, the sulfate ion, the chloride ion, the phosphate ion, etc. (This way of visualizing excess acid makes chemists shudder, but for our present purposes, it is a useful one.) These anions can be excreted in the urine, combined with cations (like sodium), but this does not rectify the *excess* of anions. Furthermore, with our type of diet, it would lead to wasteful losses of sodium potassium, magnesium, and calcium (the fixed bases). The renal tubules are able to excrete a small amount of free hydrogen ion, so that although the body fluids have a pH of 7.4, that of the urine can be reduced to 5.0 or even lower. There are however, physiological limits to the acidification of the urine and much of the acid load must first be neutralized to be excreted.

An extremely important mechanism for neutralizing excess acid to this end is the production of *ammonium ion* (NH_4^-) within the renal tissues. Ammonium hydroxide (NH_4OH) has chemical properties similar to the hydroxides of sodium and potassium and it can form relatively neutral salts with strong acids, such as ammonium chloride ($NH_4\ Cl$). Ammonia is a product of deamination of all amino acids, but is usually converted to urea. In the renal tissue it is formed from the amino groups of amino acids and from a glutamic acid derivative, *glutamine,* by the pathway shown in Fig. 10.3.

Free hydrogen ion is excreted through its exchange with the cations, typically sodium and potassium, contained in the tubular urine. The urine cations exchange with hydrogen ions from carbonic acid in the tubular cells; the intracellular sodium and potassium bicarbonate is then reabsorbed into the plasma leaving an excess of carbonic acid in the tubular urine. This in turn reacts with the phosphates in the tubular urine, changing the dibasic phosphate ions to monobasic phosphate ions according to the following equation; the potassium bicarbonate is then free to enter into the previous exchange reaction again.

$$K_2HPO_4 + H_2CO_3 \rightarrow KH_2\ PO_4 + KHCO_3$$

Therefore, the phosphate buffers are important in both the body fluid buffering and in buffering the hydrogen ion excreted in the urine.

In brief then, excess acid is excreted in relatively neutral form as ammonium salts and also as the more acid phosphate salts. These are called the *base-saving* mechanisms.

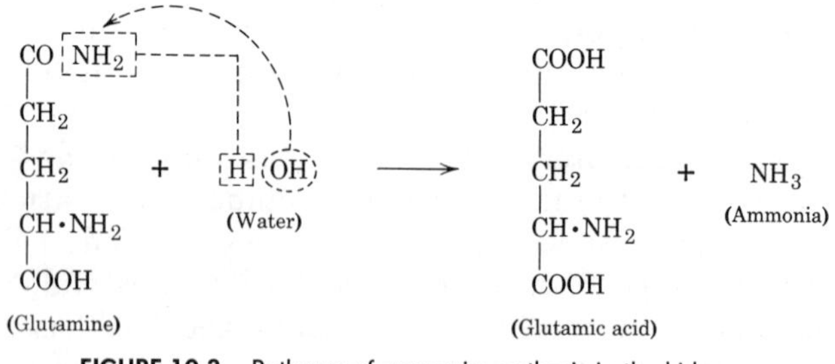

$$\begin{array}{l} CO\,NH_2 \\ | \\ CH_2 \\ | \\ CH_2 \\ | \\ CH\cdot NH_2 \\ | \\ COOH \\ \text{(Glutamine)} \end{array} + \begin{array}{c} H\ OH \\ \text{(Water)} \end{array} \longrightarrow \begin{array}{l} COOH \\ | \\ CH_2 \\ | \\ CH_2 \\ | \\ CH\cdot NH_2 \\ | \\ COOH \\ \text{(Glutamic acid)} \end{array} + \begin{array}{c} NH_3 \\ \text{(Ammonia)} \end{array}$$

FIGURE 10.3. Pathway of ammonia synthesis in the kidney.

Carbonic acid is produced whenever carbon dioxide is dissolved in water, but its formation is greatly speeded up in the body by an enzyme, *carbonic anhydrase.* This enzyme has an important role in the lungs where it catalyses the rapid conversion of plasma carbonic acid to carbon dioxide for gas exchange. In the renal tubular cells it is important in maintaining the supply of carbonic acid for the hydrogen ion exchange with cations in the tubular urine. There are several drugs which inhibit the action of this enzyme and slow down the equilibration of carbonic acid with carbon dioxide (in either direction). These drugs are useful experimental tools for investigating the mechanism of the ion exchange processes in the renal tubules. They also have considerable clinical value in patients with excessive retention of cations, notably sodium. Some of the most effective treatments of persistent edema due to sodium retention are by the combined administration of the mercury diuretics and "Diamox," one of the carbonic anhydrase inhibitors.

Base excretion. We have mentioned that a meat diet leads to a net acid production and the excretion of an acid urine. The herbivorous animals on the other hand, metabolize the organic acids in their food to carbon dioxide and water and are therefore faced with a surplus of cations (fixed bases), which must be excreted. The chemical problem here is much simpler, because the production of carbon dioxide in the tissues provides a never-ending source of carbonic acid which can react with the excess bases. The bases are then excreted as their bicarbonate salts, giving the alkaline urine characteristic of these animals.

Derangements of Acid-Base Balance

pH derangements can occur in either direction, resulting in acidosis or alkalosis. Acidosis may be due to an excess of the acids ordinarily excreted by the kidney or to an excess of carbon dioxide because of reduced pulmonary gas exchange. Similarly, alkalosis can result from renal or pulmonary disfunction. Therefore, pH derangements can be considered in four categories: 1. respiratory acidosis, 2. respiratory alkalosis, 3. metabolic acidosis, and 4. metabolic alkalosis.

Respiratory acidosis can follow the reduction of the area of the lung surface available for gas exchange, as in emphysema or pneumonia. Similarly, partial paralysis of the respiratory muscles and the consequently inadequate ventilation can also lead to the accumulation of carbon dioxide in the alveoli and an elevated carbon dioxide tension in the body fluid. This is a hazard in curarised patients undergoing operations.

An increased carbonic acid concentration lowers the pH of the body fluid. Over a period of hours, a patient can adjust to this by the renal excretion of acid and by retaining an excess of fixed base in the blood as bicarbonate. Since the pH of the internal environment largely depends on the ratio of carbonic acid to bicarbonate ion, the effect of the increased carbonic acid concentration can be offset by increasing the bicarbonate concentration. Therefore, the patient in *compensated* respiratory acidosis has an elevated bicarbonate concentration in the plasma. The compensation may well be so effective that the pH of the plasma is near normal.

Respiratory alkalosis is encountered in the hyperventilation at high altitudes already referred to, or in the overenthusiastic artificial respiration of a curarized or open-chest patient on the operating table. We have also mentioned the temporary hyperventilation in neurotic or overexcitable people In adjusting to high altitude, the body excretes fixed base and lowers the bicarbonate concentration of the internal environment; this restores the normal ratio of bicarbonate to carbonic acid concentrations.

Metabolic acidosis can occur from a great variety of causes. Our previous example of diabetic acidosis is a useful one; the accumulation of keto-acids lowers the pH of the internal environment and leads to several well recognized symptoms. We have mentioned that the respiratory center is sensitive to pH, whether the acidity is due to carbonic acid or other acids. The ventilation rate is therefore increased, just as if to compensate for respiratory acidosis; in extreme cases, the patient displays *Kussmaul breathing,* a harsh and rapid panting. At the same time the accumulated acids react with the bicarbonates of the blood, the resultant carbon dioxide is excreted through the lungs, and the plasma bicarbonate concentration is lowered. The kidneys continue to excrete acid, but the base-saving mechanisms are overloaded so that sodium and potassium are lost in the urine as salts of the keto-acids. In compensated metabolic acidosis, then, the patient displays hyperventilation, a reduced plasma bicarbonate concentration, and an extremely acid urine. Once more, the pH of the plasma may not be very low, but the continued production of acid, as in uncontrolled diabetes, inevitably overloads these temporary control mechanisms and the compensation fails.

Diarrhea is a common cause of metabolic acidosis in children.

Since the intestinal secretions are alkaline, their continued loss in the stools leads to a depletion of fixed base, or a relative excess of acid. In both diabetic acidosis and diarrhea, dehydration is an inevitable complication. Patients in renal shutdown also become acidotic because of accumulation of the acids derived from the breakdown of protein. *Dietary* protein is not the only source of fixed acid, because even though the patient is not eating, the breakdown of body protein also produces acids.

Renal shutdown is a sequel to many conditions including renal disease *per se* and the plugging of the tubules by hemoglobin after transfusion reactions or crushing injuries. In consequence, acidosis is the final cause of death in a great many disease conditions.

Metabolic alkalosis can be caused by excessive vomiting from any cause. The loss of gastric acid (which is usually neutralized and reabsorbed in the intestine), leads to a relative deficit of the chloride anion. Metabolic alkalosis is sometimes encountered in patients with renal defects if they take large amounts of sodium bicarbonate or other gastric antacids for "indigestion." The inability to excrete the excess base causes the alkalosis. For reasons beyond the scope of the present discussion, a chronic loss of potassium in lower bowel disorders also leads to alkalosis.

Metabolic alkalosis is not as readily compensated by ventilation as the other pH derangements, but some degree of hypoventilation increases the carbonic acid content of the internal environment and helps return the pH toward normal. The respiratory centers in the medulla are not as responsive to an alkaline condition of the circulating plasma as they are to acid. In other words, a depression of ventilation is less readily evoked than an increase.

Consequences of pH derangements. When the pH of the internal environment becomes very different from the normal, the pH of the *intracellular* fluid is also changed, leading to redistribution of electrolytes across the cell membranes. In particular, hydrogen ion diffuses into the cells where it is buffered by the cell proteins. This buffering causes displacement of potassium ions, which then leave the cells and increase the extracellular potassium at the expense of the intracellular concentration. The concentration gradient of potassium across the cell membrane, it will be recalled, is the principal determinant of the resting potential, which in turn determines the excitability of the cell. Therefore, these electrolyte changes partly account for the neuromuscular symptoms associated with acidosis.

In alkalosis, the extracellular calcium is less ionized. Since it is the ionized calcium concentration which largely determines the permeability characteristics of the cells, alkalosis leads to a hyperexcitability which is almost indistinguishable from hypocalcemia.

In metabolic acidosis, besides the redistribution of electrolytes due to the diffusion of hydrogen ions into the cell, the loss of fixed base in neutralizing the excess acid for excretion leads to a total body depletion of potassium, which further complicates the electrolyte derangement. An extra hazard is encountered even when the metabolic acidosis is corrected by therapy, because as the intracellular fluid returns to its normal pH, potassium re-enters the cells from the extracellular fluid which is already depleted of potassium by the renal losses. This further depresses the extracellular potassium concentration, causing changes in excitability which, if uncorrected by replacement therapy, can lead to neuromuscular disorders culminating in paralysis of the respiratory muscles and death.

The chemical determination of pH derangements. The measurement of the actual pH of the blood or plasma is useful, but somewhat difficult to achieve under routine conditions. If the pH determination is to mean anything, the blood must be equilibrated with carbon dioxide at the level encountered in the body, the pH electrode apparatus must be gas-tight so that there is no exchange of blood gases with the atmosphere during the measurement, and finally, the electrode's temperature must be controlled at the body temperature. The prior equilibration of blood with an appropriate gas mixture can be avoided by drawing the blood from the patient in an oiled syringe and by keeping the blood from contact with the atmosphere. An accurate pH determination is not a particularly good guide to the degree of *clinical* acidosis or alkalosis because, as we have mentioned, the temporary compensating mechanisms can minimize the actual pH change.

Chief among these compensating mechanisms are the changes in bicarbonate concentrations. Therefore, a measurement of the blood or plasma bicarbonate concentration is a useful index of acidosis or alkalosis, especially if combined with a pH determination.

It is difficult to measure the bicarbonate concentration *per se,* but if some strong acid is added to a sample of blood, the carbon dioxide evolved is largely from the reaction of the blood bicarbonate with the strong acid. In the *Van Slyke* blood-gas apparatus, lactic acid is added to a 1 milliliter sample of blood which is then subjected to a vacuum and agitated. This brings out of solution, not only the carbon dioxide derived from bicarbonate, but also that dissolved in the plasma and present as carbonic acid. The total gas extracted from the blood contains carbon dioxide, nitrogen, oxygen, and water vapor. After an appropriate extraction time, the mercury reservoir of the apparatus is manipulated so that the gas extracted from the 1 milliliter blood sample is contained in a known volume and the pressure it exerts can be measured by a mercury manometer. Some strong base is then admitted anaerobically, completely absorbing the carbon dioxide. The pressure exerted by the gas mixture when contained in the same original volume (but now without its carbon dioxide), is again determined, and the difference between this pressure and the first is the partial pressure of carbon dioxide in the original gas mixture. One mole of an ideal gas contained in 22.4 liters exerts 1 atmosphere of pressure, and from this (with appropriate corrections for the departure of carbon dioxide from the characteristics of an ideal gas) the amount of carbon dioxide exerting that partial pressure at the known temperature can be calculated.

The result is usually expressed in millimoles per liter of blood, but the older expression of *volumes per cent* of extracted carbon dioxide is still used in some centers. The normal values of total CO_2 are given in Table 10.1, together with some typical values encountered in the clinical derangements of pH.

TABLE 10.1 Some Typical Plasma Bicarbonate Values

	meq/1
Normal plasma	26
In compensated metabolic acidosis	15
In compensated respiratory acidosis	40
In compensated metabolic alkalosis	50
In compensated respiratory alkalosis	12

NUTRITION

11

The composition of the internal environment, and indeed the composition of the whole animal, must depend ultimately on what it eats and drinks. Everything required by the cells must be provided for them dissolved in the interstitial fluid; an adequate nutrition is accomplished when the internal environment contains all the materials the cell requires in the right proportions and concentrations. It is profitable therefore, to consider what are the requirements of the cells.

Water. The absolute necessity of water is self-evident. It is consumed as such or in beverages, but is also present in most "solid" foods. Furthermore, the oxidation of other nutrients releases "metabolic water" discussed in a later section.

An energy source. The very existence of a cell as an island of highly organized material requires energy. Any form of work, such as muscular contraction or secretion, needs even more energy. The most important source of energy for the cells is the oxidation of glucose or its intermediate products of metabolism. Therefore, the diet should contain materials which will yield glucose. These materials are the *carbohydrates.* Although glucose is nevertheless essential, much of the energy requirement of the cell can be met by fats, and these too are a normal constituent of food stuffs.

Structural material for growth and repair. Much of the cell's structure is made of protein. Because proteins have extremely large molecules they cannot readily diffuse out of the blood capillaries or across cell membranes. Therefore, the internal environment must contain units from which proteins can be constructed by the cell. These building blocks, the *amino acids,* are obtained by the digestion of dietary protein in the gut, whence the readily diffusible amino acids are absorbed and carried to the cells. There are about twenty-two amino acids commonly found in proteins and ten of these cannot be synthesized in the mammalian body; they must be present in the original dietary protein. Cells undergo wear and tear all the time and the need for repair and replacement makes a constant demand for amino acids. In the growing animal the demand is, of course, even greater.

The cells' structures include mineral elements and water as well as protein. The great importance of suitable concentrations of salts in the body fluids was described in Chapter 5. Although the fine regulation of salts and water is achieved by the kidney, it is clear that they must be consumed in quantities at least sufficient to replace wastage.

Calcium and phosphorus are not only important constituents of intracellular and extracellular fluids, but also comprise much of the hard parts of the bones and teeth. Therefore, an adequate intake of these elements is necessary for maintaining the skeleton; and again, in the growing animal the requirement is very large. Similarly, iron is essential for manufacture of hemoglobin, despite the economy with which iron from wornout erythrocytes is re-used.

Reproduction may be considered as a special form of growth in this respect. Although the nutrient requirements for the manufacture of spermatozoa and ova are small, the growing fetus makes great demands on the materials in the body fluids of the mother.

Enzyme components. Many chemical reactions which proceed quietly in living cells can be duplicated in the laboratory only by the use of high temperatures or strong reagents. The facility of the reactions in the cells is due to the biochemical catalysts, or *enzymes,* already discussed in Chapter 4. Enzymes themselves are always protein in nature, but many of them require the presence of other organic compounds combined or associated with them before they can perform their catalytic tasks. These compounds are called *co-enzymes.* The body is able to manufacture many of the coenzymes from simple materials, but in many cases, part or all of the molecule must be provided intact in the diet. The essential dietary constituents from which many of the coenzymes are replenished are the *vitamins.* They are required only in minute quantities as compared with the proteins and carbohydrates because their necessary concentration in the tissues is very small.

In addition to the organic coenzymes, most enzymes require the presence of a metal ion for their activity. For some enzymes, sodium, potassium, magnesium, or iron are required; all metals which are necessarily parts of the cell's structure. Other enzymes require manganese, for example, which is not present in large concentrations in the tissues. Since manganese is essential for the activity of many enzymes, however, it must be present in the diet, even if only in minute amounts. Because of the very small dietary requirement, manganese is referred to as a *trace element.* Iodine is also a trace element but is required as a constituent of the thyroid *hormone* and not of an enzyme.

From this brief review it is apparent that an adequate nutrition requires water, carbohydrates, fats, proteins (of suitable amino acid composition), minerals, and vitamins. These dietary constituents and their metabolism in the body will now be described in more detail.

CARBOHYDRATES

This name originally referred to carbon compounds containing hydrogen and oxygen in the same proportion as in water. A simple and representative example is the sugar *glucose,* with the empirical formula $C_6H_{12}O_6$. The *structural* formula of glucose is shown in Fig. 11.1, along with two different compounds with the same number of carbons, but with the hydrogens and oxygens differently arranged. The many different spatial arrangements of the atoms combined with carbon make great variety and complexity in organic compounds. A discussion of this type of *isomerism* is beyond the scope of this text but it should be appreciated that there can be several distinct compounds with the same atomic content. Furthermore, the possibilities of asymmetry in carbon chain compounds also make "left-handed" or "right-handed" compounds which are otherwise identical. Such "mirror-image" compounds are called *stereoisomers.* The stereoisomers of glucose, for example, are very similar chemically, but the enzymes of the body are able

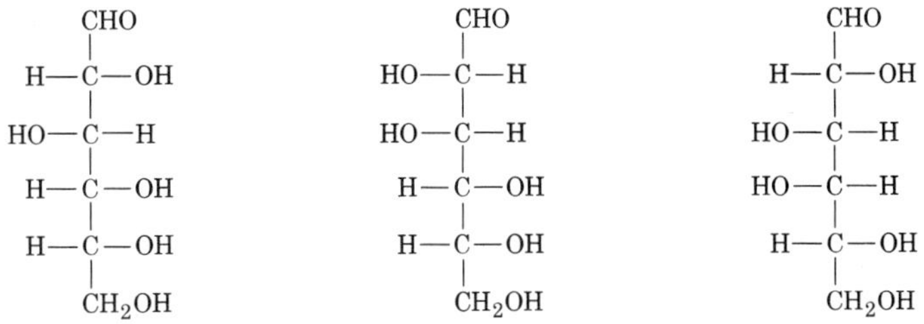

(+) Glucose	(+) Mannose	(+) Galactose
CHO	CHO	CHO
H—C—OH	HO—C—H	H—C—OH
HO—C—H	HO—C—H	HO—C—H
H—C—OH	H—C—OH	HO—C—H
H—C—OH	H—C—OH	H—C—OH
CH_2OH	CH_2OH	CH_2OH

FIGURE 11.1. Structure of three hexoses.

to use only one of them. Generally, where there are two stereoisomers possible, only one is found in nature.

The simple sugars with six carbon atoms, such as glucose, are called *hexoses;* other sugars with five or seven carbons are called *pentoses* and *heptoses,* respectively.

Monosaccharides and disaccharides. Glucose is the most important sugar in the internal environment, but it is seldom found free in foodstuffs; it is usually a component of a more complex carbohydrate molecule. Two of the simpler combinations found in food are *sucrose* (which is cane or beet sugar) and *lactose* (which is milk sugar). Both of these are called *disaccharides* because they are made of two conjugated monosaccharides, or simple sugars like glucose.

The structure of sucrose is shown in Fig. 1.12. Sucrose has a glucose unit joined to a fructose unit with one molecule of water removed, and lactose has a glucose and a galactose unit. *Maltose* is a disaccharide formed of two glucose units. Any disaccharide may be split into two monosaccharides if the missing water molecule is reintroduced under appropriate conditions. This kind of splitting is therefore called *hydrolysis.* It can be accomplished in the laboratory by boiling with hydrochloric acid whereas in the digestive tract it is quietly accomplished by *hydrolytic* enzymes.

Polysaccarides. Monosaccharide units can be joined into large molecules of endless complexity called *polysaccharides.* One of the commonest in foods is *starch,* because this is the form in which most plants store their reserves of food. Starch is made up of glucose units entirely, and on exposure to the enzymes of the digestive system it is split into successively smaller units and finally into glucose molecules, which can be absorbed from the gut into the plasma.

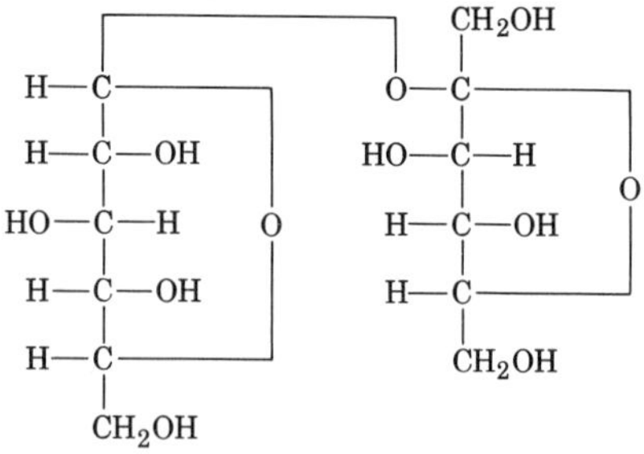

(Glucose unit) (Fructose unit)

FIGURE 11.2. Sucrose structure.

Animals store their carbohydrate reserves as *glycogen,* which has been aptly described as "animal starch." The glucose units are somewhat differently arranged but there is considerable resemblance to plant starch. Glycogen is found in liver and other meats but it does not compare with plant starch as a dietary source of glucose.

The strong, rigid structures of plants are partly made of *cellulose,* which is yet another polysaccharide. However, mammals are unable to digest it to simpler units as they do not have suitable enzymes. In consequence, the greater part of leafy food passes through the digestive tract unchanged, and its value lies in the bulk, or *roughage,* it provides for intestinal motility. The herbivorous animals, especially the ruminants (cattle, deer, sheep, etc.), are able to utilize cellulose because their digestive tracts contain swarming microorganisms, which break it down to simpler compounds the "host" animals are able to utilize. If it were not for these *symbiotic* microorganisms, cows would not thrive so well on an exclusively grass diet.

Inulin is a relatively unimportant polysaccharide found in the tubers of dahlias and Jerusalem artichokes. It is made of fructose units rather than glucose, as is potato starch.

None of the disaccharides and polysaccharides can cross cell membranes so they must be digested to monosaccharides before they can cross the gut wall. The necessary enzymes are present only in the gut. Therefore, if a complex carbohydrate is *injected* into an animal, it is not split to its simpler components and therefore does not enter the cells, but merely distributes itself in the extracellular fluid until it is excreted by the kidney. For this reason, inulin is useful for extracellular fluid volume determinations as described in Chapter 5.

The most important dietary origins of the glucose in the internal environment are those foods containing starch. In western man, these are chiefly bread, potatoes, oatmeal, cornmeal, and other cereals. Some sugar is also available from fruits, and often the consumption of table sugar or candy is significantly large. Infants get their carbohydrate as lactose in milk. In the orient, rice and beans are the most important sources of starch, and sugar consumption is low. It is interesting to note that the carbohydrate intake of Eskimos and other meat-eating hunters is small, despite the liver glycogen they consume. The energy requirements of their cells are partly met by the fat of the diet, and the glucose in their body fluids is derived from the metabolism of their excess protein intake, by pathways discussed in a later section.

PROTEINS

Just as the polysaccharides are combinations of simpler units (monosaccharides), proteins are combinations of amino acids. In a protein however, the twenty-two natural amino acids may occur in any number or combination so that there is almost an infinity of possible structures. Within an organism there are hundreds of different proteins with different functions and characteristics. Plant proteins are noticeably different from animal proteins, the proteins in one species of animal are different from those in another, and there are many minor differences between the proteins of individuals within a species.

The *amino acids* normally found in proteins have structures like those in Fig. 11.3. In each natural amino acid, an amino group ($-NH_2$) is attached to the α-carbon, that is, the carbon next to the carboxyl group ($-COOH$). Some amino acids have more than one amino group, but in naturally occurring amino acids, one of them is always attached to the α-carbon. If we consider the general formula of the amino acids, R-CH(NH_2)COOH, the variation between them is in the nature of the R-group. It may be very simple, as in glycine and alanine, or complex as in tyrosine and tryptophane. In cysteine and methionine, the R-group includes sulfur, and in lysine, it includes another amino group. The structures of the amino acids may be found in any textbook of biochemistry.

Peptide bonds. The amino acids are linked together to form complex protein molecules through anhydride bonds between adjacent

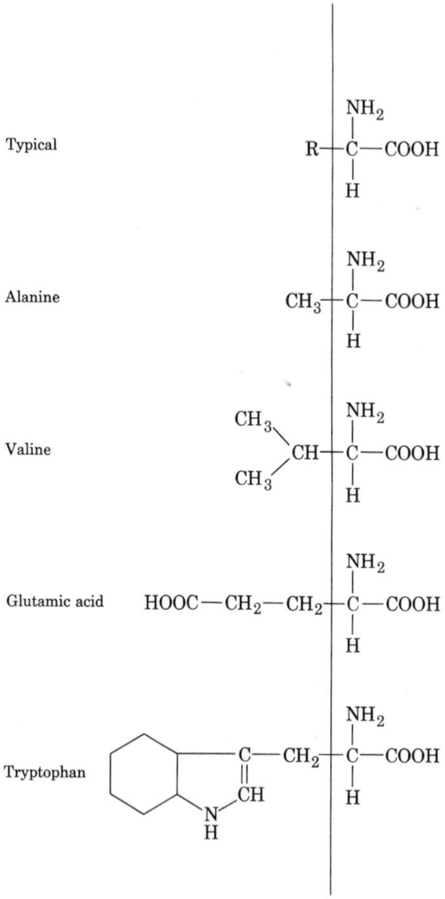

Typical	$R-C(NH_2)(H)-COOH$
Alanine	$CH_3-C(NH_2)(H)-COOH$
Valine	$(CH_3)_2CH-C(NH_2)(H)-COOH$
Glutamic acid	$HOOC-CH_2-CH_2-C(NH_2)(H)-COOH$
Tryptophan	(indole ring: C, CH, NH) $-CH_2-C(NH_2)(H)-COOH$

FIGURE 11.3. Some amino acid structures.

amino and carboxyl groups. A molecule of water is removed in the formation of this *peptide bond* as indicated in Fig. 11.4. A compound of two amino acids is known as a *dipeptide,* three amino acids form a *tripeptide* and so on. Proteins themselves are formed of many such linkages, since their molecular weights are all greater than 30,000. When they are broken down by boiling with strong acids or alkalis, or by the even more efficient digestive enzymes, they are *hydrolysed* to polypeptide units, which are in turn further broken down into smaller and smaller units, leaving ultimately the single amino acids which can be absorbed across the gut wall and distributed to the tissues via the internal environment. The similarity to the breakdown of starch is apparent except that there is great variety in the end products of protein hydrolysis.

Within the tissues, some of the amino acids are reconstituted to form the structural proteins required and the remainder are used for fuel, in much the same way as is glucose.

Essential amino acids. As mentioned in the introduction, ten of the amino acids cannot be synthesized from other compounds in man, and these *essential* amino acids must be present in the diet for proper growth and repair. On the whole, there is a poorer supply of the essential amino acids in plant proteins, and an entirely vegetable diet can easily be deficient in this respect. Most animal tissues, including fish, have approximately the right composition for human needs, which is another way of stating the harsh fact that the best food for an animal, is another animal. In addition to meat however, the other animal products including eggs, milk and cheese are good dietary sources of the essential amino acids.

In recent years, the development of chromatography has made it possible to analyze proteins for their constituent amino acids without the extremely tedious methods previously necessary. At the same time, a great deal of careful work has been done to determine the absolute requirements of the various amino acids in man, chickens, hogs, and rats. We may therefore predict the "completeness" of the protein in a diet from an analysis of its amino acid content. Before this was practical, however, nutritionists had devised a measure of the "biological value" of a protein by feeding experiments. Egg white will support adequate growth in the rat even if it is the sole source of dietary protein and may therefore be given a biological value of 100.

Other proteins may be scored on this basis by comparing the growth of rats fed the test protein with those fed the same weight of egg white. It should be remembered, though, that a diet seldom contains only one protein source. A diet may well contain two proteins of low biological value when eaten separately, but which

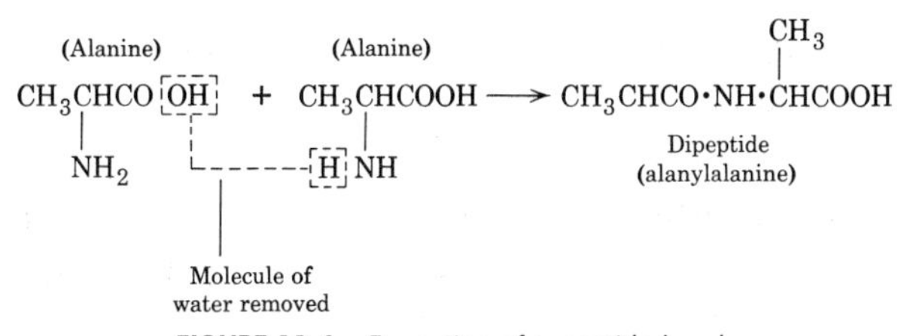

FIGURE 11.4. Formation of a peptide bond.

have essential amino-acid contents which *complement* each other to yield a completely adequate mixture. It is necessary, however, to eat the two incomplete proteins together; the value of the mixture is lost if they are eaten several hours apart. Furthermore, the essential amino acid requirements vary with the intake of nonprotein foodstuffs and with the metabolic rate of the individual. Although the question of the biological value of proteins is extremely complicated, it is safe to make the general statement that diets containing animal protein are superior to all-vegetable diets.

It is clear that in most economies, the animal protein foods like fish, meat, eggs, and milk are the expensive foods. For pork and eggs, for example, it takes many pounds of grain fed to the pigs or chickens to produce a pound of protein. The grain is potential human food to start with and therefore, such foods are always more expensive, and in the countries constantly on the verge of famine, it is positively absurd to feed grain to livestock. Sheep, cows, and goats are less wasteful converters since they can graze on land unsuited to the growing of grain and other human food, and through the activities of their symbiotic microorganisms, convert grass protein into milk and meat. The American practice of feeding grain to steers in feed lots makes excellent steaks, but from the point of view of the best utilization of the world's resources, it is very wasteful.

In many of the underdeveloped countries of Asia, the crop production is barely adequate to support the human population; there is certainly no surplus of cereals which can be fed to livestock. Consequently, these populations live on an almost entirely vegetarian diet (rice, beans, etc.) and their intake of animal protein is inadequate. In many countries the situation is further complicated by religious objections to the consumption of animal products.

Because of its good protein (and calcium) content, milk is an excellent food, especially for infants. The size and health of modern American children may be rightly attributed in part to their large intake of milk. This has so caught the public imagination however, that almost magical qualities are assigned to this food which is provided in nature for the nourishment of *infants*. The need of an adult for milk has been exaggerated in some quarters.

Composition of proteins. Since each amino acid contains an amino group, proteins are characterized by containing *nitrogen*. There are indeed nonprotein nitrogen compounds in animal tissues (and foods generally), but an analysis for nitrogen gives an approximate index of the protein content. Most proteins contain about 16% nitrogen, so if the nitrogen content is multiplied by 6.25, an approximate protein content is arrived at.

Although chromatographic analysis can give us the amino acid composition of a hydrolysed protein, the arrangement of the amino acids within the intact protein can be determined only with the utmost difficulty. This has been attempted in only a few simple proteins, and of course the synthesis of a protein from its component amino acids is as yet in the future.

Much of the difference between animals or between individuals may be attributed to the structure of their protein molecules. There is good evidence that the consistent similarity of the proteins made during an animal's lifetime is due to patterns, or "templates," carried by the genes in the chromosomes. These patterns (or half of them) are passed on to the animal's offspring.

Simple and conjugated proteins. Simple proteins are those containing only amino acids, such as those discussed previously. Conjugated proteins are those comprising a simple protein combined with another complex molecule and many important structural components of cells fall into this category. *Nucleoproteins* are those combined with nucleic acids, and *glycoproteins* contain carbohydrate units and *lipoproteins* lipid units. Hemoglobin consists of a simple protein, globin, combined with the iron-containing *heme* molecule.

LIPIDS

Lipids form a large class of organic compounds including fats and oils, their constituent *fatty acids,* and other still more complex combinations of the fatty acids with various materials. They are all soluble in solvents, such as ether and chloroform.

Fatty acids are the chief units of the lipids in much the same way that amino acids are the units of proteins, except that no lipids are as complex as even the simpler proteins. The fatty acids form a *homologous series* of the general structure R-COOH, in which R may be a chain of carbons from one to more than twenty in length. The short-chain fatty acids such as acetic and propionic acid are not normal constituents of lipids. In general, the C-10 to C-20 fatty acids are found in the lipids, with C-14, C-16, and C-18 as the most common. These numbers each *include* the carboxyl-carbon, and it will be noted that the natural fatty acids contain an *even* number of carbon atoms.

Some of the saturated fatty acids are listed below:

Butyric acid: $—CH_3\ (CH_2)_2COOH$. Found in butter fat to a small extent.
Myristic acid: $—CH_3\ (CH_2)_{12}COOH$
Palmitic acid: $—CH_3\ (CH_2)_{14}COOH$
Stearic acid: $—CH_3\ (CH_2)_{16}COOH$

Myristic acid, palmitic acid, and stearic acid are common in animal fats and vegetable oils. A greater proportion of the longer chain fatty acids is partly responsible for a higher melting point of the animal fats compared with the vegetable oils. (The solidity or fluidity at room temperature is the only distinction between fats and oils.) Another factor affecting the melting point is the proportion of *unsaturated fatty acids,* which are similar to those described previously, except that between one or more pairs of carbon atoms there is a *double bond* rather than the full complement of hydrogen atoms to complete the covalence of the carbons. Important examples are:

Oleic acid: $C_{17}H_{33}COOH$
Linoleic acid: $C_{17}H_{31}COOH$

From the numbers of hydrogens, it is apparent that these molecules contain one and two double bonds respectively. The unsaturated fatty acids have a lower melting point than saturated fatty acids of the same chain length. They are commoner in plant oils and contribute to the generally lower melting points. Plant oils may be treated with hydrogen under pressure, which saturates some of the fatty acids with double bonds and raises the melting point of the oil. This hydrogenation process is important in the manufacture of margarine and vegetable shortening, which custom demands to be solid at room temperature.

The content of unsaturated fatty acids in a fat may be determined by treating it with a known amount of iodine. The unsaturated carbon bonds combine with the iodine, and the extent of iodine uptake is an index of the number of double bonds. This is expressed as the *iodine number* of the fat.

It is now known that some of the highly unsaturated fatty acids are important structural constituents which cannot be synthesized by the human body. These *essential* fatty acids must be present in the diet in much the same way as is true of the essential amino acids, but only in minute amounts.

Fats and oils. The simplest compound lipids are the fats and oils, which are the esters of long chain fatty acids with the alcohol, *glycerol.* Each molecule of glycerol can combine with three fatty acid molecules to give a *triglyceride.* (The structure of a typical triglyceride is given in Fig. 11.5.) Triglycerides may be simple, with all three fatty acids the same, or they may be mixed. Animal fat comprising only triglycerides of these types is called *neutral fat* to distinguish it from other lipids to be discussed.

It is common knowledge that fats are insoluble in water. They may be rendered soluble by hydrolysis, which yields glycerol and the fatty acids. This hydrolysis is accomplished by enzymes in the gut, but as will be discussed in the section on digestion, many authorities believe that some fat is absorbed in an emulsified state without prior hydrolysis. An emulsion is a suspension of a fat in which the droplets are too small to separate out. An everyday example is homogenized milk.

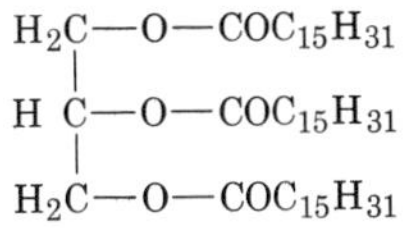

$$\begin{array}{l} H_2C-O-COC_{15}H_{31} \\ \quad | \\ H\ C-O-COC_{15}H_{31} \\ \quad | \\ H_2C-O-COC_{15}H_{31} \end{array}$$

A triglyceride
(tripalmitin)

FIGURE 11.5. Structure of a triglyceride.

Compound lipids. In addition to the simple fats and the fatty acids, there are many compound lipids, in which a typical fat molecule is combined with some other structure. For example, the *phospholipids* are triglycerides in which one of the fatty acid molecules is replaced by a complex phosphoric acid and nitrogen-containing group. The phospholipids are constituents of cell membranes and are believed to be important in the discriminatory permeability of the cells. Much of the plasma lipid content is as phospholipids too. Three important phospholipids are lecithin, sphingomyelin, and cephalin. Other compound lipids are *glycolipids* (containing a carbohydrate unit) and *sulfolipids.*

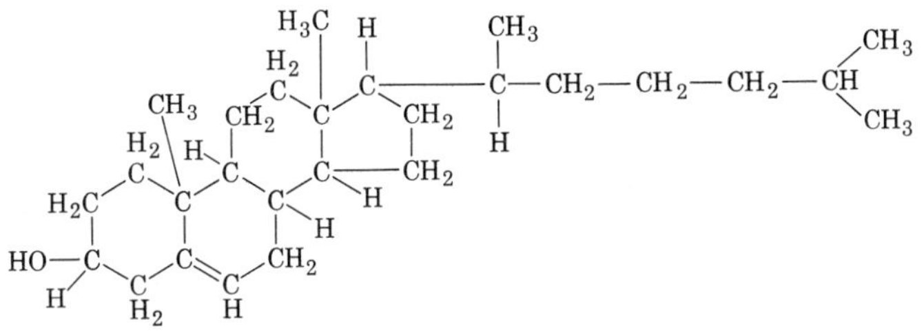

FIGURE 11.6. Structure of a sterol (cholesterol).

A further group of lipids which are rather different from the others are the *sterols*. The four-ringed skeleton of the structure of a typical sterol shown in Fig. 11.6 is characteristic. Many hormones are sterols, and in consequence, the chemistry of these compounds is subject to very intensive research (see Chapter 16.)

VITAMINS

The vitamins are often called *accessory food factors* and this describes them quite well. Many nutritional studies towards the end of the nineteenth century showed that animals did not thrive on a diet composed of purified materials, which apparently met all the then-known requirements for carbohydrate, fat, and protein. The addition of only 2 ml. of milk to the diet of rats had a growth-stimulating effect out of all proportion to the known nutrient content of the added milk. From this and many similar experiments it was concluded that milk and other foods contain "accessory factors" which are essential to life, but which are required in only the most minute quantities.

Fifty years of intensive research have led to the identification of many of these factors we now call the *vitamins*. The chemical structures and specific roles in metabolism of most of them are at least partially understood. In the early days, when their chemical structure was unknown, they were named as vitamins A, B, C, D., etc., and this nomenclature has persisted to a large extent, especially in the popular literature. From the chemical studies of the vitamins, we now know that several of them comprise groups of similar but slightly different compounds and that vitamin B is actually a whole complex of *unrelated* substances.

The vitamins fall into two main groups; the fat-soluble and the water-soluble. The distinction is of more than academic chemical interest because their distribution in foodstuffs is partly determined by their solubility. Vitamins A, D, E, and K are fat soluble, and those of the B complex and C are water soluble.

Vitamin A. We now realize that the growth response on adding milk to the purified rat diets in the early experiments was mostly due to the vitamin A in the milk fat. Vitamin A is also found in the fats of eggs and fish, and in extremely large amounts in fish livers. The structure of vitamin A is known to be one-half of the molecule of carotene, a yellow pigment found in squash and carrots, etc. and also in the leaves of green plants. Carotene is readily split by enzymes in the body to yield vitamin A, and so a plentiful supply of green or yellow vegetables is a good source of the *precursors* of vitamin A. In this special sense therefore, the diet does not need to contain vitamin A itself provided the carotene intake is adequate. If fresh vegetables or other suitable foods are not available, supplemental vitamin A can be obtained from small quantities of fish-liver oil.

A deficiency of Vitamin A leads to several separate consequences, and one of these is night blindness. Night vision depends on a retinal pigment, *visual purple* (see Chapter 22), which is bleached by exposure to light. The regeneration of visual purple uses up vitamin A, and, of course, the regeneration is diminished in a deficiency of the vitamin. A decrease in the ability of the eyes to adapt to dim light has been proposed as a sensitive test for vitamin A deficiency but its usefulness in adults is in question. More serious symptoms are seen at a later stage of deficiency in the *keratinisation,* or *cornification,* of delicate epithelial tissues in the body. The ciliated membranes which line the upper respiratory tract become dry and flaky. In consequence, foreign material is not efficiently swept out by the beating of the cilia in the fluid film on their surface and the membranes are therefore more vulnerable to bacterial invasion. This is the factual basis for the popular belief that vitamin A is an "anti-infective" agent.

The delicate membranes of the eyes are also affected by a deficiency in vitamin A leading to the condition called *xerophthalmia.* This literally means "dry eyes," and well describes the dry, crusty, and inflamed condition. Keratinisation of epithelia in the reproductive organs leads to sterility and in the urinary system it leads to the formation of calculi (stones). Vitamin A deficiency also causes poor bone growth but in this respect its effects are overshadowed by vitamin D, to be discussed. Occasionally, people have suffered from excessive bone growth through taking absurdly large amounts of vitamin A. This is an outcome of the assigning of near-magical qualities to the vitamins by popular imagination. Although a diet with an adequate supply of vitamins is good, it does not follow that a diet with even more is better. Ordinary excesses of vitamin A over immediate requirements are stored in the liver and are available during subsequent periods of deficiency. This is clearly of survival value to wild animals or to primitive man whose winter diet might well be deficient in vitamin A. Similarly, the stored vitamin A is one of the factors which make liver such an excellent food.

Vitamin D (calciferol). Closely related in structure to the *sterols* discussed in the section on the lipids is vitamin D. This vitamin is essential for normal bone growth, and children with vitamin D deficiency develop *rickets,* even if their intake of calcium and phosphorus is otherwise adequate. Conversely, children with low intakes of these minerals usually have normal bones as long as they receive plenty of vitamin D. In rickets, the bones do not grow to normal size and the mineral deposition is subnormal, so that they are both soft and pliable. Rickety children are usually stunted and have characteristically bowlegs, and the defective growth of the facial bones leads to impaired eruption of the teeth. An early sign of rickets is a bead-like enlargement of the costochondral junction on each rib, which gives a row of bumps on the child's thorax called the "Rachitic Rosary." In adults, a vitamin D deficiency leads to one form of *osteomalacia,* in which the bones become light and fragile.

Vitamin D has a similar distribution to vitamin A in foods of animal origin, and again, some fish-liver oils are especially potent sources. It is also similar in that it may be synthesized in the body from dietary precusors—sterols common to ordinary diets. Their conversion to vitamin D is accomplished in the skin under the effect of ultra violet irradiation, normally in sunlight. For this reason, rickets and comparable dental problems are seldom seen in tropical regions, even in those countries with notoriously poor nutrition. On the other hand, rickets was so common in England during the nineteenth century that on the continent of Europe it was called "Der Englische Krankheit—the English disease."

The diet of the British industrial population was probably no worse than that of the West Indian slaves, but the combination of working inside factories and the cloudy English weather precluded exposure to the sun and an adequate synthesis of vitamin D. Even as late as World War II, the teeth of the poorer people in England were noticeably worse than those of the middle classes. During that war however, the rations available for children had been designed by some of the world's leading nutritionists to meet all vitamin requirements. Consequently, despite the severe shortages of food imposed by the submarine warfare, the general level of health was quite remarkable. It is only right to add, however, that the unavailability of candy during those years probably had a lot to do with the dental improvement.

The physiological role of vitamin D is a topic of active research interest. The concensus of modern work suggests that it mediates mineralization of bone by some action at the actual site of bone formation and also by modifying the uptake of calcium from the gut.

Vitamin E (Tocopherol). No symptoms of vitamin E deficiency have ever been described in man. Male rats fed a diet deficient in vitamin E, however, do not produce sperm; and in the females which do become pregnant, the embryos are reabsorbed. Furthermore, a definite wasting of skeletal muscle is observed. From this evidence it is extremely likely that vitamin E is essential to man but that the requirements are so small compared with the content of ordinary diets that deficiencies never arise. The vitamin is widely distributed in foods of plant and animal origin and is particularly abundant in wheat germ oil.

Vitamin K. The section relating to blood clotting described the liver's requirement for vitamin K in prothrombin synthesis. In a deficiency of the vitamin, prothrombin synthesis is diminished, which leads to impaired clotting mechanisms and the consequent dangers of uncontrolled hemorrhage.

Dicumarol is a compound found in sweet clover which so closely resembles vitamin K that the enzyme systems of the liver utilize it as if in error. Because dicumarol is not truly a substitute for the vitamin, the prothrombin synthesis is inhibited by its presence. This is a good example of an *antimetabolite,* of which several are known. Dicumarol is therapeutically useful since it can lower the prothrombin levels in patients subject to intravascular clotting.

Vitamin B complex. The vitamin B complex was originally thought to be one substance because its members are found in close association in foods. As it was discovered to comprise more than one, the components were named B_1, B_2, etc., and to some extent these names are still used.

Thiamine is vitamin B_1. This substance is known to form part of an enzyme important in carbohydrate metabolism and the effects of its deficiency are far-reaching. The main symptoms of deficiency are a peripheral neuritis leading to paralysis, loss of appetite, general debility, and, in the later stages, enlargement of the heart and edema of the limbs. An important source in a normal diet is the outer layers of cereal grains. In the Far East many people live on a diet of polished rice, from which the outer layers have been removed. The deficiency syndrome they develop is known as *beriberi,* but because their diet is also deficient in other B complex vitamins, notably niacin and riboflavin, beriberi should be thought of as a multiple deficiency disease rather than a simple thiamine deficiency.

Beriberi can be prevented by exchanging the polished rice for unpolished rice or by supplementation with other foods. This was done in the Japanese Navy with a dramatic improvement in the health of the sailors as early as 1883. The nature of the accessory food factor in the rice polishings was of course unknown at that time. American, British, and Dutch soldiers in Japanese prison camps during World War II suffered extensively from beriberi; as much through the incredibly inefficient Japanese supply service as from calculated brutality.

In addition to whole cereal grains; dried brewer's yeast, ham, peas, beans, and nuts are excellent dietary sources of thiamine. It is difficult to conceive of anyone's eating a normal American diet and developing a thiamine deficiency.

Riboflavin (Vitamin B_2) is a constituent of a whole series of enzymes known as the *flavoproteins,* which serve vital roles in the oxidative processes within the cells. As with thiamine, a deficiency of riboflavin alone is seldom seen in man, but sores around the mouth, nose and eyes are commonly attributed to this deficiency. In earlier times these sores on the faces of poor children were smugly attributed to lack of cleanliness—truly a most ironic injustice!

Nearly all foodstuffs contain some riboflavin, but a diet of bread, pork, and potatoes tends to be deficient and such a diet is unfortunately the lot of some low-income groups. No foods are particularly good sources of riboflavin but liver and milk are among the best, and a diet containing reasonable amounts of meat, dairy products, and green vegetables is almost bound to be adequate.

Niacin was at one time known as nicotinic acid, but this name was dropped because of the possible association with nicotine, the active drug in tobacco. The vitamin is part of the structure of co-enzymes I and II, which are important participants in as many as forty different enzymic processes.

Niacin deficiency leads to a condition known as *pellagra,* but with the same reservations as applied to beriberi concerning the multiple nature of the deficiency. The main symptoms are a rough dermatitis and soreness of the mouth and tongue, followed by serious degeneration of myelinated nerve fibers in the central nervous system and by fatty infiltration of the liver.

Because the vitamin is a constituent of the important co-enzymes

I and II, it is widely distributed in the living material eaten for food, especially liver, other meats, and leguminous vegetables. These tend to be expensive foods however, and pellagra is found in populations living on largely cereal diets. The corn meal and molasses diet formerly common among the rural poor in the Southern United States was notorious in this respect.

Other members of the water-soluble B-complex are *pantothenic acid* and *pyridoxine.* These (and yet others) are known to be components of enzyme systems, and animal studies have shown them to be essential dietary constituents. Their distribution in foods, however, is so widespread that naturally occurring deficiencies are unknown.

Vitamin B_{12}, exceptional in that its molecule includes a metal, *cobalt,* is necessary for the maturation of red cells produced in the hemopoietic tissues. Therefore, a deficiency causes anemia. Large, immature red cells are released into the circulation as described in Chapter 6. *Pernicious anemia* may be treated by injections of minute amounts of B_{12}, but the deficiency is usually due to a faulty absorption of the vitamin rather than a dietary deficiency. Patients with pernicious anemia are thought to lack an "intrinsic factor," presumably a constituent of gastric juice, which is essential for the absorption of the "extrinsic factor," Vitamin B_{12}. A potent source of Vitamin B_{12} is liver, and before purification of the vitamin, pernicious anemia patients were treated with crude liver extracts.

Folic acid is another vitamin required for normal hemopoiesis. Pure deficiencies have not been described in man, but the multiple deficiencies of pellagra and beriberi undoubtedly involve folic acid too.

Ascorbic acid (vitamin C). The history of this vitamin paradoxically begins long before its discovery. From the close of the Middle Ages on, men began to make longer and longer voyages away from land as the Spanish and Portugese discoveries in the New World were exploited. The disease, *scurvy,* was rampant during these voyages without fresh fruit, vegetables, or meat. By the eighteenth century it came to be recognized that scurvy could be prevented and cured, not only by fresh fruits, but by the stored juice of fruits. The juice of the lime fruit was particularly suitable and to this date is included in the fare of sailors in the British Navy. The nickname of "limey" for English sailors stems from this practice.

Ascorbic acid has a relatively simple structure, and most mammals are able to synthesize all their own requirements, presumably from glucose. In man, the other primates, and the guinea pig, this synthetic ability is lacking and the vitamin must be provided in the diet. Its physiological role is undoubtedly related to its properties as a reducing agent.

The chief lesion in scurvy is a diminished production of intercellular cement materials. In consequence, all tissues become more fragile, wounds heal slowly, if at all, and the fragility of the blood capillaries leads to many small hemorrhages in the tissues. These symptoms show up first in the gums which become sore and spongy, and the teeth become loose.

The specific signs of vitamin deficiencies are functional disorders or tissue degenerations, and many of the vitamins were originally studied as the *anti*pellagra, or *anti*polyneuritis factors, etc. It cannot be too strongly emphasized, however, that the deficiency signs are manifestations of derangements in the metabolism of the cells and that these are due to the absence of a vital material in the internal environment. The cells deteriorate if the fluid bathing them does not supply the raw materials for synthesizing enzymes just as surely as if it does not contain suitable amino acids for protein synthesis, or if there is an imbalance of ions.

The demonstration of the dramatic effect of small doses of the B vitamins in curing pellagra and beriberi are to be ranked among man's greatest scientific achievements. From this, however, has grown a positive cult in which vitamins are regarded as magical potions which will "soup up" flagging energies, protect us from colds, and even make us more beautiful. On a normal American diet, a case can be made for giving children supplementary Vitamins A and D during the wintertime, but some preparations of these vitamins are of such enormous potency that many people carelessly take hundreds of times the normal level of intake. There is an accumulation of evidence showing the dangers of excessive calcification of tissues due to excesses of both vitamins A and D.

MINERALS

The importance of an adequate intake of sodium, potassium, and magnesium is already clear from our consideration of the body fluids. Their widespread occurrence in all foods means that *dietary* deficiencies can only occur under the most bizarre conditions. Indeed, there is good evidence that our very large intake of sodium chloride presents a daily load on the excretory mechanisms which can become critical in relatively mild circulatory disorders.

Of the manifold minerals required by the body, those most frequently involved in nutritional disorders are calcium, phosphorus, iron, and iodine. Many other elements are also found in human tissues, but it is often not clear whether they are there by "accident" or whether they serve a vital role. These include aluminum, bromine, cobalt, copper, fluorine, manganese, nickel, and zinc.

Calcium is the most abundant mineral in the body. Salts of calcium make up the hard parts of the skeleton and teeth, and we have already referred to the importance of the plasma calcium concentration in connection with the excitability of the tissues and with the blood clotting mechanisms. The growing child requires calcium in large quantities for making bone, and pregnant women need extra for fetal bones. For pregnant women, however, the greatest calcium requirement is after the delivery of the baby during *lactation.* If her dietary intake is inadequate, a nursing mother will draw on her own skeleton to maintain the calcium content of the milk she secretes. Decalcification of the jaw bones from this cause is a possible reason for the dental troubles common in the year or two following childbirth. These difficulties can be avoided by increasing the dietary intake of calcium. Even in nonlactating adults, the calcium salts of the bone are not dead and permanent, but rather they are in a state of constant *turnover.* Old bone is continually being replaced by newly mineralized tissue, and the inevitable wastage in the exchange must be met by an adequate consumption of calcium.

The best sources of calcium are milk and its derived products. Beans, wheat, and other cereals have a good calcium content, but in the cereals a large part of it is combined with *phytic acid* and in this form it is not absorbed from the gut. Partly because of the expense of dairy products, calcium is the mineral most likely to be deficient in an ordinary diet. In wartime England, possible calcium deficiencies were avoided by adding calcium carbonate to bread flour, but this reasonable practice is not followed elsewhere. The British flour manufacturers were embarrassed by the presence outside their plants of cement company trucks unloading sacks of powdered chalk!

There are many factors affecting the absorption of calcium from the intestine, and a defective absorption may lead to a nutritional deficiency in spite of a seemingly adequate diet. We have already referred to the probable role of vitamin D in this respect. Also, any defect in fat digestion causes calcium to be lost with the undigested fat passed out with the feces. Calcium absorption is better under acid conditions, so that the availability of dietary calcium varies with the acidity of the intestinal content and this in turn depends on the diet and the amount of gastric acid secretion. Probably the most important factor in calcium absorption and its subsequent metabolism is the relative amount of phosphorus in the diet. Excess phosphorus is excreted in the feces combined with calcium.

Phosphorus is second only to calcium in the amount present in the body. Bone is made up of complex crystals of calcium phosphate and calcium hydroxide with small amounts of magnesium, fluorine and other elements. Unlike calcium, phosphorus is widely distributed *inside* the soft tissue cells and is in a relatively low concentration in the extracellular fluid. The nucleic acids in the genetic material of the cells contain phosphorus, and of course adenosine triphosphate (ATP) is the important energy "package" repeatedly made and destroyed in energy metabolism. Phosphorus is so widely distributed in foods that it is doubtful whether any *dietary* deficiencies exist in man. More important than the absolute level of phosphorus in the diet is the ratio of the calcium and phosphorus content. If these two elements are present in proportions between 1:2 and 2:1, the absorption of both is near optimal. If the ratio is very different in either direction however, the absorption of one or the other is diminished.

The concentrations of calcium and phosphorus in the plasma are largely under the hormonal control of the *parathyroid gland*. One theory is that parathormone regulates both by primarily regulating the extent of phosphorus excretion. This topic is covered in more detail in Chapter 16.

Iron is a constituent of hemoglobin and also of a whole series of respiratory enzymes which are responsible for electron transfers in the controlled oxidation within the cells. Its importance is well recognized and yet simple iron deficiencies are still common. The most obvious symptom of iron deficiency is anemia (a shortage of hemoglobin; see Chapter 6); an anemia which will respond to iron (or to iron and copper) is therefore a *simple nutritional anemia*. Eggs, meat, liver, and fruits (especially apricots) are all good sources of iron. Only a proportion of the iron in these foods is available to the body because some iron is bound in nonabsorbable complexes.

The concentration of most minerals in the internal environment is controlled principally by the kidneys, but the iron concentration is regulated by *differential absorption*. In the gut wall there is a protein called *ferritin* which combines with iron absorbed from the gut and releases it into the blood if the blood content falls below a critical level. If the blood iron concentration is optimal, the ferritin releases none of its iron. Consequently, it remains "full up" with iron and so absorbs no more. As soon as the blood-iron concentration falls a little, the ferritin gives up iron to the blood and takes up some more by absorption from the gut.

Iron is stored in the liver in sufficient quantity to tide over a deficiency of a few months. In particular, the newborn infant has a considerable supply so long as the mother's diet has been adequate in this respect. This is important because milk is deficient in iron and the baby must rely on stored iron during the first few months of life. It is now customary to give babies small amounts of supplementary iron at least until they are eating meat and other solid foods. It also follows that pregnant women should increase their iron intake to allow for the storage in the fetal liver.

Iodine is a critical constituent of thyroxine, the thyroid hormone (see Chapter 16). Iodine deficiencies were common at one time in the mountain valleys of Switzerland and in isolated communities of other countries including the United States. An iodine deficiency leads to a shortage of the thyroid hormone even though the thyroid gland itself grows very large. The consequences of *hypothyroidism* are described in detail in Chapter 16, but they may be briefly listed as dullness, sluggishness, puffiness of the face, and retardation of mental and physical development. It is now common practice to add iodine to table salt although many persons object to this on "principle." Iodine is also found in good supply in all sea foods.

REFERENCES

1. McElroy, W. D. and C. P. Swanson, "Trace elements," *Sci. American* **188** (1), 22, January 1953.
2. Nasset, Edmund S., *Food and You,* 2nd ed. Barnes and Noble, New York, 1958.

THE METABOLISM OF CARBOHYDRATES, PROTEINS, AND FATS

12

ENERGY

In the preliminary discussion of nutrition (p. 198), it was pointed out that cells need nutrients for two main functions; for combustion to release energy and as sources of building blocks for growth and repair. An adequate intake of protein of high biological value is clearly essential for growth and repair, as are supplies of vitamins and minerals. Carbohydrates and fats are also structural constituents of cells, but their greatest importance in the diet is as sources of *energy*. A discussion of the processes by which these compounds are used to supply energy belongs properly to a textbook of biochemistry and only a few essentials are reviewed here. Similarly, the student is referred to an elementary physics text for a more complete account of energy, heat, work, and entropy. These terms have highly technical meanings in thermodynamics which are far beyond the scope of this text. However, a familiarity with the philosophical concepts involved will help make clear the principles of energy metabolism.

When a piece of wood is burned, energy is released as *heat*. The energy is "available" because the atoms of the wood are in a complex *organized* structure, originally achieved by the tree's applying energy to reduce carbon dioxide (see p. 52) and by incorporating the products with water in the complex structure of wood. The tree used energy from the sun, which it "trapped" by the chlorophyll in the leaves. When the atoms comprising the wood are *disorganized,* by burning to their original state of carbon dioxide and water, some of that energy is released. It is made up of the energy of the carbon–carbon bonds which are broken—and the energy from the oxidation of carbon to carbon dioxide.

This simple example illustrates two important points. First, the organization of atoms into complex, "unlikely" arrangements requires energy, and some of this energy is released when the atoms return to the more "likely" state of a lower degree of organization. The second point illustrated is that most energy sources are derived ultimately from the sunlight energy trapped by green plants. This is true of all foods, because even meat production depends on the plants eaten by the animal. *Nuclear* energy is different from energy derived from our common *chemical* sources in that it is not derived from sunlight.

The oxidation of a complex compound is one example of changing an organized system with potential energy to a less organized system with a lower energy content, thereby releasing *free energy.* Another example would be the release of a previously stretched spring. In every such case, not all of the energy change in the system becomes available for useful purposes; a proportion is irrevocably lost with every change to a lower energy state. The proportion which *is* usefully available is called the *free energy;* the fraction which is lost goes to increase the *entropy* of the system.

The concept of entropy is a difficult one for students without a background in physical chemistry. It is perhaps easier for the beginning student to think in terms of "free energy" (some of which can be used usefully) and "bound energy," which is unavailable heat. (The bound energy is the entropy times the absolute temperature.) We can illustrate the concept of bound energy by a homely example. Anyone with a bank checking account is familiar with the legalized shakedown called, "bank charges." If a customer deposits $1,000 he cannot draw that sum out again intact. The bank makes a charge for its services, and on closing the account the customer gets only, say $999.50. The bank charge of 50¢ can be likened to that part of the energy which necessarily becomes unavailable heat in a reaction. Just as with a repetition of the bank transaction, the loss of free energy by the entropy increase is repeated with each reaction.

It should be noted that entropy increase is quite inescapable even in a machine of maximal theoretical efficiency; this has nothing to do with the mechanical imperfections of the system. Losses from mechanical imperfections are also unavoidable in a practical sense and they further reduce the useful yield of energy from the free energy of a reaction.

The entropy of a system is an index of its degree of disorganization—the more random the arrangement of a system, the higher the entropy. The entropy of an isolated part of a system is decreased when that part is made less random, as in the synthesis of a complex molecule for example, but the energy for achieving that synthesis must in turn come from a source elsewhere in the overall system, and this involves an increase in entropy. Thus, the entropy of the universe as a whole is increasing; the enormous capital reserve of energy in the sun and other stars is being gradually dissipated in "bank charges."

The living cell has been rather poetically described as an "island of negative entropy." The appropriateness of this is clear if we consider the great degree of organization of the structural proteins of the cell, and especially the maintenance of the intracellular potassium concentration and the exclusion of sodium. We have already described the necessity of a constant supply of energy to maintain the "unlikely," state of a concentration gradient across the cell membrane. If the energy supply fails, as on the death of the cell, the sodium and potassium concentrations rapidly return to those expected from their random distribution. It is believed that the energy is required for the *work* of expelling sodium ions that diffuse into the cell (see p. 65).

The very existence of a cell therefore, requires a constant supply of energy. Any work done on the external or internal environment will require still more energy. The most obvious example is muscular work, and this includes the constant beating of the heart muscle and the movements of smooth muscle in the gut walls, as well as walking, lifting, digging, etc.

Stored energy, whether of an object raised against gravity, or in the configuration of complex molecules in wood (or food), is called *potential energy.* The potential energy is released under appropriate conditions and can appear in one or more of many forms. *Kinetic* energy is energy of movement, as in a falling body. *Electrical* energy is obtained from a generator as a result of the kinetic energy of the revolving armature. *Heat* and *light* are other forms obtained by the burning of fuel. *Work* is done by applying energy to a system to impart kinetic or potential mechanical energy.

In an animal body doing no external work, all the energy used shows up ultimately as *heat.* In biological studies therefore, energy measurements are usually made in units of heat; calories and kilocalories.

A *calorie* is defined as the amount of heat required to raise the temperature of 1 gram of water from 14.5° to 15.5° centigrade. As this unit is rather small for many purposes, the kilocalorie (1000 calories) is commonly used instead. The kilocalorie is often called a "large calorie" or simply a Calorie (with a capital C), which can be confusing, and so the term "kilocalorie" is to be preferred. The possible confusion can also be avoided by remembering that in human nutrition and physiology, the term "Calorie" always refers to a kilocalorie. It is worth noting that to call a kilocalorie a "large calorie" is as illogical as calling a kilometer a "large meter."

Determination of the caloric content of foods. The heat produced by burning a compound can be determined in a *bomb calorimeter.* A weighed amount of a substance, glucose for example, is placed in a sealed chamber filled with compressed oxygen, which is immersed in water of known volume and temperature (see Fig. 12.1). The glucose is then ignited electrically and the increase in temperature of the water noted. If the apparatus is suitably insulated, the caloric content of the glucose can be calculated, knowing that 1 kilocalorie will raise the temperature of 1 kilogram of water 1°. The heat of combustion of 1 gram of glucose is 3.76 kilocalories, which is the same as that obtained from its metabolism in the body. The whole of this energy is released in the body because glucose is completely digestible and completely combustible to carbon dioxide and water, just as it is in the bomb calorimeter. Other nutrients do not yield all their *gross energy* in the body because they are incompletely digested and/or their metabolic destruction is less complete than combustion in a bomb calorimeter. This is especially true of proteins utilized for their fuel value.

Metabolism of glucose. It is obvious that burning in the usual sense is too violent a means of energy release for biological systems. The energy in nutrients is released by their more gradual destruction and oxidation. Glucose is the principal energy source in the cells and its metabolism is briefly described.

Glucose in the internal environment penetrates the cells by an

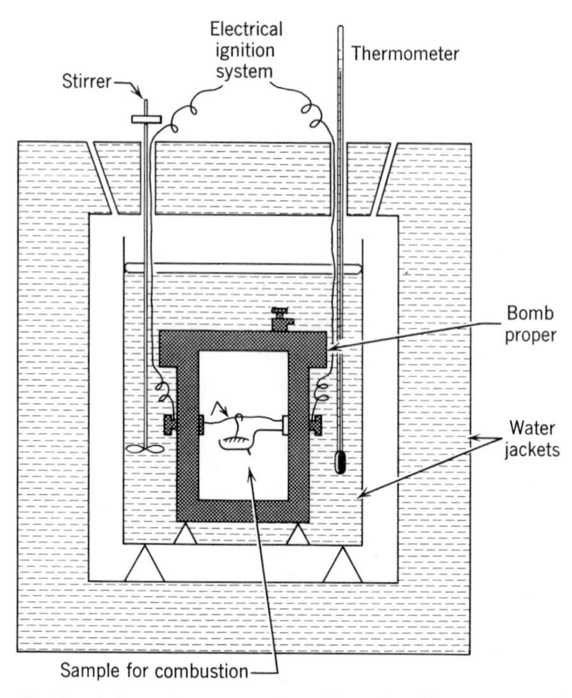

FIGURE 12.1. Bomb calorimeter in cross section. Reproduced by permission from Kleiber, in Uber, *Biophysical Research Methods,* Interscience Press, New York, 1950.

unknown mechanism, evidently facilitated by *insulin.* Within the cell (or even the membrane) the glucose is *phosphorylated.* From this more active form it undergoes the series of *glycolytic* reactions outlined in Fig. 12.2. It is noted that the compound, adenosine triphosphate (ATP) is prominent in these reactions. This complex molecule contains two "high-energy" phosphate bonds. That is, it takes considerable energy to attach the phosphate groups to adenosine monophosphate (AMP) and adenosine diphosphate (ADP) successively, but these same packets of energy can be delivered elsewhere in the cell as required when the ATP is broken down.

Thus, the "activation" of glucose by converting it to glucose-6-phosphate and further to fructose-1-6-diphosphate requires two high-energy phosphate bonds. Later, however, the fructose-1-6-diphosphate yields *two* molecules of triose (the 3-phosphoglyceric aldehyde), and in the course of their breakdown to pyruvic acid they *each* yield two high-energy phosphate bonds. Thus, glycolysis requires two moles of ATP to "prime" the glucose, but the subsequent reactions give back four moles, with a net yield of two. By this series of enzymatic reactions, some of the energy locked in the glucose is quietly converted to high energy phosphate bonds. In muscular contraction, for example, the ATP breaks down to ADP and gives up its energy. It is now known that glucose may be metabolized to pyruvic acid by pathways other than the one described, for example, by the *hexose monophosphate shunt.*

The production of pyruvic acid does not involve any oxidation, and the breakdown of the glucose molecule is only partial. Further energy is extracted by oxidation in an interesting series of reactions which proceed in a cyclic fashion (see Fig. 12.3). This series is called the *tricarboxylic acid cycle,* or the *Krebs-cycle,* in honor of the great biochemist who first proposed the scheme. Each "turn" of the cycle yields still more ATP, and the metabolites are

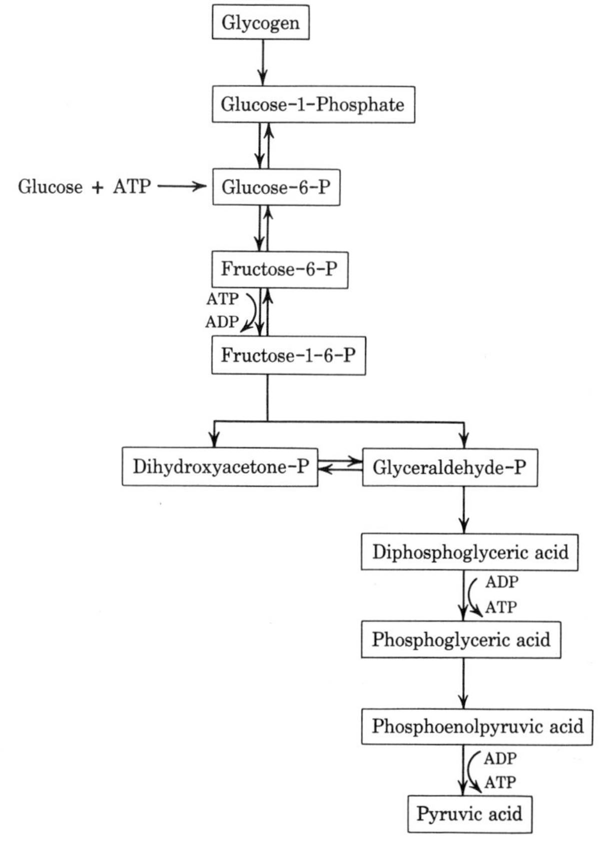

FIGURE 12.2. Glycolysis.

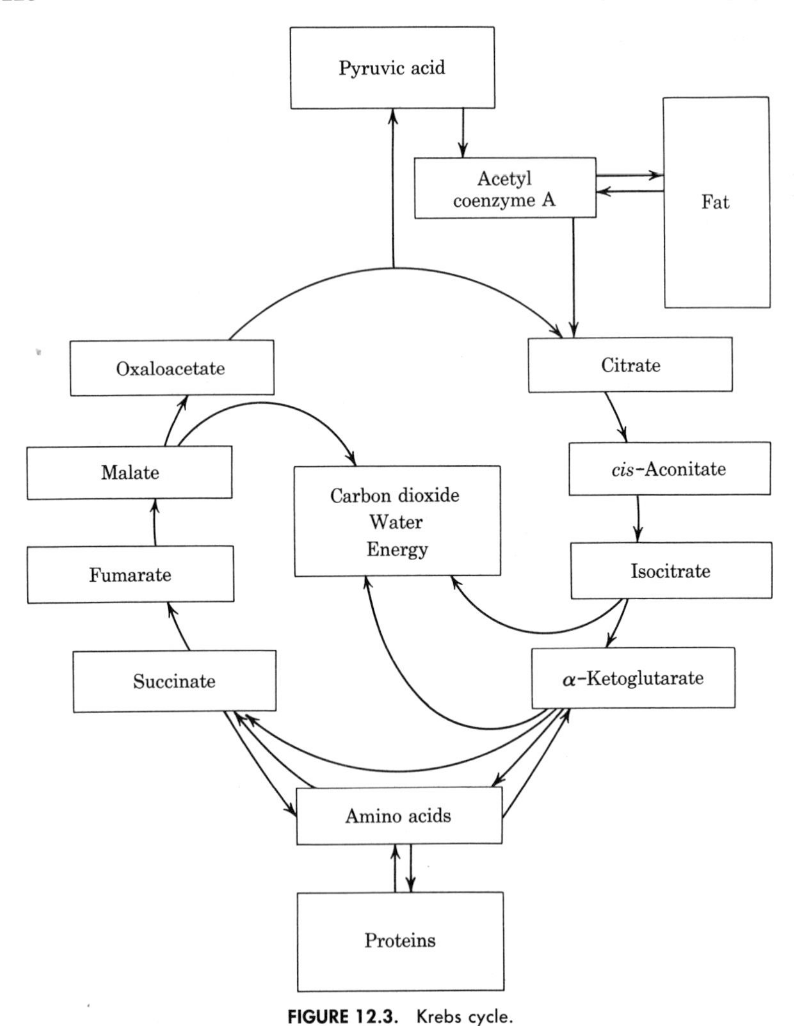

FIGURE 12.3. Krebs cycle.

completely oxidized to carbon dioxide and water. The actual transfer of electrons in the oxidations is by processes too complicated for discussion in this textbook.

Experiments have shown that the glycolytic enzymes are in the cell's cytoplasm, whereas the Krebs cycle processes take place in the mitochondria. Because the Krebs cycle reactions need oxygen, they proceed only if there is an adequate oxygen tension in the interstitial fluid to maintain a constant diffusion gradient into the cell. This depends on the blood supply to that tissue and on the oxygen content of the blood. When the oxygen supply is inadequate, however, the tissues can continue to produce ATP units by glycolysis alone. This would lead to an accumulation of pyruvic acid, except that it is rapidly converted to *lactic acid.* The accumulation of lactic acid is seen in violent muscular exercise where the energy utilization outstrips the oxygen supply to the muscle. After the exercise, during rest, the lactic acid is removed by the blood and carried to the liver. There it is reconverted to pyruvic acid, and some of this is oxidized via the Krebs cycle. The energy in the ATP units thus produced is used to convert the rest of this pyruvic acid back to glucose, and thence to glycogen, by a *reversal* of glycolysis. Of the lactic acid produced in exercise, about ⅕ is burned afterwards to reconstitute glycogen from the other ⅘ths. The increased oxygen consumption necessary for the oxidation of the lactic acid after exercise is called the *oxygen debt,* and the persistence of panting after exercise is of course the outward manifestation of this. The accumulation of lactic acid in an exercising muscle is an important factor in *fatigue.*

Before entry into the Krebs cycle, pyruvic acid is decarboxylated and combined with coenzyme A to yield *acetyl Co A* (Fig. 12.3). It is this highly active compound which combines with oxalacetic acid to yield citric acid and so recharges the cycle. As indicated, acetyl Co A is also an intermediary common to both carbohydrate and fat metabolism.

Carbohydrate storage. The concentration of glucose in the internal environment is maintained fairly constant at about 100 milligrams per cent. (No serious results follow from increased concentration, except that there is wastage of glucose in the urine when the renal threshold for glucose is exceeded see Chapter 9). A decreased concentration affects cellular function, starting with the nervous system; a depression to below 60 milligrams per cent leads to convulsions and unconsciousness.

If the absorption of glucose from the gut were the sole regulation of its plasma concentration, there would be wide fluctuations—from an excess during the digestion of a meal to a dangerous deficit between meals. Normally, excess glucose is deposited as glycogen in the liver and muscles almost as fast as it is absorbed, and the plasma glucose concentration rises only transiently after a meal. Conversely, as circulating glucose is used up by the tissues, glycogen is broken down and glucose released into the blood to maintain the optimal concentration in the internal environment. This process is called *glycogenolysis.* Although the liver contains more glycogen gram for gram than does muscle, the greater total mass of muscle means that there is actually more muscle glycogen than liver glycogen. Either way, glycogen synthesis proceeds *within* the cells, and so here again insulin exerts an important influence in mediating the entry of glucose.

Diabetes mellitus is a disease in which there is inadequate insulin production by the β-cells of the Islets of Langerhans in the pancreas. In consequence, the entry of glucose into the cells for either combustion or glycogen synthesis is impaired. The absorption of glucose from the gut continues more or less as usual, but since the glucose cannot be deposited as glycogen, the plasma concentration increases enormously and the resorbtive power of the kidney tubules is exceeded. The resultant *osmotic diuresis* causes the great thirst and polyuria characteristic of the disease (see Chapter 9).

In some diabetics, the insulin production is inadequate only in the face of a large carbohydrate load, and then it is possible to control the blood sugar level by keeping to a low-carbohydrate diet comprising several small meals per day. If the insulin deficiency is more severe, supplementary insulin must be injected nearly every day. For the comfort and convenience of diabetics there is a great deal of research work being done on preparations that can be taken by mouth with the same effect.

Although the presence of sugar in the urine is a cardinal sign of the disease, it is not the most serious aspect. Despite the high concentration of sugar in the internal environment, the cells are *functionally deficient* of glucose because it cannot enter to be metabolized. The shortage of glucose forces the body to shift to a fat metabolism, as in a starving animal. At the same time, the insulin deficiency in some way deranges the Krebs cycle reactions so that the acetyl CoA units resulting from β-oxidation of fat are inadequately oxidized; the accumulating two-carbon fragments react together to yield *ketone bodies.* One of these is acetone, which is volatile and can therefore be smelled on the breath and in the urine of diabetics. Other ketone bodies are acetoacetic acid and β-hydroxybutyric acid which, because they are strong acids, lower the pH of the internal environment and cause ion shifts as the intracellular fluid becomes more acid too. The shift of potassium in particular causes reduced excitability of tissues and leads eventually to paralysis and death.

The uncontrolled diabetic therefore suffers mainly from *acidosis,* which causes a characteristic hyperventilation in an effort to raise the pH of the internal environment by blowing off carbon dioxide (see p. 191). The acidosis itself inhibits the action of whatever insulin is present, and so the condition gets worse. For treatment the patient must receive not only insulin but also bicarbonate to correct the acidosis.

A further important point is that, in some mysterious way, potassium is laid down with glycogen and is released into the extracellular fluid in glycogenolysis. During the onset of a diabetic episode, the potassium released in glycogenolysis is added to the extracellular fluid and none is returned to the liver because there is no glycogen synthesis. The excess potassium is excreted during the osmotic diuresis and the plasma level is returned to normal. Then, when the insulin deficiency and acidosis are corrected, glycogen synthesis begins again and the extracellular fluid potassium may be depleted to a dangerous extent as it is deposited with the new glycogen. Now that this is recognized, it is standard procedure to provide supplementary potassium for such patients.

Glucose tolerance. Following the consumption of a carbohydrate meal, the rise in blood glucose is quickly corrected by glycogen synthesis. The restoration to normal is slower if the insulin production is marginal or sub-minimal; this is the basis of a test for diabetes. The subject drinks a large amount of glucose solution, or better, receives a glucose injection, and the return of the plasma glucose level is followed in serial blood samples. Glucose tolerance curves obtained in this way are shown in Fig. 12.4. The presence of urinary glucose is a sign of diabetes, but as it can result from other less sinister causes, a glucose tolerance test is always carried out for confirmation.

Glycogen Storage Disease

This disease is a rare condition encountered in children. Glycogen deposition occurs to bizarre extremes leading to enormous hypertrophy of the liver and failure of its other functions.

Fat Synthesis from Carbohydrate

It is evident that only a limited excess of glucose can be accommodated as glycogen. If the intake is larger than that required for day-to-day energy demands, the excess is stored as fat by pathways to be described.

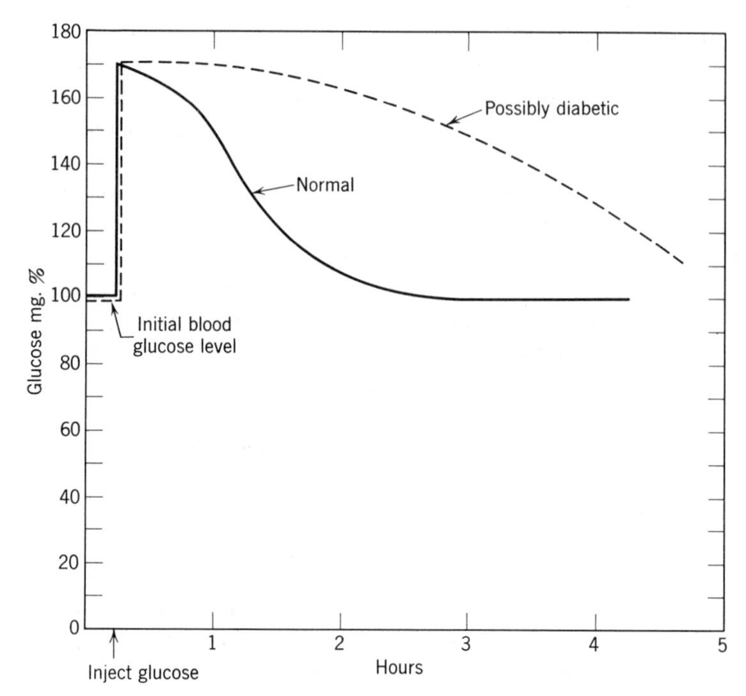

FIGURE 12.4. Glucose tolerance curves.

Blood Glucose in Exercise

During steady work, the increased glucose consumption by the cells is made good by increased glycogenolysis. In sudden activities or alarm, epinephrine stimulates the liver to a sudden release of glucose (and potassium). The rise in blood sugar may be enough to cause a transient glycosuria. This is an interesting example of a homeostatic mechanism which actually *anticipates* a change in the cells' demands on the internal environment (see Chapter 16).

FAT METABOLISM

Fat is present in the plasma and the rest of the internal environment as very small droplets called *chylomicrons.* It is possible that these may enter the cell as such, but in any case, the metabolism of fat must start with its hydrolysis to glycerol and fatty acids. This goes on in the liver and presumably in other tissues too.

The glycerol is metabolized via dihydroxyacetone phosphate which is, as we have seen, one of the intermediates in glycolysis.

The long-chain fatty acids are metabolized step-wise by a process called *β-oxidation.* The fatty acid is activated by combining with Co-enzyme A, and then enzymes successively split off two-carbon fragments combined with Co A. These are acetyl-Co A units which are identical with those formed by the decarboxylation of pyruvic acid referred to previously. It is clear therefore, that the metabolism of both carbohydrate and fat leads to acetyl-Co A, which is oxidized via the Krebs cycle.

This common pathway is important in fat *synthesis* as well as in its breakdown. Long-chain fatty acids are built up of active acetyl-Co A units, and by this route excess glucose intake is converted into body fat. The glucose is subjected to anaerobic metabolism (and the energy made available to the body), and the excess two-carbon units are stored as fat. In nature, these can be drawn on in times of food-shortage, but in our affluent society, the accumulation of excess fat is a public health problem of grave importance (see Chapter 15). Similarly, the glycerol part of fat is synthesized from glucose by reversal of the pathway described for its breakdown.

The oxidation of fat yields more energy on a weight or molar basis than the oxidation of carbohydrates, because the fat molecule is less oxidized to start with. In the typical carbohydrate, the general formula $(C_6H_{10}O_5)n$ shows that there is enough oxygen present to account for all the hydrogens as water. In a fatty acid molecule $[CH_3(CH_2)_nCOOH]$, however, there are many hydrogens as well as carbons to be oxidized, and they all yield energy. For this reason, a high fat diet may readily exceed the energy requirements of the body (see Chapter 15).

METABOLISM OF AMINO ACIDS

The synthesis of structural proteins from amino acids is thought to be controlled by the configuration of the nucleoproteins within the cells. This configuration is partly determined by heredity, as is shown by the inheritance of blood types for example, but it is also well recognized that the synthetic processes may be adapted to the body's needs, as in antibody production. Protein synthesis is carried on in all the tissues, but the liver is especially active in this respect. The liver can shrink or grow according to the protein intake and may be thought of as "storing" protein to some extent.

In an ordinary American diet, the protein intake is substantially greater than that required for growth and repair; the excess is utilized as fuel in much the same way as are carbohydrates and fats. In addition, the structural proteins of the body are constantly being replaced and repaired by new material and the old is broken down and used as fuel. This process, of course, starts with hydrolysis to yield the constituent amino acids. For further breakdown, the amino acids are first *deaminated,* that is, the amino group is removed by an enzymatic reaction similar to that in Fig. 10.3. The deaminated "skeletons" of the amino acids then undergo further reactions to yield pyruvic acid or some of the Krebs cycle intermediates, which are metabolized by processes already described. Protein is quite acceptable as an energy source and the body's requirements can be met from a diet containing only protein, but such a diet is usually expensive. For greatest economy, a diet need only contain a modest proportion of protein of high biological value, with cheap carbohydrate foods for fuel.

Deamination yields *ammonia* and this must be immediately converted to something else because it is highly toxic to the cells. The *detoxification* processes occur mainly in the liver, where the ammonia enters a cyclic series of reactions to yield *urea,* which is relatively nontoxic. In mammals, most nitrogenous waste is excreted as urea. Urine also contains some ammonia but this is actually made in the kidney as part of its acid-base control and is not itself derived from the general protein metabolism (see p. 192). In birds and reptiles the detoxified end product of nitrogen metabolism is uric acid rather than urea.

The excretion of waste nitrogen is one of the principal functions of the kidney, and, as described in Chapter 9, an accumulation of nitrogenous end-products in the blood is often an indication of failing kidneys. Temporarily this in itself is not too serious so long as the liver continues to convert ammonia to urea, because the tissues can tolerate large concentrations of urea.

The tissue proteins are in a constant state of replacement and repair, so that even in the fasting animal there is always a minimum breakdown of protein and consequent excretion of nitrogen. This is known as the *endogenous* nitrogen excretion. The adequacy of a protein diet may be determined by *nitrogen balance studies.* If a subject is fed a diet of known nitrogen content, and at the same time the urinary and fecal losses of nitrogen are determined, the overall retention or loss of nitrogen can be determined. If the protein in the diet is inadequate in quantity or essential amino acid content, the synthesis of protein for repair is less than the endogenous breakdown. Consequently, more nitrogen is lost than is taken in and the subject is said to be in *negative nitrogen balance.* If this state persists, there is a visible wastage of tissues.

Conversely, an adequate protein intake results in nitrogen balance. Positive nitrogen balance is achieved by a *growing* animal on an adequate diet, because the animal's tissue protein is increasing, and the excretion of nitrogen is less than the intake.

REFERENCES

1. Adams, Elijah, "Poisons," *Sci. American* **201** (5), 76, November 1959.
2. Dole, Vincent P., "Body fat," *Sci. American* **201** (6), 70, December 1959.
3. Doty, Paul, "Proteins," *Sci. American* **197** (3), 173, September 1957.
4. Fruton, Joseph S., "Proteins," *Sci. American* **182** (6), 32, June 1950.
5. Gale, Ernest F., "Experiments in protein synthesis," *Sci. American* **194** (3), 42, March 1956.
6. Green, David E., "Enzymes in teams," *Sci. American* **181** (3), 48, September 1949.
7. Green, David E., "Biological oxidation," *Sci. American* **199** (1), 56, July 1958.
8. Green, David E., "The synthesis of fat," *Sci. American* **202** (2), 46, February 1960.
9. Green, David E., "The metabolism of fats," *Sci. American* **190** (1), 32, January 1954.
10. Greenberg, Leon A., "Alcohol in the body," *Sci. American* **189** (6), 86, December 1953.
11. Hoagland, Mahlon B., "Nucleic acids and proteins," *Sci. American* **201** (6), 55, December 1959.
12. Kamen, Martin D., "Tracers," *Sci. American* **180** (2), 30, February 1949.
13. Linderstrom-Lang, K. V., "How is a protein made," *Sci. American* **189** (3), 100, September 1953.
14. Pauling, Linus *et al.,* "The structure of protein molecules," *Sci. American* **191** (1), 51, July 1954.
15. Pfeiffer, John E., "Enzymes," *Sci. American* **179** (6), 28, December 1948.
16. Schmitt, Francis O., "Giant molecules in cells and tissues," *Sci. American* **197** (3), 204, September 1957.
17. Stumpf, Paul K., "ATP," *Sci. American* **188** (4), 84, April 1953.
18. Stetten, De Witt, Jr., "Gout and metabolism," *Sci. American* **198** (6), 73, June 1958.
19. Trowell, Hugh C., "Kwashiorkor," *Sci. American* **191** (6), 46, December 1954.

METABOLIC RATE AND TEMPERATURE CONTROL

13

DIRECT CALORIMETRY

All the energy used by the body except that for outside work appears finally as *heat*. In mammals and birds the body temperature is kept constant, and this is only possible if heat production and heat loss are equal. Therefore, the resting rate of energy conversion can be determined by measuring the *heat loss*.

Calorimeters have been devised in which water is circulated through pipes in the walls of an animal chamber, and the animal's heat production is calculated from the rise in temperature of the water. Even more than with the bomb calorimeter though, the practical difficulties of insulation, etc., are enormous, and such methods are used in only a few centers and entirely for research purposes.

INDIRECT CALORIMETRY

For all routine determinations of metabolic rate, and for much investigative work, the heat production can be calculated *indirectly*. The energy derived from the metabolism of a mole of glucose in the body is known and we can also calculate how much oxygen is required to oxidize one mole of glucose.

$$C_6H_{12}O_6 + 6\ O_2 \rightarrow 6\ CO_2 + 6\ H_2O + 673\ \text{kilocals}$$

Therefore, if it is assumed that all heat production in a subject is from the metabolism of glucose, a determination of the oxygen consumption permits the calculation of the heat production. Actually animal heat is seldom derived from glucose alone, and a further assumption has to be made about the nutrients being oxidized. In the *fasting* subject, glucose, protein, and fat are being metabolized simultaneously in proportions such that 1 liter of oxygen is used in the release of 4.7 kilocalories (as opposed to 5.05 kilocalories if only glucose were being oxidized). In calculating the metabolic rate from oxygen consumption it is important therefore that the subject is indeed in the fasting state.

BASAL METABOLIC RATE

The metabolic rate is the heat production of the animal expressed in kilocalories per hour per kilogram animal, or more frequently, in kilocalories per hour per square meter of body surface. The significance of the surface area is clarified later. The metabolic rate of a walking or working man is greater than when he is at rest. The resting metabolic rate is sometimes called the *basal metabolic rate* (or BMR) because it is assumed to be the minimum for keeping the animal functioning. It might be thought of as roughly comparable to the rent paid for business premises; whether sales are good or bad, the rent is one expense that must always be paid.

Through its secretion, *thyroxin,* the thyroid gland acts as a "governor" on the metabolic rate. In an excess of the hormone, or *hyperthyroidism,* the metabolic rate is elevated, and in *hypothyroidism* it is lowered. The mechanism is not understood, but one attractive theory proposes that the hormone modifies the permeability of the cells to glucose and other nutrients.

A thyroid disfunction may be detected by a BMR determination which can be compared with some arbitrary "normal" rate. The resting metabolism, however, varies enormously with factors other than the thyroid status. For example, metabolism is speeded up in a cold room in order to maintain body temperature; the digestion of a meal causes extra heat production; the emotional state of the subject causes variations; and even the time of day has its effect. Therefore, to be clinically useful, BMR determinations are made under rigidly defined conditions. The subject should be in the fasted state, relaxed after a good night's sleep, and in a room at a thermoneutral temperature (about 80° F).

The oxygen consumption is measured with a recording spirometer like the one diagrammatized in Fig. 8.8, p. 151. The bell is filled with oxygen which the subject inspires through the mouth tube. On expiration, the subject's breath returns to the bell through the second tube, but the carbon dioxide content is removed by the soda-lime cartridge. Therefore, at each breath the volume of the bell's contents diminishes by the amount of oxygen removed. The scale is calibrated to give the diminution in volume of the bell from which the oxygen consumption in a unit time can be measured. Because gas volumes vary with temperature and atmospheric pressure, oxygen consumption figures are corrected to standard temperature and pressure. From the oxygen consumption, the heat production in cal/hr can be calculated.

The BMR machines commercially available for clinical use are arranged so that the determinations may be made without any knowledge of the theoretical aspects. The movements of the bell are recorded on a chart fastened to the rotating drum. The slope of the record gives a measure of the rate of decrease of volume in the bell, that is, the rate of oxygen consumption. The instructions on the chart give some rule of thumb calculations and corrections to be applied to this slope measurement. The resulting value is then compared with a "normal" BMR for the subject, calculated from tables of height, weight, sex, and age. Finally, the subject's BMR is expressed as plus or minus a percentage of the "normal." Thus a BMR of +10 is 10% higher than that expected for the subject. Because of the limitations of the method, variations of less than 5% in either direction are not considered significant.

METABOLIC RATE AND BODY SIZE

Large animals produce more heat per day than do small animals, but about 100 years ago physiologists observed that larger animals produce less heat per day *per kilogram weight.* The increased metabolic rate in larger animals is not proportional to their increased size.

The rate of heat loss of a warm body depends on, among other variables, the surface area of the body and the difference between its temperature and the temperature of the environment. The lesser metabolic rate per kilogram in larger animals was therefore explained by the *surface law.* Although larger bodies have larger surface areas than do smaller bodies of similar shape, their surface areas per unit volume are less. This can be shown by the simple example of two cubes. A cube with a length of side of 2 centimeters has a total surface area of 24 square centimeters. Its volume is 8 cubic centimeters, and assuming a specific gravity of 1.0, it weighs 8 grams. A cube with double the side length (4 centimeters) will weigh 64 grams, eight times as much. The surface area, however, is 96 square centimeters, only four times as much. On the basis of considerations like this, it was concluded that the metabolic rate is proportional to the surface area.

The difficulty encountered is that the surface area of an animal is ill-defined; it is hard to say what belongs to the surface and what does not. Should, for example, the great surface of the ears of a rabbit be regarded as body surface or not? Since it is difficult to define what does and does not constitute the surface of an animal, it is not surprising that research workers disagreed about the methods of measuring it. Surface areas can be *estimated* from more reliable parameters, such as body weight. The best known estimate of human surface area is the formula of DuBois, which is satisfactory and widely used for clinical purposes.

$$\text{Surface Area (m}^2) = 71.84 \times W^{0.425} \times L^{0.725}$$

where W = weight in kilograms
L = height in centimeters.

Several physiologists, among them August Krogh, suggested that metabolic rate might be best expressed as a function of body weight rather than of the ill-defined surface area. They did not suggest making metabolic rate proportional to body weight, but to a *power* of body weight. The surface area of bodies with similar shapes is proportional to the ⅔ power of their volume (or in objects with a constant density, to the ⅔ power of their weight). For example, the surface area of a cube is

$$S = 6.V^{2/3}$$

Therefore, instead of stating that the metabolic rates of large and small animals are proportional to their body surface areas, we can say that they are proportional to the ⅔ power of their body weights.

After analyzing many results on the heat production of large and small animals (obtained in various laboratories), a biophysicist called Kleiber concluded that the metabolic rate is more nearly proportional to the ¾ rather than the ⅔ power of body weight. He formulated, therefore,

$$B = 70.W^{3/4}$$

He expressed a prediction formula for human metabolic rate on this basis.

For men,

$$B = 71.W^{3/4}[1 + 0.004(30 - a) + 0.01(S - 43.4)]$$

For women,

$$B = 66.W^{3/4}[1 + 0.004(30 - a) + 0.018(S - 42.1)]$$

where B = basal metabolic rate in kcal per day, W = body weight in kg, a = age in years, and S = specific stature = height (cms)/$W^{1/3}$.

BODY TEMPERATURE REGULATION

The "cold-blooded" animals such as frogs, fish, and snakes are not strictly cold-blooded, but rather their tissues have almost the same temperature as their environment. Up to a limit, they function better in the warm than in the cold, and at very low temperatures they cannot function at all. In the course of doing muscular work, the heat derived from the extra metabolism helps to raise their temperature above that of the environment. These animals are called *poikilotherms.* The "warm-blooded" animals are called *homeotherms.* The hibernating animals are intermediate in that they are homeothermic ordinarily, but when environmental conditions become too severe, they suspend the struggle to maintain their body temperature and go into a state of dormancy with a much lower body temperature.

In man and the other mammals, the maintenance of the body temperature at or near 37° C means that heat production must equal heat loss and *vice versa.* It will be recalled that the calorimetric determination of metabolic rate depends on this fact. The heat production of an animal depends first on the basal metabolic rate, that is the resting rate of the body as determined by the metabolic body size, the thyroid status, and other variables, and also on the amount of activity over and above that of the basal state. The greater the muscular activity, the greater will be the heat production.

Heat loss. A homeothermic animal *loses heat* but does not *cool,* because its temperature does not fall. The routes of heat loss are radiation, conduction, and evaporation.

Radiation

A body loses heat by the radiation of infrared heat waves to nearby objects or to the sky. If the nearby objects are at a higher temperature, the body absorbs radiant heat from them. The radiation of heat to or from other objects is independent of the intervening air temperature. It is well known that skiers can be quite comfortable with their shirts off in bright sunshine, because even though the air temperature is very low, the radiant heat from the sun warms them quite adequately.

Conduction

The body transmits heat to the air around it by conduction, but since air is a poor conductor, the loss by this route in still air is minimal. We can envisage a layer of warm air on the skin, and if there is a breeze continually removing or disturbing this warm layer, the rate of loss by conduction to the air will be increased. This is one of the reasons why it is harder to stay warm in a wind. Heat loss by conduction to materials other than air (or water, if immersed in water) is seldom important, but an animal standing on snow or ice (or a sitting skier) can presumably lose considerable heat by this route.

Evaporation

The conversion of liquid water to water vapor at 30° C requires a *latent heat* of evaporation of 0.58 kilocalories per gram of water. Therefore the continual evaporation of water from the skin and lung membranes is a route of heat loss. The loss by this route can be increased by sweating. The sweat is secreted on to the skin where it evaporates and thereby dissipates 0.58 kilocalories for each gram evaporated. Evaporation takes place more rapidly in a stream of air, and so once more, the rate of heat loss is much greater in a wind than in still air.

Dissipation of excess heat. The *hypothalamus,* a part of the midbrain, is the main center for integrating and regulating the *visceral* functions of the body (see Chapter 21). In general, these are the automatic functions which regulate the internal environment (such as the action of the heart) as opposed to the *somatic* functions (such as the limb muscles). The nerves transmitting impulses to the visceral organs from the hypothalamus (and other portions of the brain) form the *autonomic nervous system,* which is discussed in a later chapter.

The hypothalamus itself contains specialized nerve cells which will respond to slight increases in blood temperature, and other cells which are sensitive to a fall in temperature. It also receives impulses over fibers from temperature-sensitive receptors in the skin. In a warm environment, or in violent muscular exercise, the beginning of an increase in body temperature evokes discharges from the hypothalamus causing sweating and peripheral vasodilation which we recognize as the flushed and damp appearance of someone who is hot. It will be recalled that sweating can enhance the evaporative loss of heat, but its extent is subject to other factors to be discussed. The extra blood flow to the skin brings heat from the body core to the surface where it can be dissipated. This increases heat loss by all routes, although there is no loss by conduction to the air if the air is hotter than the skin.

Clothes ordinarily tend to retain heat, and therefore an individual is usually better off with fewer clothes in a hot environment. White clothes, however, reflect some of the heat from hot sunshine better than the nude body does.

The extent of evaporation, and therefore the extent of the heat loss by this route, depends on the relative humidity of the ambient air. If the air is already well saturated with water vapor, evaporation is slower than in dry air. For this reason, a temperature of 90° F in air of 90% humidity, as frequently encountered in summer on the eastern seaboard of the United States, is much more uncomfortable than a temperature of even 100° F in dry desert air, as encountered in Arizona or New Mexico.

Some species, notably domestic cattle and dogs, sweat hardly at all, and thus one mechanism for the dissipation of excess heat is not available. Dogs and cattle can increase the evaporative loss from the lungs by a shallow panting which does not lead to hyperventilation and the consequent respiratory alkalosis. Despite this mechanism, domestic cattle do not thrive in very hot climates.

It will be recalled that one calorie is required to raise the temperature of water through 1° C, and therefore it will take 70 kilocalories of heat to raise the temperature of a 70 kilogram man from 37° to 38°, assuming the specific heat of tissues is the same as that of water. If the rise is not more than a degree or two, excess heat can be temporarily "stored" in this fashion with little detriment to the animal, especially if soon after the rise of temperature there is a period of lower ambient temperature in which the excess heat can be dissipated more readily. The camel is relatively unaffected by body temperature increases of 3° or even 4° C., which means it can store considerable excess heat during the daytime and can lose this heat readily during the cold desert night.

Temperature regulation in a cold environment. When the ambient temperature and the wind conditions are such that an animal's environment is below thermoneutrality, the peripheral vasoconstriction helps to maintain the core temperature of the animal and decreases the skin temperature. The temperature difference between the skin and the cold environment is thus decreased and there is a smaller heat loss. The vasomotor mechanisms in temperature control are regulated through the hypothalamus. Also, when the sensory input to the thalamus indicates a cold environment, the animal's heat production is increased. This is accomplished mainly by shivering, which is really a rapid succession of aimless muscle movements that generate heat, but also in part by a poorly defined *nonshivering thermogenesis.* Studies have shown that laboratory animals exposed to a cold environment for a period of one or more weeks show hormonal and enzymic adaptation.

In most mammals, the peripheral vasoconstriction is accompanied by the erection of the hairs of their fur, which increases the insulating layer of air trapped in and between the fibers and helps to reduce heat loss. Furthermore in long-term adaptation to cold, cattle and other animals grow thicker coats than in temperate conditions. Since man is without a covering of fur and the vasomotor and piloerector responses are seldom adequate for really severe conditions, he must depend on clothing and shelter. The advantages of constructing sound houses with artificial heating are obvious and we can restrict our discussion to the insulating value of clothing. Still air is not a good conductor, and the principle of clothing is that layers of air are trapped between successive layers of fabric and give an insulating value like the air in a fur-bearing animal's pelt.

Primarily, the insulation of clothing protects against heat loss by *conduction.* Also, the air trapped next to the skin achieves a high humidity and therefore decreases evaporation and the consequent *evaporative heat loss.* The heat radiated from the body to the innermost layer of clothes must be conducted across the poorly-conducting successive layers of air and fabric before it can be lost by *radiation* from the outer layer of clothes. Finally, good clothing protects against the increase in conduction and evaporation in a wind.

It is popularly supposed that men living in the arctic must have a large food intake for extra energy to keep warm. With the standard of clothing worn by the Eskimos and the relative newcomers to the arctic, the body lives in a microclimate comparable to tropical conditions. There is relatively little difficulty in providing enough layers of clothing for a man to stay warm under even the coldest conditions, but clothing design must also take into account the practical considerations of bulk. A further difficulty with arctic clothing, which is still not overcome, is that its insulating value is so good that the least extra exertion causes great discomfort from sweating. If this continues to the point where the sweat soaks several layers of clothing, these wet layers afterwards lose their insulating value (because of the good conductance of heat by water) and the clothing is inadequate for quieter levels of activity.

Gagge and his co-workers have studied the insulating value of clothing and have developed the measuring unit, the *clo*. A man dressed in an ordinary business suit wears about one clo. Four or five clos are needed for moderate activity in severe arctic weather, and to sleep comfortably in an ambient temperature of −40° to −50° F, a man must have clothing and a sleeping bag providing a total insulation of up to twelve clo. *Frostbite* results from freezing damage to tissues, usually the ears, nose, fingers, or toes. The extreme vasoconstriction on exposure to the cold can reduce the blood flow below the point at which the tissue can be kept warm and prevented from freezing. At temperatures of −40° F and below, especially if there is a wind, frostbite damage can occur in a very few minutes. It is especially insidious in that the general numbness of the extremities in the cold prevent any warning of its onset. Even when thawed out, frozen cells die and the tissue sloughs off to a depth depending on the depth of the freezing damage.

The folklore of New England and the prairie states is full of accounts of people who took deep breaths on very cold nights and fell dead immediately because "their lungs froze." These stories are complete rubbish.

REFERENCES

1. Barker, M. E., "Warm clothes," *Sci. American* **184** (3), 56, March 1951.
2. Bogert, C. M., "How reptiles regulate body temperature," *Sci. American* **200** (4), 105, April 1959.
3. Heilbrunn, L. V., "Heat death," *Sci. American* **190** (4), 70, April 1954.
4. Hock, Raymond J. and Benjamin G. Covine, "Hypothermia," *Sci. American* **198** (3), 104, March 1958.
5. Johnson, Frank H., "Heat and life," *Sci. American* **191** (3), 64, September 1954.
6. Kelley, James B., "Heat, cold and clothing," *Sci. American* **194** (2), 109, February 1956.
7. Mayer, Jean, "Appetite and obesity," *Sci. American* **195** (5), 108, November 1956.
8. Scholander, P. F., "The wonderful net," *Sci. American* **196** (4), 96, April 1957.

DIGESTION

14

Simple unicellular animals can absorb some of their nutrients in solution directly from the water they live in. The greater part of their requirements are met, however, by the ingestion of relatively complex solid particles which are broken down by *intracellular digestion.* In higher animals, some of the white blood cells and the cells of the *reticulo endothelial systems* (see Chapter 3) retain this property of phagocytosis, but it is merely a scavenging arrangement for the internal economy of the animal and is not related to nutrition. Most of the cells of the higher animals require all their nutrients to be presented already dissolved in the interstitial fluid. The problem then is to break down complicated foodstuffs into soluble nutrients which can be delivered to the cells. This is accomplished in the *digestive tract.*

There are simple, bag-like "stomachs" in the coelenterates and other primitive forms, but for our purposes we can consider the digestive tract of the earthworm as the ultimate in simplicity. It is a nearly straight tube in which the lining is slightly convoluted to increase surface area. Food enters the mouth and is pushed along the tube in which enzymes break down some of the organic matter into soluble components. These components are absorbed through the wall into the blood, and the undigested remainder is voided through the anus.

This is the essence of digestion; all the complexities of the digestive tract in the mammal result from its great length and the specialization of its parts for certain aspects of the overall processes. The general design of the human digestive tract is shown in diagrammatic form in figure 14.1 and is outlined in the following annotated list. A familiarity with this outline will make the subsequent detailed discussion more clear.

1. *Mouth.* In the mouth, food is chewed, mixed with saliva and swallowed.
2. The *Esophagus* is a muscular tube which conveys the food to the stomach.
3. The *Stomach* is a bag-like organ in which some phases of digestion are begun by juices secreted by the walls. The stomach also has a storage function following a large meal.
4. *Small Intestine.* Liquefied food (chyme) from the stomach passes to the small intestine. In the first section, the breakdown processes are completed by enzymes contained in the pancreatic and intestinal juices. In the lower sections, the breakdown processes continue, but the absorption of the soluble products is the chief activity. Chyme is moved along the intestine by muscular contractions of a type called *peristalsis.*
5. *Large Intestine.* The main function here is the reabsorption of water and salts from the very liquid chyme. The partially dried, undigested residue is compacted in the *rectum* as *feces* and is periodically voided in *defecation.*

We may now consider the digestive processes in more detail. For man, it might be said that digestion begins in the kitchen, because cooking renders many foodstuffs more digestible. In the physiological sense, however, digestion begins when food is placed in the mouth.

MOUTH

In chewing, the teeth tear and grind the food into manageable pieces while the secretion of *saliva* moistens and lubricates the food. The most important constituent of saliva is *mucus,* which is the lubricant encountered throughout the gastro-intestinal tract. There are three pairs of salivary glands, the *parotid, submaxillary* and *sublingual* glands. (Fig. 14.2). They secrete constantly to maintain the moistness of the mouth, but their production is increased manyfold in eating, and indeed, in the anticipation of eating.

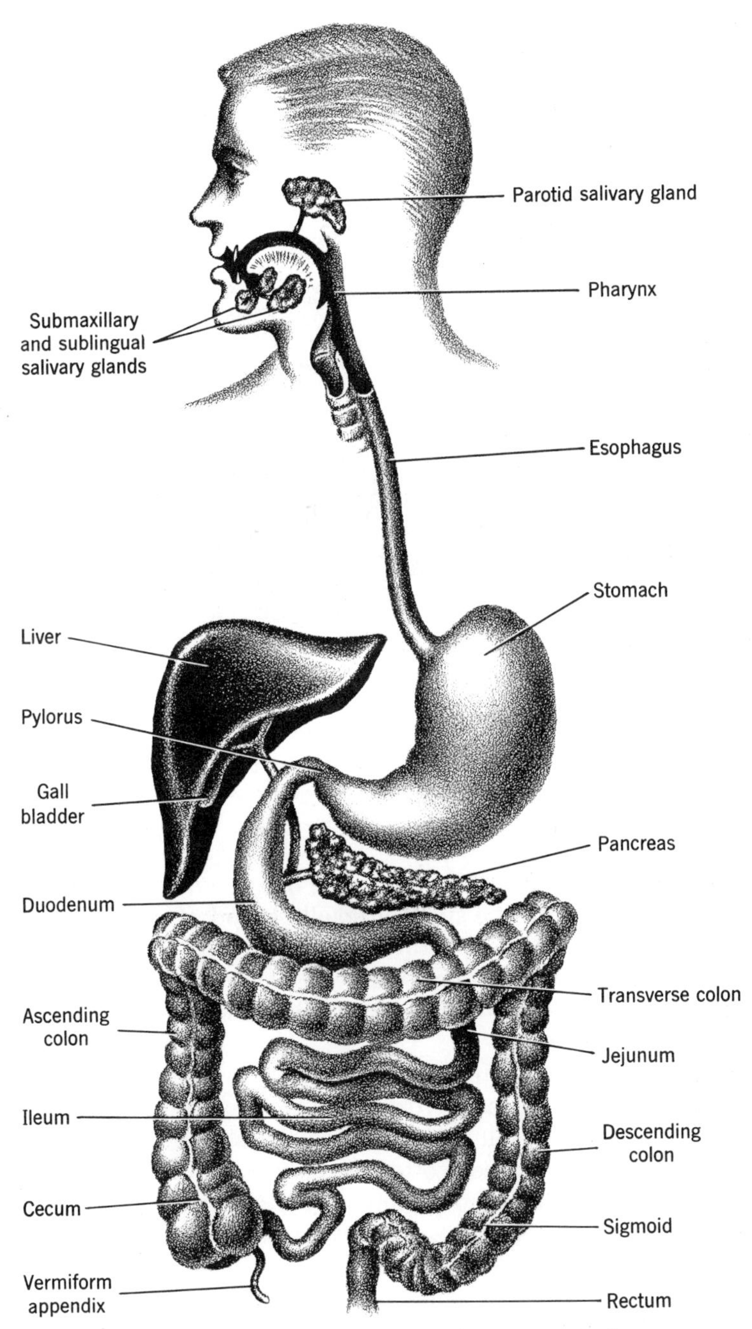

FIGURE 14.1. Diagram of the organs of the gastro-intestinal tract.

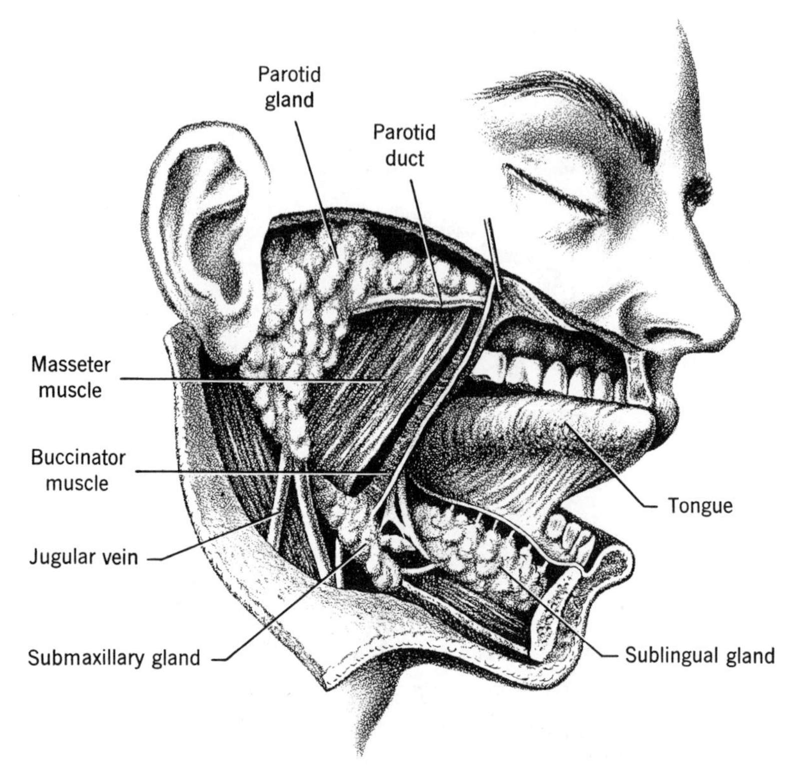

FIGURE 14.2. The salivary glands, as revealed by dissecting the side of the face. Reproduced by permission from Millard, King, and Showers, *Human Anatomy and Physiology*, W. B. Saunders Co., 1956.

The salivary flow is a classic example of a *conditioned reflex*. It does not require even the sight or smell of food; if we are hungry, the mere noises of setting a table are enough to start salivation. Copious salivation in chewing is elicited by the *taste* sensation. The taste buds, located on the tongue, are discussed in a later section on the special senses.

The saliva of man is quite unusual in that it contains a digestive enzyme, *ptyalin,* which can hydrolyse starch to yield maltose units. In most mammals the saliva functions solely as a lubricant, and even in man, its digestive value is often overrated. Food usually spends so short a time in the mouth that very little hydrolysis can take place before the acid contents of the stomach inactivate the ptyalin. For this reason, many near-cranks insist on chewing each mouthful of food twenty (or fifty) times. If a large carbohydrate meal is consumed, however, the mass in the stomach may be dense enough for ptyalin to continue to act in the center of it during the twenty minutes or so required for the acid gastric juices to permeate the whole.

After adequate chewing, the tongue pushes a *bolus* of food between the *fauces* at the back of the mouth into the *pharynx*. With the completion of this conscious part of swallowing, a complicated reflex is invoked by which the pharyngeal muscles seize the bolus and force it downwards into the esophagus. At the same time, the epiglottis closes to prevent food entering the trachea and the soft palate closes the nasopharynx. The mechanical events in swallowing have been studied in X-ray movies of subjects eating meals mixed with radio-opaque substances.

ESOPHAGUS

The *esophagus* is the tube connecting the pharynx and the stomach. The upper part is striated muscle and the lower is smooth muscle, but the peristaltic contractions which move the food down are automatic throughout. At the junction of the esophagus and the stomach is a muscular valve, the *cardia*. Ordinarily, the cardia is closed, but it opens in response to the arrival of peristaltic waves passing down the esophagus in swallowing. The closure of the cardia usually prevents regurgitation of the stomach contents into the esophagus.

STOMACH

This organ has three interrelated functions; storage, the secretion of digestive fluids, and mixing. It has two main parts, the highly distensible *fundus* which is mainly concerned with the storage and secretory functions and the smaller, more muscular *antrum*, which churns and mixes the food.

Gastric secretions. The whole of the stomach lining, including the antral portion, has glands secreting mucus; the *digestive* juice secretion is confined to the fundus. The two principal components of the fundic secretions are hydrochloric acid and pepsin.

Hydrochloric acid is secreted in a surprisingly concentrated form by the *parietal* cells. The fresh secretion has a pH of about 1.0 but dilution of the acid with other gastric contents gives an overall stomach pH of about 2.0 (which is only one tenth as acid). This extreme acidity is important in several respects; it not only kills any live food, but also prevents bacterial putrefaction if a large meal is stored in the stomach for a considerable time. The low pH also causes the *denaturation* and precipitation of soluble proteins which will then remain longer in the stomach. The most essential role of the gastric acid, however, is in providing suitably acid conditions for *pepsin,* a hydrolytic enzyme which splits proteins into smaller units, but does not carry the digestion all the way to single amino acid units.

Pepsin is secreted in an inactive form, *pepsinogen,* by the *chief cells* of the fundic glands. Pepsinogen is "unmasked" and made active on being exposed to hydrochloric acid, or to some already activated pepsin. Several other digestive enzymes in the intestine are secreted in inactive form, which is clearly necessary if the glands are not to be digested by their own products. It should be further noted at this point that the mucus secreted throughout the gastrointestinal tract is also important in protecting the gut lining from the action of the digestive juices.

Rennin is a gastric enzyme important only in the infant animal. It causes milk protein to coagulate and therefore to stay in the stomach longer. There is also a possibility that it further promotes the digestion of casein in some as yet unknown manner. The enzyme is almost absent from adult gastric juices.

Gastric secretion is stimulated in three phases. First there is a *cephalic phase* (literally, "of the head") which compares with the secretion of saliva in anticipation of food. Through this reflex, the stomach is ready to start digesting food as soon as it arrives.

The *gastric phase* of gastric secretion follows when food is actually in the stomach. The contact of the food with the stomach wall stimulates the secreting glands either directly or via vagal reflexes. Furthermore, the contact causes the production of a hormone, *gastrin,* by the antral portion of the stomach, and the circulation of this in the blood causes copious secretion by the stomach glands.

The *intestinal phase* of gastric secretion is less important and occurs, curiously enough, *after* food has left the stomach and entered the intestine.

Motility of the stomach. The wall of the fundus contributes very little to the kneading and mixing of the food. Waves of powerful contractions pass over the antral walls, pushing the food toward the *pyloric sphincter.* If this remains closed, the inner part of the food mass in the antrum is forced back into the fundus, thereby mixing it thoroughly. Contrary to what might be expected, the antral muscles do not eject the stomach contents in vomiting. This is actually accomplished by compression of the stomach against the liver and diaphragm by the muscles of the abdominal wall.

When the stomach contents are digested enough to be largely fluid, and if the first part of the small intestine is empty (or nearly so), the pyloric sphincter relaxes and the antral contractions empty that portion of the stomach.

SMALL INTESTINE

The *small intestine* is about 20 feet long in an adult, extending from the pyloric sphincter of the stomach to the *ileocecal valve* where it joins the large intestine. The first 12 inches or so is the *duodenum,* followed by the *jejunum* (about 8 feet) and the remainder is the *ileum.*

The digestive enzymes active in the small intestine are secreted by glands in the intestinal walls themselves and by the separate gland, the *pancreas,* which discharges via a duct into the duodenum. Although it contains no enzymes, the *bile* secreted by the liver is also important in digestion.

Pancreatic juice is alkaline and partially neutralizes the hydrochloric acid emerging from the stomach. The pancreas is the major contributor of a fat-digesting enzyme, *lipase,* which hydrolyses neutral fat to soluble glycerol and fatty acids. It also secretes two proteolytic enzymes, *trypsin* and *chymotrypsin,* which, like pepsin, split proteins into smaller units but not to amino acids. Trypsin is actually secreted in its inactive form, *trypsinogen,* which is unmasked by an enzyme, *enterokinase,* or by already activated trypsin. Chymotrypsin is secreted as *chymotrypsinogen,* and this is activated by trypsin. Finally, the fourth enzyme of pancreatic juice is *amylase,* which hydrolyses starch to dissacharide units.

The secretion of pancreatic juice is chiefly controlled by a hormone, *secretin.* In the presence of chyme, this is released into the blood by the tissues in the wall of the duodenum; the arrival of the hormone at the pancreas stimulates secretion. The mechanism is clearly analogous to the gastrin-induced gastric secretion.

Bile. Among its many other activities the liver manufactures bile. This complex fluid is a vehicle for the excretion of excess calcium, breakdown products of hemoglobin from wornout erythrocytes, and several other special waste products. The importance of bile in digestion lies in its alkalinity, whch helps neutralize the acid from the stomach, and in its containing *bile salts.* These complex materials promote the emulsification of fat in the chyme. When emulsified, the fat is more easily attacked by lipase, and some authorities even suggest that emulsified fat can cross the gut membrane without prior hydrolysis.

The liver secretes bile at a fairly constant rate, but unless there is chyme in the duodenum, the duct is kept closed by the *sphincter of Oddi.* Then the bile cannot leave the duct and consequently backs up and fills the *gall bladder.* The presence of fat in the duodenum promotes the release of a hormone, *cholecystokinin,* which causes a relaxation of the sphincter of Oddi and simultaneously, a mild degree of contraction in the gall bladder.

Succus entericus. The digestive juices secreted by glands in the walls of the small intestine are lumped together under the name, *succus entericus.* A large part of this is made up of mucus, which has the lubricative and protective role already discussed. The enzyme components characteristically complete the processes begun by gastric and pancreatic enzymes. Thus, there are several *peptidases* which hydrolyse polypeptides to individual, absorbable amino acids. Disaccharides are hydrolysed respectively by *sucrase, maltase,* etc., and finally there is a small amount of a lipase. The glands secreting the succus entericus are deep pits in the intestinal wall, known as the *crypts of Liberkühn.* The cells at the bottom of each crypt secrete enzymes and those nearer the surface secrete mucus.

Secretion in the intestine is under some degree of hormonal control, but the *enterocrinin* does not have so pronounced an effect as gastrin or secretin. It appears that secretion is mainly induced by the mechanical stimulation of the gut walls by chyme.

Absorption of nutrients. The simple, soluble materials resulting from the digestion of foodstuffs are absorbed into the cells of the gut lining. From these they pass into the extracellular fluid and from there into the plasma and lymph. Undoubtedly, simple diffusion accounts for the movement of some of the nutrients, but glucose, for example, is absorbed by an active transport process. For ordinary diffusion there must be a concentration gradient, but glucose is absorbed from the gut until there is none left. It is clear then, that in the later stages of the absorption at least, glucose enters the body fluids against a concentration gradient.

The small intestine presents an enormous area for the absorption of nutrients, not only because of its great length but also because of the convolutions of its lining. The largest folds are the *plicae circulares* in the duodenum, which becomes smaller and sparser further down the gut and are quite absent in the lower ileum. In addition, the *villi* are finger-like projections of semimicroscopic proportions all over the gut lining, giving it a velvet-like texture (see Fig. 14.3). Fig. 14.4 shows that the villi are highly specialized absorptive organs; each contains a loop of blood capillary and a small lymph vessel called a *lacteal.* The capillaries drain into the *portal* circulation, which passes through the liver before returning to the heart. It will be recalled that the liver regulates the glucose and amino acid levels of the circulating blood by removing the excess during periods of absorption from the gut and by releasing these nutrients at other times.

More of the lipoid materials absorbed into the villi pass into the lacteals than pass into the blood. The very name "lacteal" refers to the milky appearance of the lymph in these vessels, which is caused by the fat droplets in suspension. The lymph drains through many *lymph nodes* studded along the intestine, and enters the thoracic duct, whence it is returned to the blood circulation. Following a fatty meal, therefore, the blood plasma has a cloudy appearance due to the yet undeposited fat. Blood in this state is quite unsuited for transfusions or storage in a blood bank.

Intestinal motility. The chyme is constantly mixed in the intestine by *segmental contractions* of the smooth muscle of the gut wall. It is squeezed into segments by rings of contraction which occur

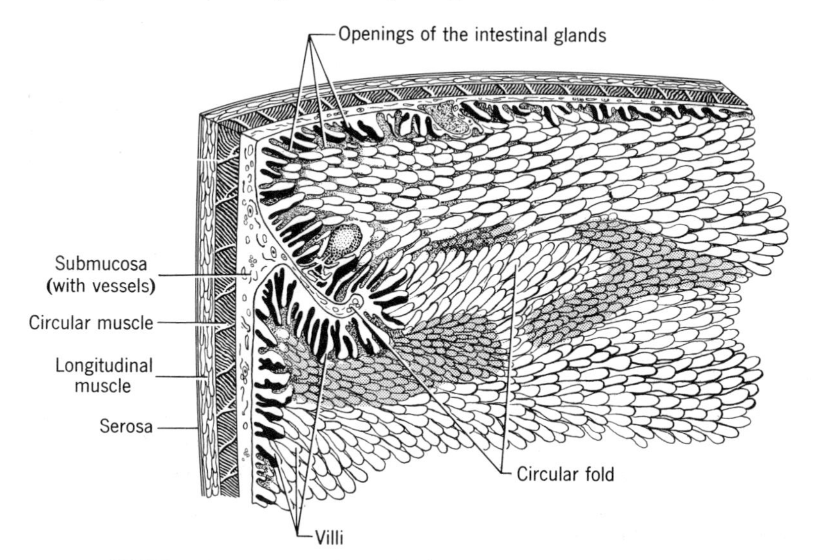

FIGURE 14.3. A portion of the wall of the small intestine. (After Braus)

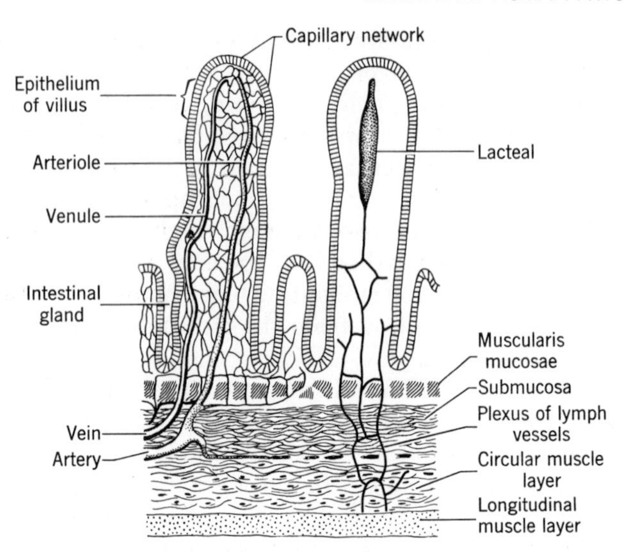

FIGURE 14.4. Diagrammatic cross section of wall of the small intestine; villus at the left shown with blood vessels only, that at the right with lacteal only, but both occur in all villi. (After F. P. Mall)

simultaneously every few inches along the intestine. These are followed by new contractions in the formerly uncontracted segments at the same time as the original contractions relax. This movement has the effect of squeezing the chyme back and forth, ensuring a thorough mixing.

The propulsion of chyme along the gut is by rings of contraction which travel preceded by a wave of complete relaxation. This *peristaltic* wave can be imitated by pinching a rubber tube full of thick liquid and sliding the fingers along it.

The chief stimulus for peristalsis is the presence of chyme in the intestine, and, in general, the larger the bulk of the gut contents the greater the gut motility. Although the structural components of plant tissues are of indigestible cellulose, they are valuable in a diet in providing bulk. Whole fruits are laxative because the undigested skins exert a mildly irritating effect on the intestine and promote even greater smooth muscle activity. Various drugs are laxative, or more violently, *purgative,* through their effect either on the gut muscle directly, or through an effect on the visceral centers which control it. Vomiting is often accompanied by reverse peristalsis which moves intestinal contents back into the stomach.

LARGE INTESTINE

The *large intestine* is, as the name implies, of larger diameter than the small intestine, but it is very much shorter. The principal parts are shown in Fig. 14.1. It is noted that the ileum does not merely expand to become the large intestine but makes an approximately T-shaped junction with it. The "blind" portion of the large intestine below the junction is the *cecum,* and from the distal portion of this extends a smaller blind structure, the *vermiform appendix.* The cecum is relatively unimportant in man. In herbivores like the horse and rabbit, however, it is very large and acts as a fermentation vat for the bacterial digestion of plant cellulose, and in this respect it is analogous to the enlarged stomach, or *rumen,* in cows, goats, and sheep. In man, the appendix is a common site of infection leading to the acute abdominal pains of *appendicitis.* If this proceeds unchecked, a rupture of the appendix allows the bacteria-laden cecal contents to contaminate the normally sterile peritoneal cavity. This is the nature of the even more sever emergency in *peritonitis.*

At the junction of the ileum and the cecum, the ileocecal valve permits chyme to pass into the cecum without regurgitation. Above the cecum, the *colon* is arranged in a curiously symmetrical fashion in the abdomen. The sections are called respectively, the *ascending, transverse,* and *descending* colons. The most distal portion of the colon which leads into the *rectum* is the *pelvic colon.*

The principal function of the colon is the absorption of water and salts from the still liquid residual chyme, which includes not only those originally contained in the food, but also the large amounts in the intestinal secretions. This absorption conserves and recycles water and electrolytes. The importance of this becomes apparent in diarrhea when the passage of chyme through the lower gut is too rapid for reabsorption. The consequent loss of water and salts can lead to serious dehydration and electrolyte imbalances unless replacement therapy is started early. This is particularly true of infants who have a more rapid fluid turnover than adults.

There is a great deal of bacterial activity in the colon; nearly 10% of the dry matter of the feces is ordinarily made up of bacterial cells. Some of this fermentation is physiologically valuable because the bacteria make vitamin K and some of the B-complex vitamins, which are then absorbed into the blood. The small amounts of undigested protein present are attacked by putrefactive bacteria, giving rise to *indole, skatole,* and *hydrogen sulfide,* which are responsible for the characteristic smell of feces. The brown color of feces is largely due to the *bile pigments* excreted by the liver; in fact, pale clay-colored feces are frequently a sign of impaired bile secretion.

The large-intestine does not show the continuous motility that the small intestine does. There are, instead, massive peristaltic waves occurring at intervals of several hours that move the colonic contents toward the rectum. This happens typically when the stomach empties, and so is called the *gastrocolic reflex.* The sensation of pressure and distension in the rectum when the feces are compacted into it gives rise to the defecation reflex.

As is the case with micturition, the defecation reflex is subject to higher control in adults and, under ordinary circumstances, can be consciously inhibited. Defecation is brought about by a relaxation of the *anal sphincter* accompanied by strong peristaltic contractions of the colon and by a respiratory maneuver that raises the intra-abdominal pressure. This familiar maneuver involves a forcible expiration against a closed glottis so that no air escapes from the lungs.

FOOD INTAKE, STARVATION, AND OBESITY

15

We have seen that even under resting conditions, the cells of the body have a relentless demand for energy-yielding nutrients. Any form of muscular work increases the combustion of nutrients as does also the increased metabolism necessary to keep warm in a cold environment. If there is no food intake, the glucose concentration in the internal environment can be maintained for a few hours by the mobilization of the glycogen stores in the liver. When these are exhausted, the blood glucose level actually falls (by about 20–30 milligrams per cent in man) and in the next 24–48 hours, the body adapts to an energy metabolism based on the combustion of depot fat and the products of gluconeogenesis.

This combustion of the body substance can proceed for some time if the animal is previously well nourished. The fat depots provide energy only; the gluconeogenesis provides compounds for combustion and also provides essential amino acids which can be used for repair of other tissues. It is, however, clearly uneconomic to break down the protein of one tissue to obtain amino acids for the repair of another, and in the long run, there is wastage of the musculature. From the diminution in the size of the liver during starvation, it can be presumed that liver protein is mobilized preferentially, but beyond that there seems to be no discrimination in the breakdown of tissues. The more vital organs, the heart muscle for example, seem to be just as vulnerable as the less vital organs such as the skeletal muscles.

In starvation, therefore, the reduction of body fat is not in itself serious, but the wastage of muscles causes a progressive weakness and eventually an inability to cope with the environment. The blood-glucose level rises somewhat after the low level encountered soon after the beginning of the fast, but it never reaches its normal level. This also contributes to a general weakness and lassitude. The emotional irritability at the beginning of a fast can be reasonably attributed to the low blood sugar.

Further serious consequences of total starvation are the multiple vitamin deficiencies which develop. Previously well-nourished individuals probably have adequate stores of vitamins A and D in their livers which will last for considerable time, but deficiencies of vitamin C and the vitamins of the B complex can develop in the space of a month or so. Starvation is therefore rendered even more dismal by the bleeding gums and the slow healing characteristic of scurvy, and by the neuropathy of a multiple B vitamin deficiency. In man especially, the adrenal cortical response to starvation promotes the retention of sodium and a consequent edema.

Examination of the carcasses of animals which have died of starvation shows that there is still some abdominal fat, suggesting that the cause of death is not the final failure of energy-yielding nutrients. It is likely that the cause is the cardiovascular collapse consequent to the wastage of the myocardium.

An intake of foodstuffs above that needed to meet the energy requirements of the animal results in the storage, first of glycogen, and then of body depot fat. As indicated previously, this is a useful survival factor in wild animals since they can store fat during a season of plentiful food as a reserve against the relative starvation of the winter season. Regardless of how much extra food they are given, laboratory rats of mature size will ordinarily regulate their food so that their body weight remains about constant. Their day-to-day intake may vary considerably, but over a period of one or more weeks it is remarkably constant.

The mechanism by which food intake is regulated so exactly has been a topic of great interest. Despite the fact that the world's population increase seems continually to draw ahead of the current food production, the relatively advanced nations of the West have a potential surplus of food. For a large part of the world's population, the limit on food intake is rigidly set by the amount of food which can be grown or purchased by the family. In richer societies the food intake is only restricted by the appetite of the individual, and economic factors are secondary.

The sensations which prompt a man or animal to consume food are, by definition, almost entirely subjective. However, there is certainly a distinction between *hunger* and *appetite,* although it is difficult to describe the difference exactly. Hunger may be considered entirely physiological; a response to the changes in blood sugar, etc. described in the previous discussion of fasting and starvation. There is good evidence that centers in the hypothalamus are stimulated by a lowering of blood glucose concentrations and cause the stomach contractions associated with extreme hunger. In these terms, there is little difficulty in explaining why an animal begins to eat, but we still do not have any good explanation for why it stops eating when it has consumed just about enough food. The problem of *satiety* is therefore still a mystery.

Appetite in man is a much more complicated sensation. We have a "good appetite" for a meal, even when our period of fasting since the last meal is short enough that the physiological changes consequent to fasting are barely apparent. A mere listing of psychological factors in appetite is enough to indicate its complexity; the kind of foodstuffs served, the standard of cooking, the temperature, the setting (solitary, domestic, or social), and finally, the well-known fact that we often do not have an appetite until actually sitting down to the meal.

A widely accepted theory postulates that, subject to the kind of psychological conditioning listed above, satiety results from a complex interrelation between blood glucose level, its rate of utilization, and its rate of storage. Specific and minute lesions in the hypothalami of rats lead to tremendous overeating (hyperphagia), and in other close-by points in the hypothalami, to no eating (hypophagia).

OBESITY

In the United States and other Western countries, obesity, due to excessive food intake, is a major public health problem. There is a very high statistical correlation between obesity and heart disease, probably due to a great many physiological consequences of obesity which it is not appropriate to discuss in the present chapter.

TABLE 15.1 Estimated Daily Caloric Requirements for Various Occupations

Logger or miner	5,000 and up
Farmer (pitching hay, etc.)	5,000
Farmer (driving tractor)	3,000
Carpenter	3,500
Milling machine operator	3,200
Assembly-line worker	3,000
Clerk	2,500

True obesity, as opposed to the puffiness of edema, can only follow excessive food intake. By excessive, we mean food intake over and above that required for the energy requirements of the individual. A logger on a mountain slope in the State of Washington may consume and utilize up to 7,000 calories per day, a food intake which would probably be excessive for two office workers in Manhattan. Tables 15.1 and 15.2 give the caloric costs of various activities. It sometimes happens that hypothyroidism depresses the basal metabolism and activity of an individual so that even a modest food intake is in excess of his requirements. This is the origin of the often-stated "glandular" cause of obesity. In the U.S., obesity is usually the result of simple overeating, especially of the fatty foods like ice cream, butter, french fried potatoes, and similar materials with a high caloric density. Our society in America places great value on a trim appearance both in men and women, but despite this strong motivation, many obese people find it extremely difficult to restrict their intake and lose weight. Even when they do manage to lose weight by rigid dieting, they frequently regain their original weight quite soon after relaxing their dietary restriction. Causes of this overeating are complex, but four factors can be considered here.

TABLE 15.2 Energy Expenditure for Various Activities

	kcal/kg/hr.
Lying (awake)	0.1
Sitting	0.4
Driving automobile	0.9
Walking	2.0–4.0
Bicycling	2.5
Dancing (not jive)	3.0
Skating	3.5
Swimming	8.0

Habit. During their twenties and early thirties, most people adapt to a food intake which maintains a constant body weight. As they get older they frequently abandon athletic pursuits and at the same time, promotion in their work causes them to be more sedentary than they were in junior positions. It then happens that they continue with a food intake commensurate with their previous level of activity, but which is now excessive. This leads to obesity which can often be seen in the "softening" of college football players in later years.

Compensation. In the most general way we can consider that life offers various areas of satisfaction or gratification. The number and importance of the various areas varies with the individual, but for most of us, important gratification is gained from the pleasures of eating and drinking, from professional competence and the regard of our colleagues, from a serene and affectionate domestic life, from a modest degree of social success and prestige, and from sexual activity. Many workers believe that the pattern of obesity suggests that if an individual is receiving inadequate satisfaction in one or more of these areas, he tends to compensate by increased indulgence in one of the others. In our society, the most accessible gratification is that of eating. J. R. Brobeck, one of the leading physiologists in the field of food intake regulation, has summarized this point: "Food is, without any doubt, the oldest and most widely used 'tranquilizer.'"

Hereditary predisposition. It is a matter of common observation that obesity "runs in families." There are probably unknown physiological characteristics making some individuals more prone to obesity than others, and these may well be hereditary. A tendency to hypothyroidism is a good example, but as we have seen, this is not a common kind of obesity. Similarly, we can postulate the inheritance of an emotional makeup leading to compensatory overeating as described, but this is even more vague. Many psychiatric workers point out that obese parents often instill their own attitudes toward food in their children, thereby passing on habits of overeating, which are more properly considered as conditioned rather than hereditary. They report cases in which obese mothers are terrified that their children are not getting enough to eat and urge them to eat fantastic amounts of food, even when the children are clearly obese too.

Alcohol. Alcohol has a high caloric value, and although it cannot itself be converted to stored body substance, the "sparing" of other foodstuffs by an alcohol intake can lead to excessive storage. Since the normal food intake seems to be geared to the energy requirements of the individual, the extra calories taken as alcohol are usually adjusted for by a proportionately smaller intake of other nutrients. One of the characteristics of "hard liquor" however, is that it promotes appetite, even if it does dull the sense of taste. Therefore, individuals with a tendency to overeat are more likely to do so if they drink liquor before meals. This results in a greater caloric intake from the other nutrients as well as that from the alcohol itself. As a rule, beer does not stimulate appetite as much as do the hard liquors, but the extraordinary amounts of beer consumed in some circles represents a large caloric intake from the alcohol and sugar in the beverage.

The contribution of alcohol to obesity is seen at its extreme in individuals who use alcohol as well as food to compensate for real (or imagined) lack of gratification in other areas. The alcoholic is not necessarily the thin, chain-smoking individual of fiction, but can frequently be extremely obese.

The obesity which is unfortunately all-too-common in young married women after they have had a baby is explicable in some of these terms. First, there is the habit of increased food intake acquired quite reasonably during pregnancy and lactation. Secondly, it often follows that the care of an infant restricts the social life of a young mother, particularly during the daytime when her husband is away and she is alone in a small house or apartment. The ensuing "loneliness" can be all too readily assuaged by frequent snacks and the consumption of candy.

From the sociological point of view, the sexual life of a woman may be interpreted to include the whole spectrum of raising a family and making a home, as well as the more specific aspects of sexuality. Therefore, we may consider sexual dissatisfaction as a contributant to obesity, not only in terms of unsatisfactory sexual intercourse, but also in terms of the necessity of living in crowded, graceless surroundings with a constant battle to maintain the decencies of life. It is an economic fact that young married people frequently have to occupy unsatisfactory quarters, often in a portion of the home of one or other parent. Many moderately obese young married women find little difficulty in regaining their trim figures when their social life improves. It should also be noted, however, that meeting new and interesting people provides a strong motivation in this respect.

In men there is no doubt that sexual dissatisfaction can be a contributing cause to obesity. A further very important factor here is professional frustration or disappointment. The obesity developing in many men in their fifties can be attributed to their increased sedentary habit, a possible dissatisfaction with their declining sexual activity, and also, the realization that they have achieved their greatest professional potential and that they must be reconciled to see younger men promoted over their heads.

ANOREXIA

Anorexia is a rare condition in which the food intake is inadequate to meet the energy requirements of the body; it results in a state of "voluntary" starvation. The causes are usually just as obscure as are those of overeating. A patient with a stomach ulcer or other condition leading to profound discomfort after eating will tend to restrict his food intake, but will usually manage to find some foods which are less irritant and which can help meet his energy requirements. Disaffected prisoners and soldiers have periodically gone on hunger strikes, but this represents an even more special case.

Psychiatrists recognize a condition called *anorexia nervosa* in which the patient refuses to eat, frequently for the most obscure reasons not even apparent to the patient himself. In a few cases that have been successfully treated it was found that the patients regarded most foodstuffs as symbols for extremely disagreeable associations. The interesting generalization has been made that neurotically depressed individuals tend to hyperphagia and obesity whereas psychotically depressed individuals tend to hypophagia and inanition.

HORMONAL CONTROL OF THE INTERNAL ENVIRONMENT

16

In the preceding chapters, we have referred several times to the *hormones,* the complex compounds secreted in small quantities by the *ductless* or *endocrine* glands. The glands are so named because they have no ducts leading from their secretory cells, as opposed to the *exocrine* glands such as the salivary glands, etc. Instead, their secretory products diffuse into the blood, whence they are distributed throughout the extracellular fluid. The discussion of circulation and renal physiology illustrated that higher animals have evolved two complementary systems for controlling the organs concerned with homeostasis. One is the nervous system, chiefly, the two divisions of the autonomic system, and the other is an endocrine system.

For example, the minute-to-minute regulation of the arterial blood pressure is through neural feedback mechanisms, whereas the slower and less dramatic regulation of the body fluid volume is through the endocrine control of water reabsorption in the distal tubule (ADH), and the reabsorption of sodium in the proximal tubule (aldosterone). These two kinds of control epitomize the differences between neural and endocrine homeostatic mechanisms; with only a few exceptions, the endocrine effects are slower and longer lasting.

It should be noted that there are endocrine functions *not* associated with the regulatory organs, just as the nervous system has important functions other than its role in homeostasis. One such function of the nervous system is the control of the skeletal muscles by which the animal responds to its external environment; we might say that there is a neural control of the minute-to-minute *behavioral* response of the animal. At the same time, some of the endocrine glands exert a slower, longer-lasting, and over-riding control, as for example, the hormones secreted by the reproductive organs that determine the pattern of sexual (and maternal) behavior. This presents an interesting analogy with the two kinds of homeostatic control. The far-reaching effects of the hormonal status on behavior are plainly seen at puberty in the human. The endocrine changes affect not only the overtly sexual behavior of the individual, but also the whole range of his attitudes and values. Without going into detail, we might note the suddenly increased interest in poetry, music, and creativity generally, which is usually accompanied by rebelliousness and frequent collisions with authority.

Finally, the growth of the animal and the differentiation of the tissues is under an all-pervading endocrine control, which extends even to the development of the endocrine glands themselves.

In summary, then, there are three main areas of endocrine control:

1. *Homeostasis* of the internal environment.
2. *Behavioral* response to the external environment.
3. *Morphogenesis* and growth.

Because this section of the book is devoted to the organs regulating and distributing the internal environment, the principal emphasis will be on the first category, the functions in homeostasis. The sex hormones will also be listed and their interrelationships indicated, subject to further discussion in the later chapter on reproduction.

In the interests of clarity, we must discuss the components of the endocrine control singly, but as with all physiological regulations, we should bear in mind that there is a complex and harmonious integration of all of them in the normal body. Figure 16.1 shows the locations of the endocrine glands.

THE THYROID GLAND

The *thyroid gland* is suitable for first consideration (Fig. 16.2). Through its secretion, it controls the metabolic rate of the tissues

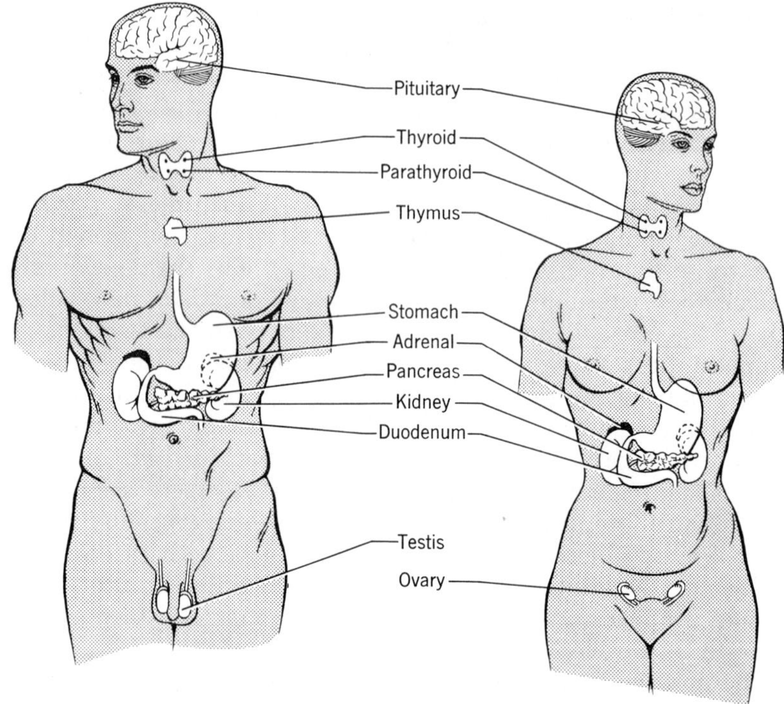

FIGURE 16.1. The endocrine glands.

much as the governor controls the speed of a tractor engine. Anatomically, it comprises two lobes, one on each side of the midline in the anterior portion of the neck, just below the larynx. The lobes are joined across the front of the trachea by an *isthmus* of tissue. Within the same connective tissue capsule, or even embedded in the lobes, are four pea-sized structures, the *parathyroid glands.* Despite their name and location, they have a completely separate function which is discussed later.

The thyroid hormone is mainly *thyroxin,* but the gland contains at least two other compounds of very similar chemical structure. The gland synthesizes the hormone from the amino acid, *tyrosine,* and from *iodine,* both of which must be present in the diet. Food does not contain free iodine, but rather *iodide,* which is oxidized to iodine by the thyroid gland itself. Thyroxin is a relatively small, diffusible molecule, and after synthesis it is stored in the gland as *thyroglobulin,* a protein complex with a molecular weight of about 700,000. The gland has a connective tissue framework

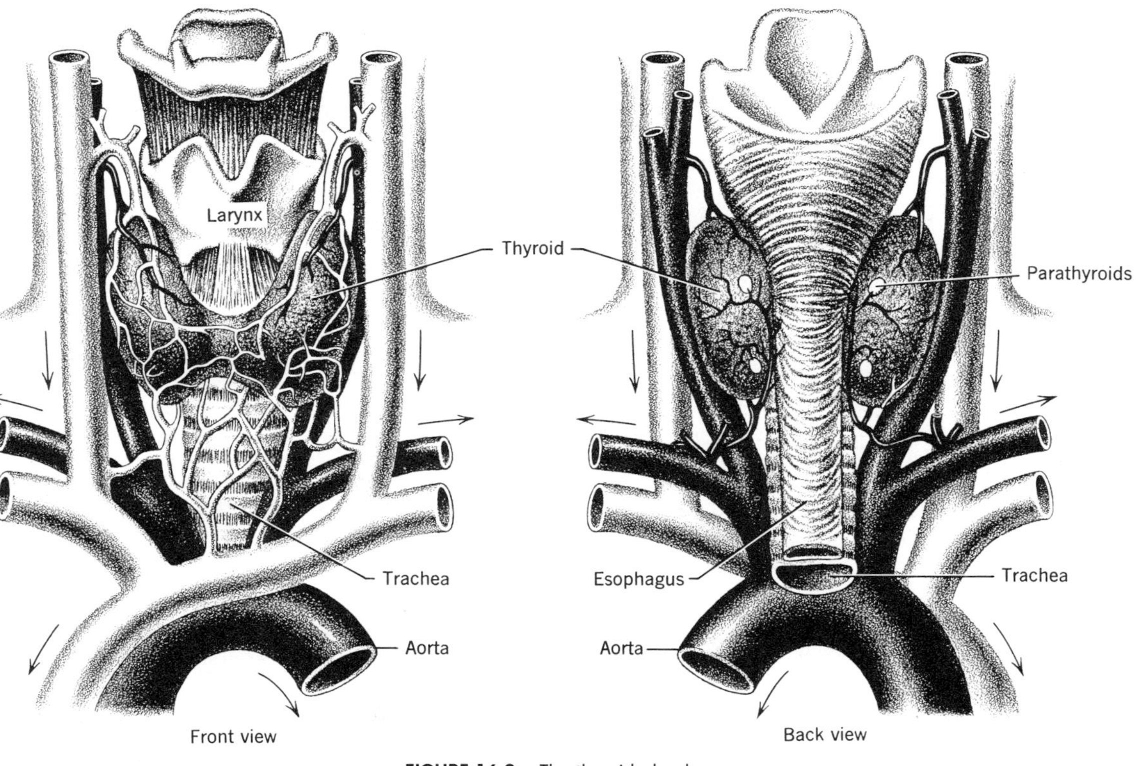

FIGURE 16.2. The thyroid gland.

containing *follicles,* or sacs, in which the thyroglobulin is stored as *colloid.*

The thyroxin level in the blood is maintained by the enzymatic splitting of thyroxin from the protein complex so that it can diffuse out of the follicles. In the blood it immediately recombines with the albumin and α-globulin of the plasma proteins and circulates in this form. Since its characteristic constituent is iodine, a determination of the *protein bound iodine* is an index of the circulating thyroid hormone and, therefore, of the thyroid status of the individual. This is now an important diagnostic tool where thyroid malfunction is suspected.

Our knowledge of the functions of the thyroid hormone, and of all the other hormones for that matter, is largely based on experiments in which the gland has been removed from laboratory animals, or on the clinical observations of patients with defective glands. The symptoms that appear in these cases indicate the role the hormone plays in the normal body, and they can usually be reversed by administering *extracts* of the gland. Additional information comes from the effects of an excess of the hormone, seen in patients with *overactive* glands, or in normal animals treated with appropriate extracts. From such studies, we are able to conclude that the thyroid has two general functions; the regulation of energy metabolism and a role in the growth process. They may well be linked by an underlying principle which we do not yet understand.

Energy metabolism. The thyroid hormone regulates the resting level of oxidative metabolism of the tissues. In its absence, the oxygen consumption falls to about half of its normal resting value; conversely an excess of the hormone speeds up the energy production in the cells. The effects at the cellular level show up in many ways in the intact body. In an excess, or *hyperthyroidism,* there is, in addition to an increased resting metabolism, an increased muscular activity, with a consequent enhancement of heat production. The heart rate is also speeded up, not only to meet the increased demands of the tissues for nutrients and oxygen, but also because of a direct effect on the myocardium. Similarly, there is increased excitability in the nervous system. The central effects in particular, often include emotional instability, a short attention span, and compulsive, rapid-fire talking. Even the activity of the renal tubular cells seems to be stepped up, which is indicated by a higher threshold for glucose excretion.

Conversely, a deficiency of the hormone, or *hypothyroidism,* makes for an overall sluggishness. Muscular activity is reduced, the pulse rate is subnormal, the reflexes are slowed, and, in extreme cases, emotional and intellectual activity is diminished to a cow-like stupor.

Growth, maturation, and tissue differentiation. An infant animal with the thyroid gland removed never attains normal growth and development. This applies to growth not only in the sense of increasing stature, but also in the more general sense which includes the development of the genital organs at puberty, the rate of hemopoiesis, the maturation of bone, and even the rate of healing in wounds. A thyroid deficiency in an adult animal has less spectacular effects on these growth and differentiation processes, but there is, nevertheless, an anemia, reduced growth of hair and nails, and slow healing.

A possibly related effect of a thyroid deficiency is a relative increase in the entire extracellular phase of the body. There is a fluid retention, which can amount to gross edema and an increased deposition of the mucoprotein cement normally found between the cells.

Feedback control of thyroxin synthesis and secretion. The thyroid hormone enhances metabolism in all tissues of the body *except* the thyroid gland. A plenitude of circulating hormone depresses the gland and a deficiency stimulates it. Of even greater importance though, is the overriding control exerted by the *adenohypophysis* in secreting thyroid stimulating hormone (TSH), also called thyrotrophic hormone (TTH). The adenohypophysis, the anterior portion of the pituitary gland, is appropriately called the "master-gland" of the body, because its secretions control the activities of several endocrine glands, the thyroid included. It is sensitive to the level of circulating thyroxin and in a deficiency, the TSH production is increased, stimulating the thyroid to greater production. The consequent rise in thyroxin inhibits the TSH production and so reduces the stimulation of the thyroid. In engineering terms, this is a *negative feedback mechanism.* By this reciprocal stimulation and inhibition of each other, the adenohypophysis and the thyroid together maintain a steady level of thyroxin synthesis and secretion.

Since the thyroxin level determines the resting metabolic rate of the tissues, the regulation of its level is vital to homeostasis. As in cardiovascular control mechanisms, however, homeostasis can be "unreasonable." When a thyroxin deficiency results from impaired synthesis, due to an inherent defect in the gland or to a dietary deficiency, the adenohypophysis still "blindly" stimulates the thyroid by greater and greater TSH secretion. This may be effective in increasing the output or not, depending on the lesion or dietary deficiency, but in either case, the gland increases in size. This thyroid *hypertrophy* is therefore commonly associated with a thyroid deficiency. In *goiter* (to be discussed) the enlargement is plainly visible.

The thyroid, through its control of the metabolic rate, is a prime determinant of heat production. There is good evidence that part of the adaptation to a cold environment (in terms of days and months rather than minutes) lies in an increased thyroid activity. It is highly probable that the *hypothalamus,* the part of the brain concerned with body temperature regulation, promotes increased TSH secretion by the adenohyphysis.

Mechanism of action of thyroid hormone. Despite the volume of empirical knowledge obtained from depletion and supplementation studies, and despite our knowledge of the chemical structure of thyroxin and its active analogs, we still do not understand how it exerts its control. For the regulation of metabolic rate, we might postulate that thyroxin improved the permeability of the cell membranes to glucose and other nutrients, but this merely poses another question, "How does it alter the permeability?"

An endocrinologist, Byrom, has an ingenious theory which embraces both the energy metabolism and growth functions. Fetal tissues have more of the mucoprotein intercellular cement, and he proposes that development can only proceed by the partial removal of this cement, making the cell surfaces more accessible to the nutrients and growth hormone in the extracellular fluid. Thyroid hormone certainly has such an effect on the intercellular substance. Furthermore, variations in the adult intercellular cement-content would similarly affect the permeability of the cells, and consequently, their metabolic rate. To date, there have been no critical experiments to substantiate the idea.

Thyroid diseases all result from an excess or deficiency of the thyroid hormone. In man, several clinical entities are recognized, depending on the underlying cause and the severity of the symptoms. *Goiter* is a term applied to any thyroid enlargement, and can occur in hypothyroidism or hyperthyroidism. As described previously, a deficiency of circulating hormone can cause the adenohypophysis to lash the thyroid gland to greater and greater activity. On the other hand, defects in the gland itself, or in the TSH secretion, can cause overactivity and enlargement, regardless of an elevated hormone level in the blood.

A common cause of goiter is a simple iodine deficiency of the kind that used to be rife in the high valleys of Switzerland and other inland areas where the soil and water are deficient in the element. The incidence is now much reduced by the use of iodized table salt. The rapid transportation and increased trading in foodstuffs, in the western world at least, has also improved the situation. Fewer and fewer communities today depend entirely on their own local food production; the importation of food from other areas tends to cancel out deficiencies in foodstuffs resulting from regional peculiarities of the soil.

Less well understood are the *antithyroid,* or goiterogenic, substances occurring in some foods, typically cabbage. These inhibit thyroxin synthesis in some way, and they may be critical if other factors are marginal. *Thiourea* and *thiouracil* are antithyroid drugs with a similar effect. In carefully adjusted dosages they are useful in controlling hyperthyroidism.

Goiter can also occur independently of the iodine supply when there are defects in the gland itself, or in the TSH control. This kind tends to run in families.

Cretinism and Myxedema

In a child, a profound hypothyroidism from any cause leads to *cretinism.* The symptoms are those to be expected from the preceeding discussion; there is very slow, stunted growth and even if the child survives to the normal age of adolescence, there is no sexual development. The mental development is extremely retarded and in the interests of everyone, cretins are usually institutionalized.

There are all degrees of hypothyroidism and if diagnosed early enough, a partial deficiency can be dramatically improved by administering thyroid extracts. One of the dangers in mild hypothyroidism is that when it occurs in one child of several in a family, it often goes unnoticed by the parents until it is surprisingly well advanced. Children develop relatively slowly, and busy parents may not notice that one child is exceptionally slow, especially since such children are quiet and "good."

A profound hypothyroidism developing in an adult is called *myxedema.* The mental processes become sluggish and confused, the skin becomes coarse and dry, hair falls out, there is a puffiness from both true obesity and edema, the metabolic rate is greatly depressed, and there is a failure of sexual function and libido. As in children, the hypothyroidism can occur over a wide range of severity. A mild degree is fairly common; it is marked by lethargy attributable to the depressed metabolic rate and frequently by some degree of obesity as well. In young women, common signs are an erratic menstrual cycle and lack of sex drive.

It is difficult to distinguish between the lethargy and obesity due to hypothyroidism and that which can only be related to the personality of the individual or to psychiatric disorders. One useful guide is the basal metabolic rate of the patient, which may be determined by the method outlined on page 235. In recent years, this has been supplemented by, if not entirely replaced by, protein-bound iodine determinations, especially since the rather difficult chemical assay has been supplanted by an isotope dilution method. This involves adding a small aliquot of I^{131}, a radioactive isotope of iodine, to a known volume of a plasma sample. It is not necessary to inject the radioactive iodine into the patient.

In an even more sensitive test of thyroid function, a minute amount of I^{131} is injected into the patient and the rate at which the thyroid gland takes up the isotope is measured. This is accomplished by placing a *scintillation rate counter* over the gland itself. Frequently, the urinary excretion of I^{131} is also followed as a double check. The measured rate is compared with normal values; the greater uptake rates are associated with greater thyroid activity. Although the radiation hazard is small, this test is only performed when there is good reason to suspect thyroid malfunction. It is not conducted on a semiroutine basis as are metabolic rate and protein-bound iodine tests.

Replacement therapy is the standard treatment for hypothyroidism not due to a simple deficiency of iodine; the patient consumes pills made of extracts of thyroid from slaughtered farm livestock. If continued for several years, it is often necessary to increase the dosage gradually, since the extra thyroxin tends to depress whatever thyroid activity the patient originally had. Sometimes the enlargement of the gland persists even after the hypothyroidism has been corrected. If prominent enough, some of the extra colloid tissue is surgically removed for cosmetic reasons.

Finally, it should be noted that hypothyroidism can occur as part of the catastrophic degeneration resulting from hypofunction of the adenohypophysis; this situation is much more complex.

Hyperthyroidism

At the other end of the spectrum of thyroid functional disorders are various degrees of oversecretion. The most severe is *Graves disease,* or exophthalmic goiter. It is believed that the primary defect in this disease is in the adenohypophysis, which produces too much TSH. The symptoms are those described earlier in this chapter, plus a characteristic protrusion of the eyeballs, which is responsible for the name "exophthalmic goiter." The basal metabolic rate may be double the normal in some cases, and the great heat production causes continuous sweating and peripheral vasodilation. Less serious and slower to develop is *adenomatous goiter,* in which the defect is an intrinsically overactive thyroid gland.

In "sophisticated" conversation, we frequently hear the term "hyperthyroid" applied to anyone of bounding energy. This is sometimes the case, but truly hyperthyroid individuals are usually erratic and their energy is jittery and ill-directed. The energy of many hard driving people can be better explained as an honest enthusiasm for their occupation, or as a psychologically deeper compensation for an abiding sense of inferiority and the need to prove themselves to others (and themselves)—not necessarily a bad thing.

Hyperthyroidism may be treated on a temporary basis with the antithyroid drugs. Since they have toxic side effects, however, they cannot be used continuously. The more usual treatment is to reduce the amount of tissue synthesizing thyroxin by the partial removal of the gland. This was formerly done surgically, but the controlled irradiation of the gland is being more widely practiced. Although most therapeutic irradiation is by X-rays, the specificity of iodine uptake by the thyroid gland means that it can be irradiated from *within* by the gamma-rays of I^{131}. The isotope is injected in carefully calculated amounts, which are of course much larger than those used in the function test previously described. Since I^{131} is a short-lived isotope (its radioactivity is halved every eight days), the radiation does not continue to damage the gland indefinitely.

THE PARATHYROID GLANDS

The parathyroid glands are usually found closely associated with the thyroid (Fig. 16.1). Early attempts to correct hyperthyroidism by surgical means led to bizarre neuromuscular consequences. We now know that these were due to the inadvertent removal of the parathyroid glands.

The secretion of the parathyroid glands, *parathormone,* is not a simple compound like thyroxin, but is a complex protein of undetermined structure. Its role is the regulation of inorganic calcium and phosphorus levels in the internal environment. The ionized calcium concentration in the extracellular fluid, it will be recalled, is an important factor in the excitability of tissues. In hypoparathyroidism, the ionized calcium concentration is reduced and causes the excessive excitability in muscle and nerve known as *tetany,* a tonic contraction of the muscles (see p. 300). The hands (and feet) of a tetanic patient are drawn into a characteristically claw-like appearance.

Even in the adult animal the mineral portions of bone are in a constant state of turnover, with new material being laid down and old mineral being resorbed. The deposition of bone mineral depends on, among other factors, the product of the concentrations of calcium and phosphorus in the fluid around the bone. This *solubility product* is normally high enough for bone deposition only if *both* elements are present in adequate amounts. A low concentration of either element gives a low solubility product and leads to net bone resorption, even if the other is present at normal concentration.

Feedback control and mechanism of parathormone function. It is almost certain that the parathyroid glands secrete parathormone in response to a decrease in serum inorganic calcium and that this is not subject to any master gland control. A widely accepted theory postulates that the hormone promotes the excretion of phosphates by the renal tubular cells, thereby lowering the extracellular phosphate concentration. The product of the calcium and phosphorus concentrations therefore falls below the solubility product constant and a net dissolution of bone results; the dissolved bone mineral raises the calcium level back to normal and replaces the excreted phosphate. The correction of serum calcium is thus achieved at the expense of a small amount of bone mineral, but the bone may be regarded as a store of calcium and phosphorus, just as much as a structural support. Such losses are only serious if protracted indefinitely, as on a diet deficient in calcium; the depletion of the bone mineral then leads to *osteoporosis.*

An increased serum calcium level leads to greater urinary excretion of calcium, but also causes decreased parathormone secretion. This decrease diminishes the excretion of phosphorus, raises the product of the ion concentrations until it exceeds their solubility product constant and causes a net deposition of bone. Without adequate renal excretion of calcium, an increased serum calcium level (as from a diet high in calcium) leads to an indefinite accumulation of bone minerals. A hyperfunction of the parathyroid glands leads to an elevated serum calcium level and indiscriminate calcification in various tissues. Kidney stones, in particular, are frequent.

Another theory of parathyroid function is that the hormone controls the activities of the *osteoclasts,* the bone-dissolving cells. Yet a third theory proposes that both the renal tubules and the osteoclasts are the site of action.

Parathyroid defects are much scarcer than thyroid diseases, but it should be noted that severe derangements in calcium and phosphorus metabolism can occur even if parathyroid function is normal. They are particularly likely on diets with inadequate vitamin D, inadequate calcium or phosphorus, or with a large imbalance between the intake of these two elements (see p. 218).

THE ADRENAL CORTEX

The two *adrenal,* or *suprarenal,* glands lie one on each side, just above the kidneys, and are usually covered by the perirenal fat. Each is an apparently compact anatomical unit, but actually comprises *two* endocrine glands. The outer part, the *cortex,* secretes a whole range of *steroid* hormones to be discussed in this section; the core of the gland, the *medulla,* has an entirely separate function in the secretion of epinephrine and nor-epinephrine.

The adrenal cortex secretes so many hormones, with so wide a range of functions, that the classical experimental approach of removing the glands gave a bewildering series of results. In general, though, it was noted that animals with the cortex removed showed poor work performance and a low tolerance to infection, temperature changes, or to any form of stress. They also suffered profound water and electrolyte disturbances. The animals usually died after two weeks, but life could be extended by a further two or three weeks by giving them unlimited access to salt and water.

Studies of this nature are now supplemented by great strides in the chemical identification of the steroid hormones secreted by the cortex. The structure of a typical steroid molecule and that of cholesterol, a metabolic precursor of the hormonal steroids, is shown in Fig. 11.5.

The array of steroids secreted fall into three main categories:

Sex Hormones

These comprise *androgens* (male hormones), *estrogens,* and *progesterone* (female hormones). Their effects are similar to those of the same hormones produced by the testes and ovaries, except that the adrenal cortex secretes both male and female hormones in each sex. It is likely that the androgen secretion in the female promotes the growth of pubic and axillary hair. Adrenal tumors in women, leading to hypersecretion of androgens, suppress libido and even promote an aggressive homosexuality. In normal men, the adrenocortical androgens are probably unimportant in view of the large amounts secreted by the testes. However, hypersecretion in prepubertal boys accelerates adult muscular development, the growth of pubic hair, and sexual enterprise.

Mineralocorticoids

Mineralocorticoids are the steroid hormones regulating the electrolyte metabolism (we have previously discussed the role of aldosterone in the control of sodium excretion). The absence of these hormones causes a urinary loss of sodium and consequent electrolyte imbalance in adrenalectomized animals; the palliative effects of a high salt intake are due to replacement of the salt wastage. Aldosterone is the most potent mineralocorticoid and it has *glucocorticoid* properties as well.

Glucocorticoids

Glucocorticoids have far-reaching effects, but as the name implies, their principal role is in carbohydrate metabolism. In general, they promote the deposition of liver glycogen and inhibit the utilization of glucose by the cells. They also promote *gluconeogenesis,* that is, the conversion of body protein to sugar, leading to a wastage of muscle and a rise in plasma glucose concentration. At face value, then, gluconeogenesis looks a very undesirable process, but it is really a valuable response in starvation and other stressful situations. The lost resilience of adrenalectomized animals is chiefly attributable to their lack of glucocorticoids and the consequent depression of their gluconeogenic potential. Our present knowledge of the effects of the glucocorticoids in other aspects of metabolism is too confused for presentation at this level. It may be noted, however, that they diminish allergic responses; especially the more serious inflammatory types like rheumatoid arthritis.

Nomenclature. Although all three groups of hormones are steroids, the useful term, *corticosteroids,* is usually restricted to the mineralo- and glucocorticoids.

Feedback control of adrenocortical secretion. As with the thyroid, the adrenal cortex is controlled by the *adrenocorticotrophic* hormone (*ACTH*), also called *corticotrophin.* In man, the principal effect of ACTH is on the glucocorticoid secretion; the mineralocorticoid production is less affected. A diminished glucocorticoid concentration in the blood stimulates the secretion of ACTH. This in turn stimulates the adrenal cortex and, by the familiar feedback mechanism, the rising glucocorticoid level reciprocally inhibits further ACTH secretion by the adenohypophysis.

We have referred to the value of the adrenocortical response in stress and there is interesting evidence that ACTH secretion is increased in adapting to stressful conditions. Three contributing mechanisms have been proposed:

i. The extra utilization of the corticosteroids lowers their concentration in the blood and so promotes ACTH secretion by the normal feed-back control.

ii. The epinephrine released by the adrenal medulla in an emergency stimulates ACTH secretion.

iii. The hypothalamus can stimulate the adenohypophysis directly, either by neural pathways or by an as yet unidentified hormone of its own.

Diseases of the adrenal cortex. *Hypofunction* of the adrenal cortex in man is called *Addison's disease.* The deficiency of mineralocorticoids leads to decreased plasma sodium levels and increased potassium. Because of the salt wastage, the extracellular fluid volume cannot be maintained and the mean arterial pressure falls. This in turn causes inadequate filtration in the kidney and an accumulation of waste products. The glucocorticoid deficiency causes a lowered blood sugar concentration and renders the patients very susceptible to stress and infection. A further distinctive feature is a "bronzing" of the skin due to the accumulation of a pigment *melanin.* The disease is treated by injections of the cortical hormones and by maintaining a high intake of sodium chloride to counter the urinary wastage.

Hyperfunction of the adrenal cortex can result from tumors in the gland. As mentioned previously, it can lead to masculinization in adult women, excessive masculinization in men, or sexual precocity in children.

Cushing's syndrome is a hyperfunction largely restricted to the corticosteroids. The principal effect is a salt retention leading to edema and cardiovascular difficulties. Genital atrophy is common and also a curious obesity restricted to the neck and trunk, while the limbs remain thin. The most hopeful treatment in these cases is the location and removal of the tumors. In some cases it is necessary to remove all the cortical tissue and to maintain the patient subsequently with cortical extracts, as in Addisonism.

THE ADRENAL MEDULLA

This gland produces two very similar hormones, *epinephrine* and *nor-epinephrine,* also known as *adrenaline* and *nor-adrenaline* respectively. Their molecular structures are relatively simple and closely similar; both are amines. They are sometimes referred to collectively as the *catecholamines.*

These hormones are among the few exerting a rapid and short-lasting effect. In a way, the adrenal medulla might as well be regarded as part of the sympathetic nervous system as part of the endocrine system. Unlike most of the other endocrine glands, it receives a direct nerve supply which can stimulate its secretion. It has the same embryonic origin as the sympathetic nervous system, and its hormones are identical with those released at the ends of sympathetic post-ganglionic fibers in the heart and smooth muscles.

Effects of epinephrine and nor-epinephrine. The most important effects of these two hormones have already been discussed in the chapter on circulation, but at that time they were treated as one substance. Although their effects are complementary, they are actually different.

Epinephrine increases the rate and amplitude of the heart beat. It also inhibits visceral muscle and promotes a vasodilation in skeletal muscle and the coronary vessels.

Nor-epinephrine has its principal effect on the vasoconstrictor muscles of the skin and visceral vessels. It has relatively little effect on myocardial contractility or on skeletal muscle blood vessels.

The net effect of the two hormones on the circulation is an increased arterial pressure and an increased flow through exercising muscles. Epinephrine increases cardiac output and nor-epinephrine induces discriminatory vasoconstriction in the skin and viscera, shunting the blood to where it is needed.

The two hormones together have a further important effect.

They promote *glycogenolysis,* the breakdown of liver glycogen to glucose, causing an increase in blood sugar concentration. Like the cardiovascular effects, this is a useful response to a burst of exercise or as preparation for combat (or for running away!)

Control of Secretion

As already indicated, the gland secretes in response to impulses over sympathetic nerves. The kind of "emergency" promoting the release may range from the mild effort of walking upstairs to profound emotional shocks or the violent struggles in asphyxiation. It is possible that there is a normal rate of nervous stimulation to the gland, giving rise to a "basal" level of catecholamine secretion. Such a resting level must indeed be small, as laboratory animals show no ill effects after removal of their adrenal medullae.

Diseases of the adrenal medulla. The gland is not essential to life, and beyond a tendency to hypoglycemia in exercise, there is no recognizable ill-effect from its absence.

Rare tumor conditions can cause hypersecretion. In cases where the hypersecretion occurs sporadically, the patient suffers bouts of tachycardia, transient hypertension of an extreme nature, and a terrifying sense of apprehension and foreboding. Where the hypersecretion is continuous, the clinical picture is very similar to essential hypertension.

The sympathetic system and adrenal catecholamine secretion in cats is much more *labile* than in dogs. For cats, cardiovascular and blood sugar responses to a given emergency tend to be greater. Similarly, some humans have a more labile response in this connection, especially under emotional stress. They suffer acute physical discomfort from tachycardia in only mildly distressing social situations. The incidence of hypertension among Negroes is greater than in corresponding Caucasian populations. Even allowing for the social pressures and frustrations affecting the American Negro, there is still reason to believe that it results from a greater sympathetic lability.

THE PANCREAS

The pancreas lies close to the posterior abdominal wall behind the stomach. It functions as both an exocrine and an endocrine gland. Its exocrine functions involve the secretion of bicarbonate and digestive enzymes into the gut, which is discussed in the chapter on digestion (see p. 249). The endocrine function, which concerns us at present, is carried on by "islets" of special tissue scattered through the *acinar* or exocrine tissue; these are the *islets of Langerhans.* Their first discovered secretion was accordingly named *insulin.*

The islet tissue contains two histologically distinct kinds of cells, the α and β cells. We now know that insulin is secreted by the β cells, and another hormone, *glucagon,* by the α cells. Both hormones are of intermediate chemical complexity; they are polypeptides with 51 and 29 amino acid components respectively. Their exact structures have recently been established.

Functions of insulin. Insulin is another hormone with a rapid action. A description of its function is inseparable from that of carbohydrate and fat metabolism, which is discussed on p. 228. The following account, however, relates only to common-sense details and should be intelligible without prior discussion of intermediary metabolism.

The hormone controls the level of glucose in the internal environment. It will be recalled that this sugar is the prime energy-yielding nutrient of the cells. By mechanisms still not understood, insulin promotes the uptake of glucose by the tissues for oxidation, for glycogen synthesis (in liver and muscle) and for fat synthesis. Whatever the fate of the glucose, the net result is its removal from the internal environment. Conversely, a diminution in blood sugar promotes the breakdown of liver glycogen, partly through the effect of the other hormone, glucagon, but also through the stimulation of the liver by epinephrine from the adrenal medulla and from sympathetic nerves in the liver itself. The balance of these mechanisms is the maintenance of a fairly constant glucose concentration. The blood glucose tends to rise in an emergency, sometimes even to the extent of exceeding the renal threshold, but the insulin effect quickly restores the normal level after the sympathetic stimulation ceases.

The actual site of insulin action, in both the topographical and biochemical sense, is of great research interest and is still unresolved. There may well be more than one. The most likely possibility is concerned with the transport of glucose into the cells. Before it can be transported, or before it can enter any synthetic or catabolic reactions, glucose is first "activated" by combination with a phosphate radical, to yield glucose-6-phosphate. The reaction is catalyzed by an enzyme, *hexokinase,* and it is proposed that insulin facilitates the enzyme action.

In a deficiency of insulin, the glucose uptake by the cells from the blood is reduced, although absorption from the gut and the breakdown of liver glycogen both continue. In consequence, blood glucose rises to high levels (hyperglycemia), exceeds the renal threshold, and causes glycosuria.

The injection of insulin into a normal animal leads to *hypoglycemia,* a lowered blood sugar. If the insulin dose is large, the blood glucose may fall low enough to seriously affect the metabolism of the central nervous system. The convulsions and unconsciousness of *insulin shock* are due to profound hypoglycemia.

Glucagon function is far less well understood. Like insulin, it facilitates the oxidation of glucose in the tissues, but as noted above, it promotes the breakdown of glycogen and so tends to *increase* blood glucose concentrations. It is sometimes called the "hyperglycemic" factor.

Control of Insulin Secretion

The islet cells respond to elevated blood glucose levels by increasing the insulin secretion. The reciprocal inhibition of secretion in hypoglycemia has not been demonstrated as conclusively in the laboratory.

Diseases involving the pancreatic islet cells. The disease, *diabetes mellitus,* is caused by hyposecretion of insulin. It can be experimentally induced by surgical removal of the pancreas, or by *alloxan* poisoning, which selectively destroys the β cells.

The prime effect of an insulin deficiency is the reduced ability of the cells to use glucose. This causes the hyperglycemia; the consequent glycosuria in turn causes the polyuria and thirst typical of the condition (see p. 173). The functional shortage of glucose in the cells disrupts their energy metabolism. In the absence of glucose, the cells resort to fat as an energy source, but the oxidation of fat as the sole nutrient does not proceed to completion. Instead there is an accumulation of 2-carbon fragments which condense to yield *keto-acids.* The acidosis caused by these strong acids, together with the diuresis, causes electrolyte derangements which eventually result in paralysis and death. The process is self-exacerbating, because the accumulating acids further inhibit the function of any small supply of insulin still available.

Diabetics may be successfully treated through a long lifetime by insulin injections and a common sense diet. In some cases, the hyposecretion of insulin may be so marginal that the patient can manage without insulin therapy if he avoids taking on large carbohydrate meals, which strain his glycogenetic capacity.

Sometimes a diabetic inadvertently gives himself too large an insulin injection. The resulting hypoglycemia can cause unconsciousness, but often produces a drowsy stupor, superficially like drunkenness. A few cubes of sugar or a candy bar can immediately correct the situation but sometimes patients have been locked up by officious policemen and have become very ill in consequence. Diabetics are encouraged to carry a card describing the first aid to be given them if they are found in a state of collapse.

Tests for Insulin Status

The classic symptoms of diabetes mellitus are glycosuria, polyuria, and thirst. The first can occur under several other conditions and the last two are subjective enough to make it difficult to get exact answers from patients. A more conclusive guide, therefore, is the *glucose tolerance test.* The patient drinks, or is injected with, a glucose load in solution. Blood samples are taken over the next two or three hours and their glucose contents determined. If the blood glucose does not return to normal within one to two hours, there is indication of insulin deficiency.

Insulin assays can be carried out by measuring the uptake of glucose in isolated rat tissues, incubated in a medium containing an aliquot of the subject's plasma. The method is cumbersome and requires extremely skilled attention, so that it is impractical as a routine diagnostic test.

THE ADENOHYPOPHYSIS

The *hypophysis,* also known as the *pituitary gland,* is about the size of a small marble. It is attached by a stalk to the floor of the third ventricle, in the base of the brain. From its location, the early anatomists thought it secreted phlegm ("pituita" in Latin), and this is the origin of its name.

Despite its superficial unity of structure, it really comprises two separate organs, in which the functions, embryonic origin, and histologic structure are quite different (see Fig. 16.3). The posterior portion is derived from the brain; it is of nervous tissue and is in direct connection with the hypothalamus. This part is the *neurohypophysis* which was previously referred to as the storage site of *antidiuretic hormone.* It will be discussed in the next section.

The anterior portion is derived from the same embryonic epithelial tissue as the naso-pharynx. It is strictly an endocrine gland

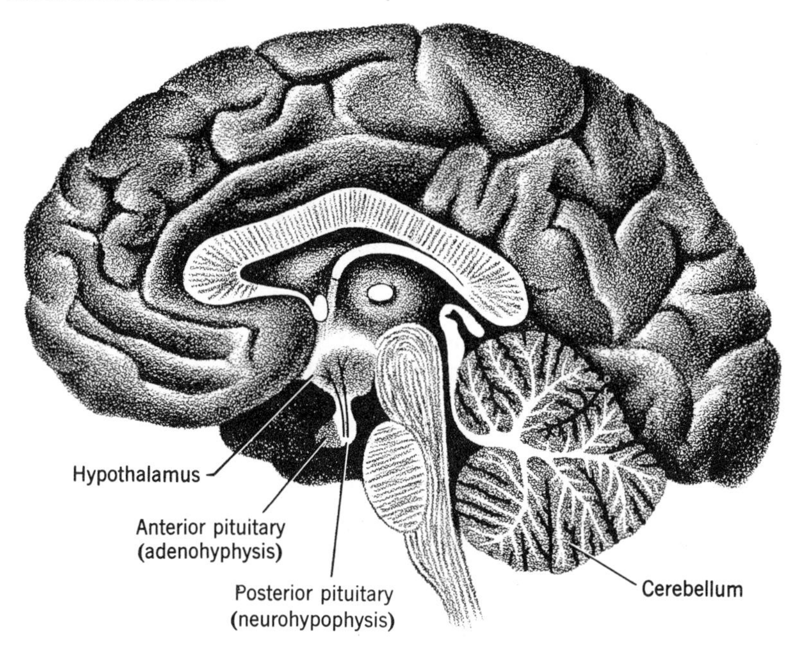

FIGURE 16.3. Section of the human brain, showing location of the hypophysis (pituitary gland).

and if it has any functional relationship with the neurohypophysis, it is extremely tenuous. The anterior pituitary is called the *adenohypophysis,* the prefix, "adeno-" meaning "glandular."

The adenohypophysis and neurohypophysis are also widely referred to as the "anterior" and "posterior pituitary glands." The former nomenclature is used in this text, not implying that the latter is wrong, but rather conforming with the agreed practice of modern endocrinologists. The older names tend to obscure the fact that the functions of the two parts are as distinct as those of the thyroid and parathyroid glands.

Functions of the adenohypophysis. We have already referred to the role of this organ as a "master-gland" in connection with its control of thyroid function through TSH secretion and its control of adrenocortical function through ACTH.

Early studies with *hypophysectomized* animals showed the symptoms we would expect from the decreased thyroid activity and glucocorticoid insufficiency (the mineralocorticoids, it will be recalled, are not so subject to ACTH control). At the same time, adult animals showed atrophy of the gonads and recession of the secondary sexual characteristics. Young animals did not grow to full size and did not develop sexually.

In man, the adenohypophysis secretes six hormones (they are all proteins except ACTH, which is a polypeptide).

1. Thyroid-stimulating hormone. TSH (Thyrotrophic hormone TTH).
2. Adrenocorticotrophic hormone, ACTH (Corticotrophin).
3. Growth hormone (Somatotrophin STH).
4. (*a*) Luteinizing hormone LH (in female).
   (*b*) Interstitial cell-stimulating hormone. ICSH (in male)
   } Identical
5. Follicle-stimulating hormone (FSH).
6. Prolactin (no known function in male) (Luteotrophic hormone LTH).

Hormones 1, 4, 5, and 6 control gonadal function. They are accordingly referred to as the *gonadotrophins.* In frogs and other cold blooded vertebrates, there is a seventh hormone, the *melanocyte-stimulating hormone (MSH),* which controls the amount and arrangement of the dark pigment (melanin) in the skin.

Attention is drawn to the suffix "trophic" in several of these names. It is derived from the Greek word for "nutrition," which is appropriate. The rather similar suffix "-tropic" means "turning toward" and is quite inappropriate in this connection.

Thyroid-Stimulating Hormone

The function of this hormone and the push-pull relationship of its secretion with that of thyroxin have already been described.

Adrenocorticotrophic Hormone

This hormone has also been discussed.

Growth Hormone

The functions of this hormone are difficult to describe beyond saying that animals do not grow without it. If an adult animal eats a diet with a caloric and protein content greater than daily needs, the amino acids of the excess protein are deaminated, and the sugar-like residues are converted to fat (see p. 232). This leads to an increase in *weight,* but not one that we ordinarily regard as *growth.* On the other hand, a young animal on a similar diet will utilize some of the excess protein for increasing its own tissue protein; this is growth in the true sense. Without growth hormone, the young animal makes fat from extra protein just as the adult does. Therefore, we can say that growth hormone promotes the building of tissue protein from dietary protein, but it may very well do it by *impeding* the conversion to fat. The relative leanness of actively growing animals is consistant with this, although subject to other interpretations.

Growth hormone also promotes and *prolongs* the activity in the epiphyseal cartilages of the bones (see p. 34). A deficiency of growth hormone causes early closure of the epiphyses and a short stature. In this respect, the sex hormones may be regarded as *antagonists* of growth hormone, because their increased production in later adolescence promotes the epiphyseal closure. As long as sex hormone secretion remains at the childish or early adolescent level, growth hormone continues to promote bone growth. An everyday observation illustrates this. Boys who develop an adult appearance (beard, voice, and muscle) at an early age, also reach their adult stature early. On the other hand, boys in whom the secondary sexual characteristics develop more slowly, continue to grow for longer. Early maturing men tend to be shorter, but not necessarily so, because clearly, many boys who look adult at fifteen or sixteen may have reached six feet in height already.

It is important to remember, furthermore, that some true growth continues after epiphyseal closure. The increased muscularity of the adult male is by growth.

Control of growth hormone secretion is a complete enigma. It is hard to see how tissue growth can exert a feedback control of the secretion. In this respect, the growth hormone differs from the others of the adenohypophysis.

The Gonadotrophins

The follicle-stimulating hormone promotes the development of a ripe ovum in each menstrual cycle in the female and promotes *spermatogenesis* in the male testes. These processes are described more fully in a later section on reproduction.

The next gonadotrophin is a single hormone, although it is called "luteinizing hormone" in the female and, "interstitial cell-stimulating hormone" in the male. It also promotes cellular changes in the gonads which will be taken up later. In addition, however, this hormone controls the endocrine activities of the gonads, which produce hormones as well as sperms and eggs.

The ovaries secrete estrogen and the testes secrete testosterone. Both are similar to the sex hormones of the adrenal cortex already described. They control the *secondary sexual characteristics;* beard, deep voice, etc. in the male, and mammary development and rounded contours in the female. As previously noted, they also affect the *libido,* or sex drive, which has behavioral repercussions far beyond the overt sexual activities.

The gonadotrophins control the sex hormone secretion through the now familiar push-pull kind of relationship of their respective concentrations. There is, however, an accumulation of evidence suggesting that the sex hormone level affects the adrenohypophysis via the hypothalamus, rather than directly. For our present purposes though, the principle is the same.

Prolactin seems to have a function only in the female. The hormone helps maintain the *corpus luteum,* a temporary endocrine organ vital in early pregnancy. It also promotes the proliferation of the breast tissue in late pregnancy and helps maintain lactation after parturition. It does not have these latter functions if injected into a nonpregnant female, because it requires the estrogen and progesterone status of pregnancy.

Diseases of the adenohypophysis. Hypofunction of the gland occurs at all degrees of severity and can involve all the hormones, or just one or two.

A profound hyposecretion of all the hormones is scarce in children, presumably because it is so fatal that the fetus would not survive. The *Lorain-Levi* type of dwarfism results from hyposecretion of the growth hormone and gonadotrophins. Such children remain as children in stature and sexual development all their lives. If the TSH and ACTH secretions are adequate, there is no illness from hypothyroidism or adrenocortical insufficiency, and there is no mental retardation. At the same time, the personalities of these dwarfs tend to be childlike, presumably because of their lack of sexual development.

A general hypofunction of the gland is called pan-hypopituitarism, or sometimes, *Simmond's disease.* In women, a cardiovascular collapse during childbirth can lead to thrombi in the blood vessels of the pituitary and subsequent necrosis of the gland. The disastrous consequences resemble a relentless premature senility. This post-partum hypopituitarism is called *Sheehan's disease.*

As would be expected, hyposecretion of the growth hormone alone leads to stunted growth. Hyposecretion of the gonadtrophins in adults leads to atrophy of the gonads and a recession of secondary sexual characteristics. In male children, it leads to *eunuchism,* in which the penis and testes remain of infantile dimensions, and the secondary sexual characteristics do not appear. The lack of sex hormones, however, allows the growth hormone to promote bone growth beyond the ordinary adolescent period. Therefore, eunuchs are characteristically tall, but without the angular muscularity associated with a good male physique. Similarly deficient female children also grow tall, have no breast development, and are frequently somewhat masculine in appearance because of the adrenocortical androgens.

The consequences of hypersecretion and hyposecretion of TSH or ACTH have already been discussed in the sections on the thyroid and the adrenal cortex. Hyposecretion of these hormones is almost always accompanied by gonadotrophic insufficiency as well.

Hypersecretion of the growth hormone in children leads to *gigantism*. Several unfortunate freaks have recorded heights of over eight feet. In adults, the epiphyses of the long bones are already closed and the extra hormone in a hypersecretion cannot promote further lengthening. Instead, there is an enlargement and thickening of the bones of the hands, feet, and face; the condition is called *acromegaly*. Apart from a distressing coarsening of the appearance, the extra bone growth in the skull compresses the brain, giving rise to headaches, blindness, and, ultimately, death.

Treatments

The availability of suitable adenohypophyseal extracts offers hope for the treatment of Simmond's disease, but to the present, the results have not been particularly successful. Defects in TSH and ACTH secretion are usually met by supplementing the thyroid or adrenocortical hormones appropriately. Hyposecretion in children has been less successfully treated, and so far, no suitable growth hormone preparation is available.

The various hypersecretions are frequently due to tumors in the gland itself or in the hypothalamus. The removal or irradiation of the tumor tissue can be spectacularly successful.

THE NEUROHYPOPHYSIS

As described in the previous section, the neurohypophysis is the posterior portion of the pituitary body. It releases two hormones, *oxytocin* and *antidiuretic hormone* (ADH). They are both polypeptides with molecular weights of about 1,000. There is now conclusive evidence that they are not actually synthesized in the neurohypophysis, but rather in the hypothalamus, and migrate as bead-like droplets along the connecting fibers to be stored in the neurohypophysis. Neural impulses from within the hypothalamus, or from the general pattern of afferent impulses reaching it, control the synthesis of the hormones and their release by the neurohypophysis. These hormones have a fairly rapid action, but not as rapid as that of epinephrine.

In the chapter on renal physiology, we have already discussed the feedback control of ADH secretion and its role in promoting the facultative reabsorption of water in the kidney. This hormone has a second less important function in that it can cause a vasoconstriction and some rise in blood pressure. For this reason it is also called *vasopressin.* In man, the vasoconstrictor effect is mainly in the coronary vessels, which tends to lower the cardiac output. There is, therefore, no dramatic increase of systemic blood pressure. The pressor response in some experimental animals is much greater.

The other neurohypophyseal hormone, *oxytocin,* promotes a generalized contraction of the smooth muscle in the body. While almost all such muscles are affected, the greatest response is from the uterine muscle, especially in late pregnancy. We might expect that oxytocin has some role in the contractions of labor, but nothing is known for sure. The hormone is frequently injected post-partum to promote the contraction of the empty uterus and thereby slow down bleeding from minor lacerations incident to expulsion of the fetus.

In the lactating female, the infant's sucking at the nipple sends impulses to the hypothalamus which cause oxytocin release. Because of the increase of the hormone in the blood, the smooth muscle of the mammary gland contracts and ejects the milk. The response to the "*sucking reflex*" is called the "letdown" of the milk. Emotional disturbances will inhibit the reflex. The hormone is sometimes used by farmers to facilitate the first milking of a recalcitrant, freshly-calved heifer.

REFERENCES

1. Constantinides, P. C. and Niall Carey, "The alarm reaction," *Sci. American* **180** (3), 20, March 1949.
2. Csapo, Arpad, "Progesterone," *Sci. American* **198** (4), 40, April 1958.
3. Fieser, Louis F., "Steroids," *Sci. American* **192** (1), 52, January 1955.
4. Funkenstein, Daniel H., "The physiology of fear and anger," *Sci. American* **192** (5), 74, May 1955.
5. Levine, Rachmiel and M. S. Goldstein, "The action of insulin," *Sci. American* **198** (5), 99, May 1958.
6. Li, Choh Hao, "The pituitary," *Sci. American* **183** (4), 18, October 1950.
7. Thompson, E. O. P., "The insulin molecule," *Sci. American* **192** (5), 36, May 1955.
8. Wilkins, Lawson, "The thyroid gland," *Sci. American* **202** (3), 119, March 1960.
9. Zuckerman, Sir Solly, "Hormones," *Sci. American* **196** (3), 76, March 1957.

PART 4

THE EFFECTORS AND RECEPTORS

MUSCLE-NERVE INTRODUCTION

As described in the introductory chapter, an animal is "aware" of the external environment through its external receptors. These organs are sensitive to various kinds of stimuli, as, for example, the eyes are sensitive to light and the ears to sound. Information from the receptors is communicated to a central clearing house (the brain and spinal cord), from which instructions are sent to the appropriate effectors (muscles) so that the animal can take some action to respond to the environment. An example is seen in a feeding animal. The image of an edible object is projected by the lens on the retina of the eye. The pattern of stimulation of the retinal receptors causes a series of signals to be sent over the optic nerves to the brain. One portion of the brain receiving the visual impulses interprets them as a picture, and in the light of experience, or of a less easily defined "instinct," the animal recognizes the edible object in its immediate vicinity. Impulses are then sent from the brain over other nervous pathways to muscles, causing them to contract and relax systematically so that the animal walks over to the food, picks it up, and eats it. The deceptive simplicity of the example is apparent when we consider the enormous complexity of antimissile missiles, which are designed to perform the relatively elementary task of locating a hostile missile and closing on it to make a simple collision.

The eyes and muscles are *external* receptors and effectors respectively, since they are concerned with response to the *external* environment. It should be noted that there are also systems of internal receptors and effectors through which the *internal* environment is controlled. For example, the pressoreceptors in the arch of the aorta keep the cardiac centers of the brain informed about the blood pressure; impulses to the pace makers of the heart (an internal effector) permit appropriate changes and corrections to be made.

The pattern of response to both internal and external environmental changes is the same in that: 1. a sensitive organ is stimulated by the change; 2. signals from that organ travel to the central nervous system where they are "interpreted"; and 3. signals are sent out over other pathways to muscles or other effectors which can do something about the change.

An essential part of these activities is *communication;* this is typically the function of the nerves. The receptors are also considered as part of the nervous system.

PROPERTIES COMMON TO BOTH MUSCLE AND NERVE

Both nerve and muscle tissue are highly *irritable.* As described in Chapter 5, a living cell has a potential across its membrane due to the unequal distribution of sodium and potassium ions between the intracellular and extracellular fluid. Stimulating the membrane by mechanical pressure, electric shock, or other means, changes the permeability of the membrane drastically. In this state, sodium ions flow into the cell and potassium ions flow out, reversing the membrane potential. This brief event is called *depolarization.* Immediately afterwards, the membrane's normal permeability is restored and the potential returns to its normal resting level. The depolarization at one point sets up *local circuits* so that the membrane on each side is depolarized and a wave of depolarization spreads out from the single site of stimulus. This wave is called an *action potential.*

In nerve cells, the action potential travels along the nerve fiber and is, in fact, the *nerve impulse* by which information is carried from one part of the body to another. The passage of the action potential along the cell membrane in muscle triggers the mysterious chemical events which lead to a shortening of the contractile fibers within the muscle cell.

MUSCLE

17

There are three kinds of muscle in the body: (1) *skeletal* muscle, also called *voluntary* muscle; (2) *smooth* muscle, also called *visceral* or *involuntary* muscle; and (3) *cardiac* muscle.

SKELETAL MUSCLE

The *skeletal muscles* comprise about 40% of the body weight and as the name indicates, they are attached to the bony skeleton and are responsible for moving the limbs, etc. All the external movements of an animal, including the maintenance of its posture, depend on the coordinated contraction and relaxation of skeletal muscles.

Because these muscles can be moved at will, they are called *voluntary* muscles, but this is apt to be confusing as there are many involuntary movements of skeletal muscle. The diaphragm, for example, is of skeletal muscle but its excursions during ordinary respiration are quite unconscious. Similarly, the thigh muscle (*quadriceps femoris*), twitches involuntarily when its tendon is tapped below the patella (knee-jerk reflex). The contraction of the skeletal muscle, however, is certainly more voluntary than that of smooth muscle, which cannot be consciously controlled at all.

The units of skeletal muscle are long,

slender, multinucleated cells called muscle fibers. They are contained in bundles by sheaths of connective tissue and groups of the bundles are again contained in yet tougher sheaths. The connective tissue sheaths extend throughout the whole mass of a muscle and they are all structurally continuous with the *tendons,* by which the muscle is attached to the bones of the joint it moves. In this way, the tension from the contraction of each individual muscle fiber is applied usefully at the points of origin and insertion of the muscle.

The muscles of a joint are usually arranged as *pairs* of *antagonists* for movement in each direction. Thus, at the elbow joint the contraction of the *biceps brachii* flexes the arm and that of the *triceps* extends it. A smooth flexion is accomplished by the *antagonism* of the triceps to the pull of the biceps, and vice versa in extension. In addition to the movement of the joints, a steady contraction of some skeletal muscles maintains the posture of the animal by a *tonic* extension (of the knee joints for example). Such muscles are aptly called the *antigravity* muscles.

The fibers of skeletal muscle are of two main types, red and white. The red fibers have a less dense intracellular structure and they contain a red pigment, *myoglobin,* which resembles hemoglobin both in chemical structure and in its ability to combine with oxygen. The myoglobin keeps a small store of oxygen in the muscle. White fibers are denser and contain no myoglobin. Both types of fibers can be found in a muscle but the red fibers predominate in the antigravity muscles of the trunk and limbs, whereas the white fibers predominate in flexor muscles which are called upon to make rapid but less protracted contractions. As might be expected, red fibers contract less rapidly than white fibers, but are less subject to fatigue. The classic example of red and white muscle is seen in the thigh and breast muscles respectively of a chicken.

Skeletal muscle is also called *striated muscle* because of the striations visible under the microscope (Fig. 17.1). The relationship of this microstructure to the function of muscle has been of great interest for many years. Modern studies with the electron microscope and X-ray diffraction techniques have provided a picture summarized in Fig. 17.2. The muscle cell contains a fluid, *sarcoplasm,* which is presumably similar to the intracellular fluid of other cells, and in addition, contains parallel fibers (*myofibrils*), which are the contractile elements of the muscles. The myofibrils have light and dark bands and it is the coincidence of the bands in adjacent fibrils which gives the whole muscle fiber its character-

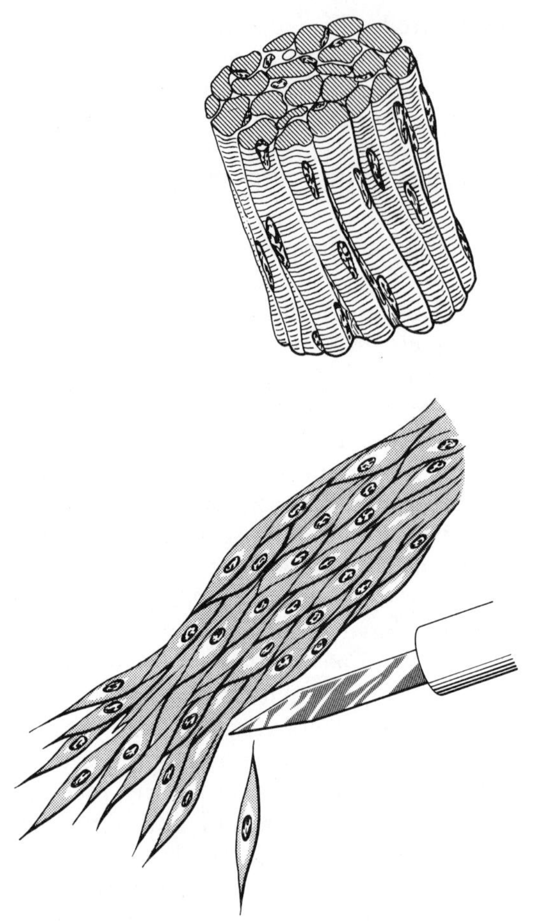

FIGURE 17.1. Differences between smooth, or involuntary, muscle fibers (lower) and striated, or voluntary, muscle fibers (upper). (Redrawn from Hall & Moog, *Life Science*, John Wiley and Sons, 1955)

istically striated appearance. The bands are arbitrarily named as in part *E* of Fig. 17.2. Each myofibril is made up of strands of two kinds of protein, *actin* and *myosin,* arranged alternately in a cross section of the myofibril. The strands are arranged as shown in the diagram in Fig. 17.2. The dark bands occur where both actin and myosin strands are superimposed; the light bands occur where there is only one protein or the other.

These structural details are well established. Less completely accepted is the theory that muscular contraction results when cross linkages between actin and myosin strands pull hand-over-hand, as it were, so that the interdigitated strands "crawl" past

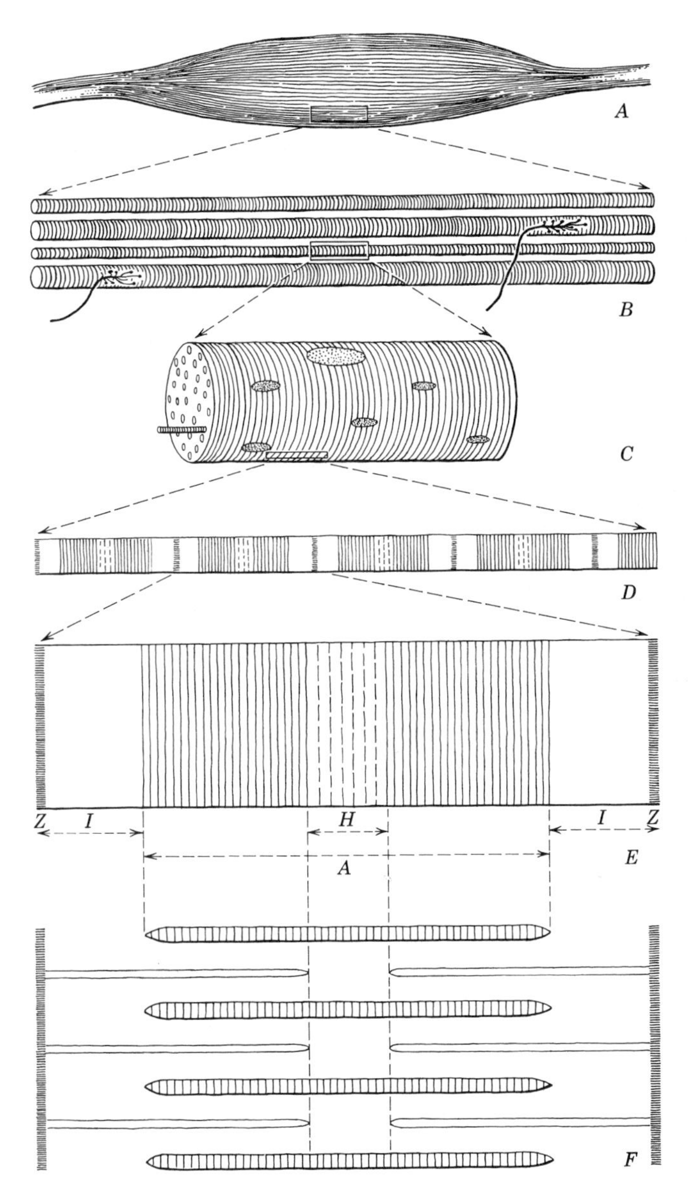

FIGURE 17.2. Striated muscle is dissected in these schematic drawings. A muscle (*A*) is made up of muscle fibers (*B*) which appear striated in the light microscope. The small branching structures at the surface of the fibers are the "end-plates" of motor nerves, which signal the fibers to contract. A single muscle fiber (C) is made up of myofibrils, beside which lie cell nuclei and mitochondria. In a single myofibril (*D*) the striations are resolved into a repeating pattern of light and dark bands. A single unit of this pattern (*E*) consists of a "Z-line," then an "*I*-band," which is interrupted by an "*H*-zone," then the next *I*-band and finally the next *Z*-line. Electron micrographs have shown that the repeating band pattern is due to the overlapping of thick and thin filaments (*F*). (From Huxley, H. E., *Sci. Am.* **199** (5) 66, 1958.)

each other. This would account for the observed disappearance of the M-band in contraction.

The energy for muscular contraction comes from the "packets" of energy available in the form of the high-energy phosphate bonds of ATP, which are generated by the combustion of glucose as described in the chapter on metabolism (p. 225). The splitting of the bonds is accomplished by the enzymic activity of the actomyosin of the muscle itself. The ATP content of muscle is small, but a stock of high energy phosphate bonds is maintained in the form of phosphocreatine. During contraction (or at some other stage of muscular activity), some of the ATP is converted to ADP, but is immediately reconverted to ATP by high energy bonds released from the breakdown of phosphocreatine. Meanwhile, the metabolism of glucose is also restoring ATP units and continues to do so after the contraction, regenerating the phosphocreatine reserve. If the demand for ATP (energy) is greater than can be met by oxidation, some extra ATP is generated by increased *anaerobic* metabolism of glucose, leading to lactic acid accumulation. Some of this is afterward fully oxidized in the liver; the extra oxygen intake this involves is the payment of an oxygen debt (see p. 227).

Superficially, no one would doubt that the energy is used in the active *contraction* of the muscle, and this is indeed the commonly accepted view. Recent experiments have shown, however, that ATP bonds do not appear to be split during the contraction phase of a muscle twitch; this observation is clearly not consistent with the accepted view. Therefore, there are good grounds for postulating that the resting muscle is like a stretched spring that delivers its potential energy when stimulated and that energy must then be applied to restore it to its resting length, which would involve an *active relaxation.*

Isolated muscle-nerve preparation. Many characteristics of muscular contraction have been worked out using a frog muscle-nerve preparation and the simple equipment shown in Fig. 17.3. The muscle is the gastrocnemius, or calf muscle. By a simple dissection, it can be removed from a pithed frog complete with a length of the sciatic nerve and a portion of the femur attached. The piece of bone may be held in a clamp, effectively securing one end of the muscle, and at the other end, a thread tied to the Achilles tendon is arranged so that when the muscle contracts, it pulls on the *muscle lever.* Weights attached to the muscle lever cause the muscle to contract against a *load,* which can be varied. A smoked drum, rotating at constant speed is placed against the pointed tip of the muscle lever, which leaves a clear scratch on the smoked surface. Movement of the muscle lever is magnified because of the length of the lever arm and shows up clearly in the record on the drum. The apparatus rotating the smoked drum is called a *kymograph.*

The muscle is conveniently stimulated by a small electric shock applied either to the muscle itself or to the nerve. The shock may be of the simple *Galvanic* kind, in which the stimulating electrodes are connected directly to the poles of a battery, or by a Faradic or *induced* voltage, which is more convenient in many respects. In the Faradic arrangement, a high voltage with a small current flow is developed in an induction coil similar to that in the ignition system of an automobile. A coil of this kind, called an *inductorium,* is shown in Fig. 17.3. The induced voltage can be varied by sliding the secondary coil over the primary, and this of course varies the strength of the stimulus applied to the muscle.

It should be noted that a current is induced in the secondary coil when the circuit is closed and again when it is opened. That is, the preparation is stimulated on both the "make" and the "break." Because the fall-off of the current on the break is faster than the build-up on the make, the break voltage is always higher than the make voltage for any given setting of the inductorium. Elaborate stimulators with vacuum tube circuits are available today and these are more versatile and more accurate than the simple inductorium.

A useful refinement in carrying out experiments of this kind is the inclusion of a signal magnet. This has a writing point applied to the kymograph drum which is depressed by an electromagnet when the stimulator circuit is closed. In this way, the exact instant of applying the stimulus is recorded on the smoked paper.

In carrying out experiments with the frog muscle-nerve preparation, it is important to distinguish between responses due to the characteristics of the muscle proper, and those of the nerve or the junction between the nerve and muscle. For this reason it is often better to stimulate the muscle itself rather than the nerve.

Single twitch. The simplest muscle response is the single twitch which follows a single stimulating shock. Its characteristics are best shown when the kymograph drum is set to turn rapidly. A typical record is shown in Fig. 17.4. Note the delay between the application of the stimulus and the onset of contraction; this is the *latent* period. Its duration may be measured by applying a tuning fork of known frequency to the rotating drum; the fork's oscillations then make a time scale for reference.

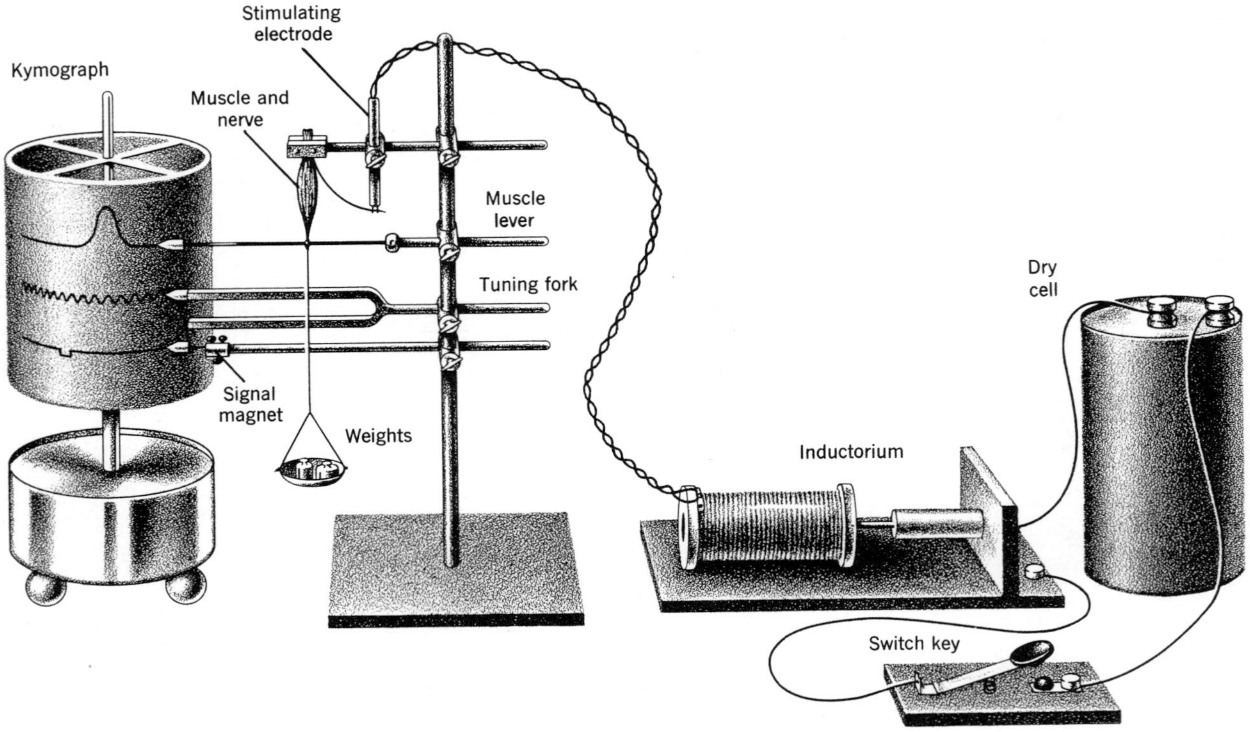

FIGURE 17.3. Frog muscle-nerve preparation.

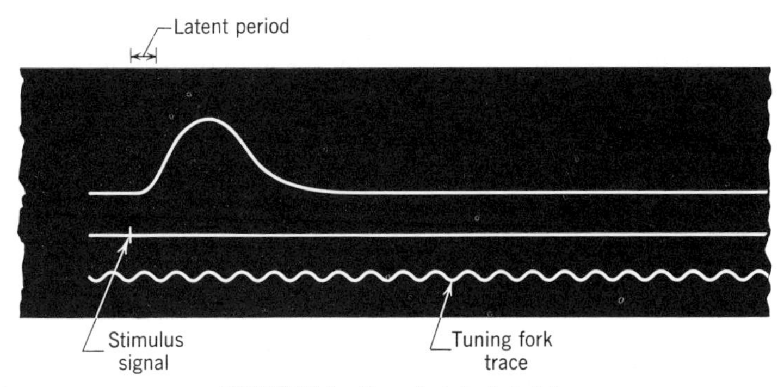

FIGURE 17.4. Record of single twitch.

Strength of stimulus. Using a stationary drum advanced by hand after each stimulus, it is possible to determine the relationship of the contraction amplitude to the strength of the stimulus. A stimulus can be so small that it is inadequate to depolarize the membrane and cause an action potential; therefore, no contraction is elicited. Such a stimulus is *subliminal.* If the strength of the stimulus is increased gradually, it eventually reaches a level at which it will barely evoke a small contraction. This strength is called the *liminal* (or minimal) stimulus. With further increments of stimulus strength, the amplitude of the contraction increases to a maximum, beyond which there is no further increase with stronger stimuli (see Fig. 17.5). The stimulus strength beyond which there is no further increase of amplitude is called the *maximal* stimulus.

Duration of stimulus. It will be appreciated that the liminal and maximal stimuli strengths are not absolute values for any particular muscle, since they depend on the metabolic state of the muscle, the nature of the stimulating electrodes, and many other variables.

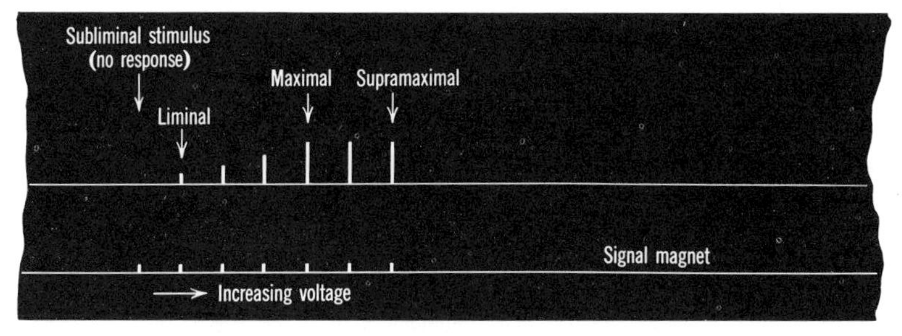

FIGURE 17.5. Kymograph record showing muscles response to stimuli of varying strength.

Important among these is the *duration* of the stimulus. In general, a small stimulus is more effective when applied for a longer time than for a short time. Nevertheless, for each preparation, there is a stimulus strength which is still just subliminal even if applied for an infinitely long time (see Fig. 17.6). This stimulus strength is known as the *rheobase,* and from it another value has been derived, the *chronaxie.* This is defined as the time for which a stimulus of twice the rheobase strength must be applied in order to elicit a contraction. At one time, the measurement of chronaxie was thought to be of great theoretical importance, but muscle physiologists today have diminished interest in it.

Summation of subliminal stimuli. It has been found that the application of a subliminal stimulus to a nerve or muscle causes a *local excitatory state* in the membrane, even though no action potential is evoked. While this state persists, the application of another subliminal stimulus, or of a succession of them, may be enough to trigger an action potential. The cumulative effect of a succession of subliminal stimuli is called *summation.*

All or None Law

At one time it was believed that the greater amplitude of contraction with increasing stimulus strength was due to the greater

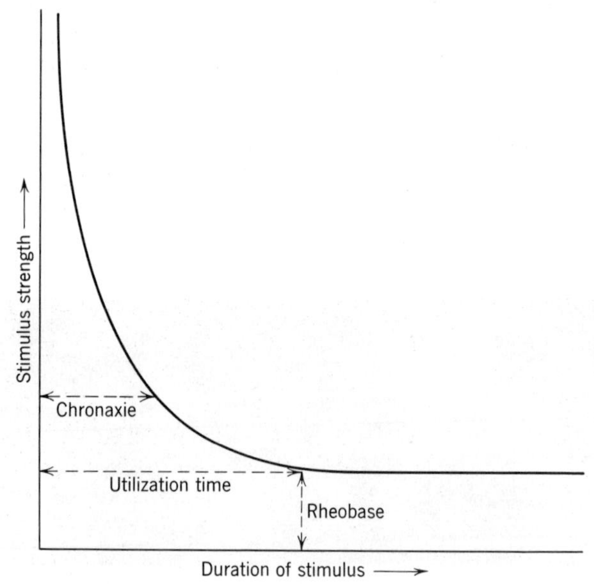

FIGURE 17.6. Graph showing strength-duration relationship.

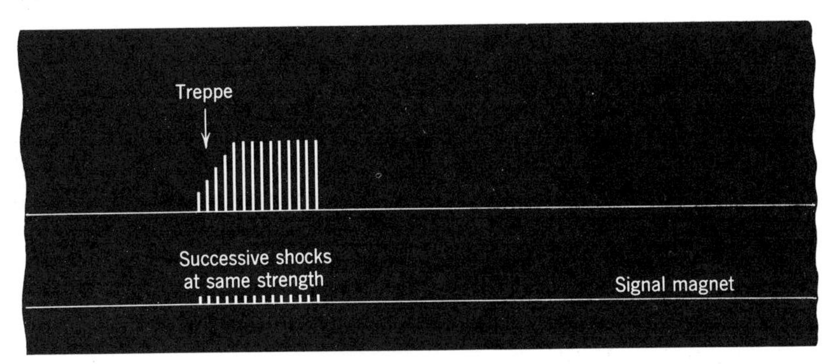

FIGURE 17.7. Treppe.

contraction of each muscle fiber. It is now known, however, that each fiber either contracts maximally, or not at all; increased contraction results only from the contraction of *more* fibers in the muscle.

In the individual muscle fiber, a stimulus of above liminal strength causes an action potential and subsequent contraction. A subliminal stimulation has no effect (by itself). The same is true of the other kinds of muscle and of nerve fibers. This characteristic is known as the *All or None Law*. It does *not* state that the strength of contraction of a muscle is invariable, because obviously it can be diminished in fatigue, for example, but rather that the strength of the stimulus has no effect providing it is above liminal strength. An analogy is the firing of a pistol; provided the pressure on the trigger is strong enough to overcome the hammer spring, the strength of the pull has no effect on the velocity of the bullet. There are many apparent exceptions to the All or None Law in neuromuscular physiology and some are to be discussed, but the general principle still stands.

Treppe

When a muscle is stimulated as rapidly as three or four times per second, the first few contractions are of increasing amplitude, even though each stimulus is of the same strength. This apparent contradiction of the All or None Law is called the *Treppe effect*. Treppe is German for "staircase," and the name refers to the step-like increments in contraction. It is presumed that the chemical events of the first few contractions cause changes in the muscle fibers, rendering them more contractile in a manner that can be likened to "warming up" an automobile motor (Fig. 17.7).

Summation of twitches. For a brief period immediately after a stimulation, a further stimulus has no effect on a muscle. This *refractory period* is very short in skeletal muscle and is over before the muscle has finished contracting. Therefore, a stimulus applied after this time will cause a further contraction superimposed on that already started. Subsequent stimuli at the right intervals will cause more and more superimposed contractions. This phenomenon, shown in Fig. 17.8*A*, is called *summation.* If the successive stimuli are applied closer together, the result is as in Fig. 17.8*B*, and finally, if they are in extremely rapid succession the summated contractions blend into a smooth and sustained *tetanic* contraction, Fig. 17.8*C*.

Muscle contractions in the intact animal are rarely of the single twitch kind; they are usually tetanic contractions produced by volleys of nerve impulses. It is important not to confuse the summation of twitches (a mechanical summation) with the summation of subliminal stimuli (which is electrical). Furthermore, the staircase effect must be clearly distinguished from the summation of twitches. The staircase effect is seen in successive contractions, each of which is completed to the point of relaxation before the next one starts.

The liminal stimulus strength for tetanic contraction is lower than for single twitches, but otherwise the pattern of response to varying strength is similar. In particular, maximal tetanic contraction results when *all* the fibers are contracted. If the increasing stimuli are applied to the nerve, a similar response is observed because the increased stimulus strength exceeds the threshold of more and more individual *nerve* fibers in the nerve trunk until they are all sending impulses to the muscle fibers they serve. Similarly, in the intact animal, the greater the number of nerve fibers carrying impulses, the greater the number of motor units which contract, and so the total contraction of the muscle is

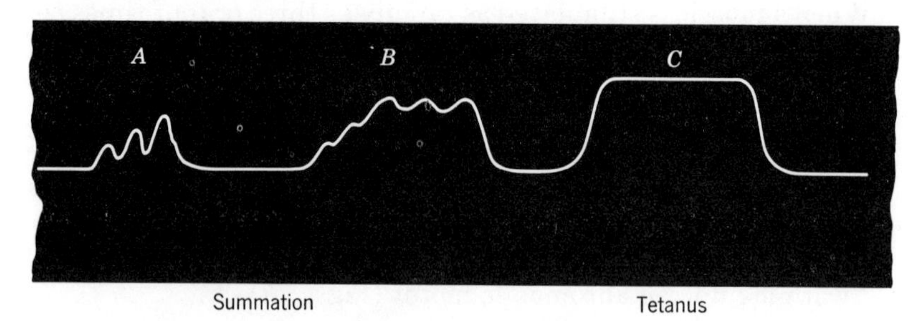

FIGURE 17.8. Effect of applying successive stimuli closer and closer together.

increased. A partial contraction is achieved by stimulating fewer nerve fibers. The antigravity muscles are in partial contraction for very long periods of time, which would cause excessive fatigue if it were always the same fibers contracting. In the living animal, the *tonic* contraction of the postural muscles is achieved by different groups of fibers briefly maintaining tension by tetanic contraction and then relaxing while other groups take over.

Isotonic and isometric contraction. In the preceding discussion we considered the contraction of the isolated frog muscle when it is shortening against a constant load. This is an *isotonic contraction.* The muscle can also be attached to a rigid support at both ends, so that although the physiological events of contraction occur, there is no actual shortening; such a contraction is *isometric.* Our own muscles contract isometrically when we try to lift a weight we cannot budge. Valuable techniques have been developed for studying isometric contraction in isolated muscles, because in such a system, the muscle's characteristics are not obscured by the inertia of all the moving parts in the simple isotonic apparatus previously described.

Length and tension. An interesting characteristic of skeletal (and cardiac) muscle is that the force of contraction can be increased by stretching the muscle before it contracts. This is conveniently shown with the frog muscle preparation. If a suitable stimulus strength is selected and maintained, the amplitude of contractions can be measured after the addition of increasing loads to the weight pan attached to the muscle lever. Increasing the load tends to stretch the muscle, and, up to a point, will increase the contraction. The results of such an experiment can be plotted as a *length-tension diagram* shown in Fig. 17.9. The most profound importance of this phenomenon in the body is the response of cardiac muscle to stretching by increased filling pressure (see p. 127).

Work and heat. When a muscle contracts and moves a load, part of the energy supplied is used to do *work* and the remainder appears as *heat.* The thermodynamics of muscular contraction is beyond the scope of this text and so this heat will be simply called the *heat of contraction.* After the muscle has relaxed, still more heat is generated while the energy reserves of the muscle are restored; this is the *heat of recovery.* If a muscle contracts isometrically (without shortening) *all* of the energy appears as heat. The heat produced by the muscles is the principal source for the maintenance of body temperature, and this is painfully obvious to

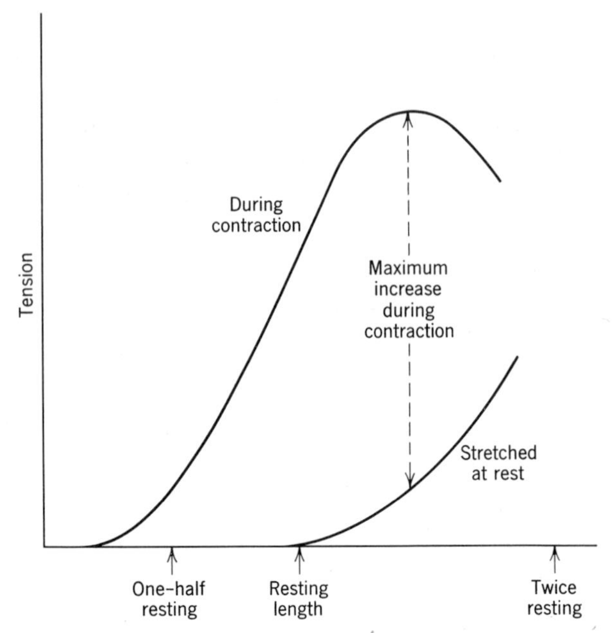

FIGURE 17.9. Length-tension diagram of skeletal muscle.

anyone doing heavy muscular work in a hot climate. Conversely, if heat production by normal and useful contractions is inadequate to maintain body temperature in a cold environment, then rapid and aimless contractions are induced to generate heat. We recognize these as *shivering*.

Injury potential. The potential which exists between the inside and outside of a muscle fiber membrane can be measured by placing a minute recording electrode inside the cell and another on the outside. This requires very specialized equipment, however, and is not easily performed in the student laboratory. A more easily demonstrated manifestation of the membrane potential is the *injury potential.* If a living muscle is cut, the ends of the severed fibers are continuous with the insides of the cells through the conductivity of the sarcoplasm. Therefore, if one electrode is placed in the wound and another on the undamaged surface of the muscle, the potential recorded is at least *related to* the membrane potential (see Fig. 17.10). The injury potential may also be demonstrated without using any equipment beyond the living tissue. One muscle of a frog is dissected out complete with its nerve. The nerve is then draped over another muscle so that it touches both

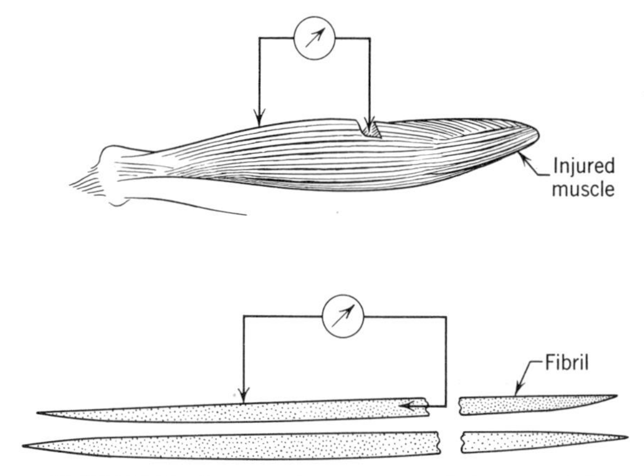

FIGURE 17.10. Diagram showing nature of injury potential.

the intact surface and an injured portion (Fig. 17.11). As the nerve makes contact, the injury potential will evoke an action potential in the nerve causing its attached muscle to twitch. This classic experiment was first performed by Galvani.

Myoneural junction and motor units. An isolated muscle can be stimulated to contract by applying either an electrical or a mechanical stimulus. In the body however, muscles normally contract in response to impulses arriving over the motor nerves. The efferent nerve fibers terminate on muscle fibers in a structure known as an *end-plate,* or a *myoneural junction* (shown diagrammatically in Fig. 17.12). A nerve fiber divides into many branches as it approaches a muscle and each end-plate serves several muscle fibers. A nerve fiber and its associated muscle fibers is called a motor unit. In the large muscles of the legs, a single motor nerve fiber may serve up to a hundred and fifty muscle fibers, but in the finer muscles of the eye, the number is between five and ten.

At one time it was thought that the nerve impulse crossed to the muscle fibers much as an electric spark jumps a gap. It is now

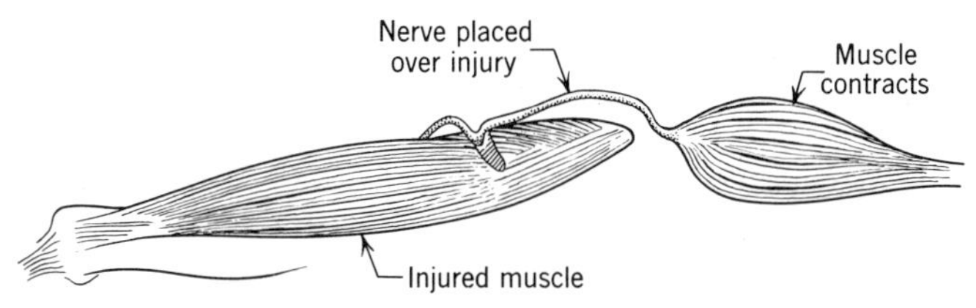

FIGURE 17.11. Rheoscopic injured muscle preparation.

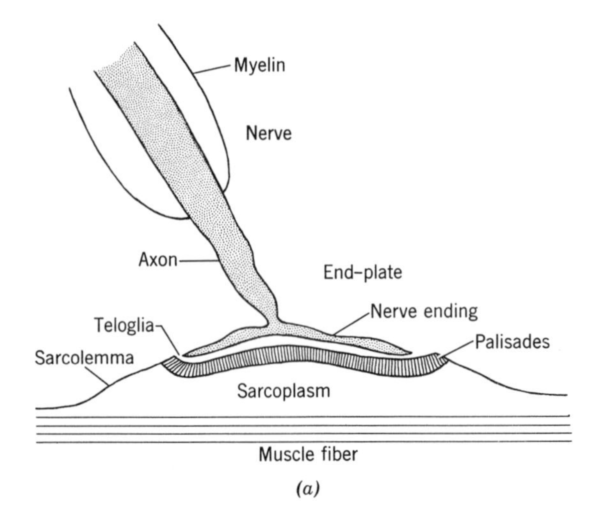

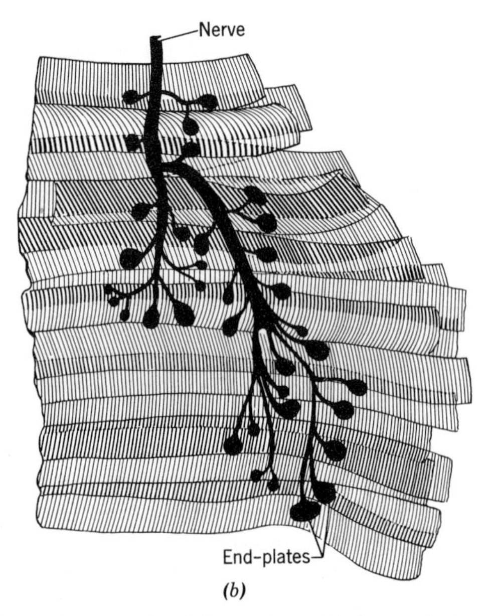

FIGURE 17.12. (*a*) Diagram of end-plate region. (*b*) Motor end-plates of neuro-muscular junctions in striated muscle.

known that the arrival of nerve impulses at the end-plate causes the production of *acetylcholine,* and that this chemical depolarizes the muscle cell membranes much as does an electric shock. The wave of depolarization then spreads along the muscle in each direction from the end-plate and the muscle contracts. A single nerve impulse will produce enough acetylcholine in the end-plate to cause a single twitch in the muscle. The acetylcholine is destroyed almost immediately by an enzyme, *cholinesterase,* and in consequence, the muscle cell membrane is then repolarized and can respond to another impulse. If the nerve impulses are sufficiently rapid, the muscle does not have time to relax between stimuli and so remains in *tetanic contraction.*

Many features of the myoneural junction can be studied with the isolated frog muscle-nerve preparation, or alternatively, the nerve and muscle may be left in the animal and the sciatic nerve stimulated *in situ.* The Achilles tendon, by which the gastrocnemius inserts on the *calcaneus,* is cut and tied to a muscle lever so that on stimulating the nerve, the contraction of the muscle records on the smoked drum (see Fig. 17.3). Prolonged tetanic stimulation via the nerve causes the amplitude of the contraction to decrease rapidly. If the muscle itself is stimulated directly, however, the decrease is much slower. The faster decrease when the nerve is stimulated at high frequencies results from a depletion of acetylcholine precursors at the end plate and so fewer muscle fibers are depolarized by each impulse. This is known as *end-plate fatigue,* which should be distinguished from the muscle fatigue caused by an accumulation of lactic acid during hard work.

The drug, *curare,* blocks the transmission of the impulse across the myoneural junctions and leads to paralysis. South American Indians used curare as a poison for their blow pipe darts and it is used now to relax the muscles of an anesthetized patient undergoing surgery. However, the respiratory muscles are paralyzed too, and the patient must be given constant artificial respiration during the operation. *Succinyl choline* also blocks transmission at the myoneural junction, but by competitive inhibition of the cholinesterase. It ties up all the cholinesterase for a few minutes, and during that time, the continued presence of undestroyed acetylcholine prevents the muscle fibers from being repolarized after they have contracted once. This drug is useful in obtaining muscular relaxation for a short time, such as that required for inserting an endotracheal tube.

Some complex organic fluorine compounds are potent inhibitors of cholinesterase and cause a paralysis similar to that with succinyl choline except that it is irreversible. These compounds are the "nerve gases" which are among the least attractive of modern weapons. Paradoxically though, many important studies have been made of the fundamental properties of the end-plate by using the nerve gases as experimental tools.

SMOOTH MUSCLE

Inasmuch as skeletal (striated) muscle is largely concerned with response to the external environment, the smooth muscles of the body are found in organs that help control the internal environment. The walls of the hollow viscera such as the stomach, intestine, bladder, etc. are partly composed of bands or layers of smooth muscle, and for this reason, it is alternatively called *visceral* muscle. Smooth muscle is also found in the walls of blood vessels where it is instrumental in controlling the blood pressure through the *vaso-motor tone* (see Chapter 7). The skin contains smooth muscle which can erect the hairs in response to cold or fright; these are the *pilomotor muscles.*

Skeletal muscle tissue is remarkably uniform in appearance and functional characteristics in all parts of the body. Smooth muscle on the other hand is extremely variable in appearance, both microscopically and grossly, depending on the organ it is found in. A few generalizations are possible, however.

Smooth muscle cells are cigar-shaped with a single nucleus and are usually much shorter than skeletal muscle fibers. Also, the fibrils in each cell are smaller and do not have bands like those of skeletal muscle. Therefore, the tissue itself does not have a striated appearance, which is of course why it is called smooth muscle (Fig. 17.1).

In general, smooth muscle is slower acting than skeletal muscle and might even be described as sluggish. It takes more to stimulate it and the contractions occur more slowly. In the smooth muscle of the gut walls there are dense masses of cells with relatively sparse innervation, whereas the vasomotor and pilomotor muscles are organized in a manner similar to the motor units of skeletal muscle.

Smooth muscle may be stretched over a considerable range without altering the tension at each end. This property is called *plasticity* (as opposed to *elasticity*) and it is apparent in the walls of the stomach or urinary bladder. These organs can become quite full before the muscular walls are under any increased tension.

Much smooth muscle receives *dual innervation* from the sympathetic and parasympathetic divisions of the autonomic nervous system (see Chapter 21). One set of fibers are *excitatory* and others are *inhibitory.* Some autonomic fibers release acetylcholine at their endings as do the motor nerve fibers serving skeletal muscle, and these are therefore called *cholinergic* fibers. Other fibers are *adrenergic,* however, since they release the hormone *adrenalin* (actually nor-epinephrine) to mediate the contraction of the smooth muscle. In the smooth muscles that are dually innervated, one set of fibers is cholinergic and the other is adrenergic. In some muscles, the cholinergic fibers are excitatory and the adrenergic are inhibitory, and in other muscles *vice versa;* there is no uniform rule. Some smooth muscle, as in the skin and eye, is singly innervated.

Smooth muscle frequently displays *inherent rhythmicity* independent of its innervation. Excised pieces of intestinal or uterine smooth muscle will contract rhythmically for a long time if placed in a nutrient solution and oxygenated.

CARDIAC MUSCLE

The heart is a large mass of muscle responsible for pumping the blood. This muscle tissue is different from the others in the body and has several unique features; it may be thought of as intermediate in properties between visceral and skeletal muscle. It possesses intrinsic rhythmicity to a high degree and yet its contractions are rapid and forceful. Under the microscope, the cells are distinctly striated and the arrangement of actin and myosin filaments is assumed to be similar to that in skeletal muscle.

The distinctive characteristic of cardiac muscle is its *syncitial* arrangement. This refers to the intercommunication between cells by their branching (see Fig. 7.5). Because of this arrangement, an impulse originating at one point in the heart spreads throughout the muscle mass and the whole heart contracts with every beat.

A further important characteristic is that of depolarization, cardiac muscle remains depolarized for a relatively long time (0.3 seconds as opposed to 0.03 seconds for skeletal muscle). The muscle remains contracted so long as it is depolarized, and this allows enough time for the heart chambers to empty of blood at each contraction. Furthermore, the heart muscle is *refractory* to further stimulation during its long period of depolarization, and this helps to protect it from stray impulses which would otherwise interrupt its rhythmic beating.

All parts of the myocardium display intrinsic rhythmicity to some degree, but the contraction is synchronized through the dominance of one special area, the *sino-auricular node,* which imposes its rhythm on the rest of the heart. The sino-auricular node is therefore called the *pacemaker.* It is composed of specialized cardiac muscle, and the impulses arising from it spread through the syncitium of the whole mass of the myocardium. The intrinsic rhythmicity of the pace maker is modified by dual autonomic innervation, which can excite or inhibit it in a manner similar to that described above for visceral muscle. This is discussed in more detail in the section on circulation, p. 108.

All or None Law

The spreading of the action potential throughout the myocardium means that any impulse strong enough to elicit a contraction causes the maximum contraction possible under the metabolic conditions prevailing.

The strength of contraction depends on the metabolic well being of the muscle and also, as in skeletal muscle, on the initial degree of stretch before contraction. The heart muscle is stretched by the inrush of blood between contractions, and the extent of this stretch determines the force of the next contraction much as was discussed under "Length and Tension" for skeletal muscle and in Chapter 7 (circulation) under "Starling's Law of the Heart" (p. 127).

HOMEOSTASIS AND MUSCLE

If it were possible to ascribe a "purpose" to any biological function, it would be said that the purpose of homeostasis is to maintain a suitable environment for the efficiency of the external receptors and effectors. The economy of the animal seems to be devoted to the care of the muscles and nerves through which it is able to collect food and adapt to the external environment. Therefore, we can regard skeletal muscles and associated motor and sensory nerves as the prime beneficiaries of homeostasis. Cardiac and visceral muscles are, however, important agents in *maintaining* homeostasis. The factors to be controlled in the internal environment are discussed in Chapter 5, which introduces the topic of homeostasis. For muscle in particular, the requirements of the tissue for oxygen and glucose and for the renewal of waste products are self-evident. The incomplete combustion of glucose during very hard work or a diminished oxygen supply leads to the accumulation of lactic acid; this is a major cause of fatigue.

REFERENCES

1. Anon. Photos by Brunokisch, "Heart muscle," *Sci. American* **185** (2), 48, August 1951.

2. Hayashi, Tern and George A. W. Boehm, "Artificial muscle," *Sci. American* **187** (6), 18, December 1952.
3. Huxley, H. E., "The contraction of muscle," *Sci. American* **199** (5), 66, November 1958.
4. Katchalsky, A. and S. Lifson, "Muscle as a machine," *Sci. American* **190** (3), 72, March 1954.
5. Steinbach, H. B., "Animal electricity," *Sci. American* **182** (2), 40, February 1950.
6. Szent-Gyorgyi, A., "Muscle research," *Sci. American* **180** (6), 22, June 1949.

NERVE

18

The unit of nervous tissue is the nerve cell or *neuron*. An idealized neuron is shown in Fig. 18.1. It comprises a *cell body* which contains the nucleus, several processes called *dendrites* and another process, the *axon*. The distinction between dendrites and axons is not always clear-cut, but dendrites carry impulses *toward* the cell body and axons carry them away. An *axon* is also called a nerve fiber, and it is bundles of such fibers that make up what we recognize as *nerves* in dissection. Axons terminate in *end-tufts* which can connect with dendrites of other neurons or end-plates on muscles. The *myelin sheath* indicated in Fig. 18.5 is not always present, but it is characteristic of large, rapidly conducting fibers. The myelin sheath is of a fatty material and gives a glistening white appearance to nerves and to the "white matter" portions of the brain and spinal cord. The cell bodies are always unmyelinated and are found in the "gray matter."

THE NERVE IMPULSE

The action potential is the nerve impulse, and it is propagated along a nerve fiber by the mechanisms discussed previously. The passage of an action potential in nerve is quite invisible unless it

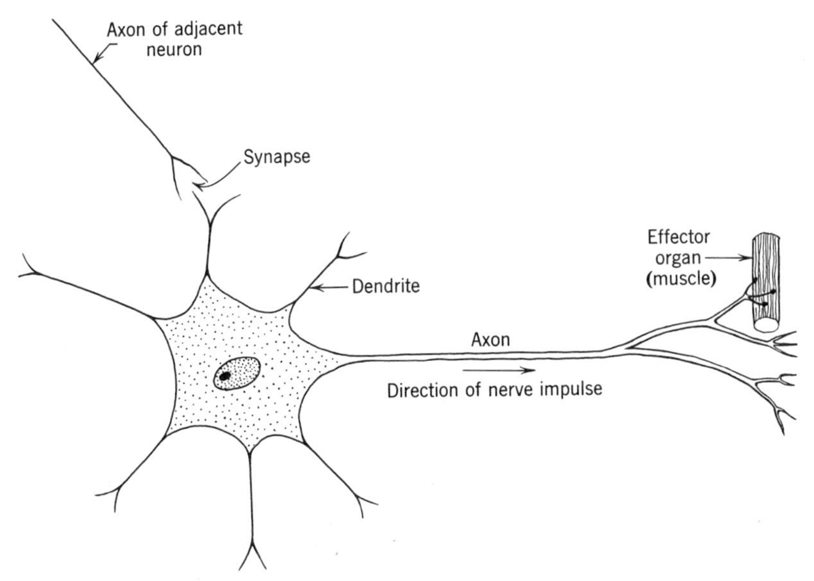

FIGURE 18.1. Idealized neuron.

stimulates a muscle to contract. Even this is of limited value in studying the properties of nerve itself, however, because it is impossible to distinguish visually between the properties of the nerve, the myoneural junction, and the muscle.

Methods of studying nerve impulses. A diagram of a nerve cell is shown in Fig. 18.2. If the axon (fiber) is stimulated by a small electric shock at *A*, an action potential will pass along the fiber and can be picked up by recording electrodes at *B* and *C*. The small amplitude of the action potential and its extreme rapidity mean that very sensitive equipment must be used to detect it.

The study of the electrical properties of nerve and muscle has received great impetus from modern developments in electronic instrumentation. Powerful amplifiers can magnify the minute changes of potential passing along the cell membrane and cause them to deflect the beam of electrons in a cathode ray oscilloscope. This instrument is so useful in neuro-physiological research that it merits a brief description here.

Oscilloscope

A stream of electrons from the emitting cathode at *A* in Fig. 18.3, is focused and accelerated by the anodes at *B*. The stream passes down the evacuated tube until it hits the back of the fluo-

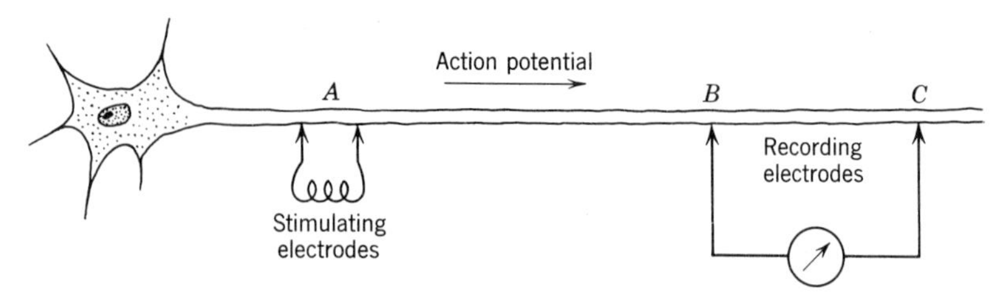

FIGURE 18.2. Diagram of stimulation of a nerve fiber and recording of the action potential.

rescent screen *C* at the other end of the tube. A device changes the potential on the *X* plates so that the stream of electrons moves quickly and regularly across the screen and then returns immediately to the beginning point. This makes a bright *base line* across the fluorescent screen. The *Y*-plates have equal potentials, but if they are connected to recording electrodes through an amplifier, the passage of an action potential past the recording electrodes briefly unbalances the potentials. A small change in potential on one or the other Y-plate causes an upward or downward deflection of the electron beam, which shows up as a *spike* on the base-line. In Fig. 18.3, the passage of an action potential past the two recording electrodes causes the electron beam to be deflected first in one direction and then in the other. The result (shown in Fig. 18.4) is a picture of a *diphasic* action potential. If the nerve is injured between the recording electrodes, a single deflection, or monophasic action potential, is recorded.

When the nerve is stimulated, some current is conducted along the surface of the fiber to the recording electrodes, giving a brief deflection or *shock artefact* at the instant the stimulus is applied.

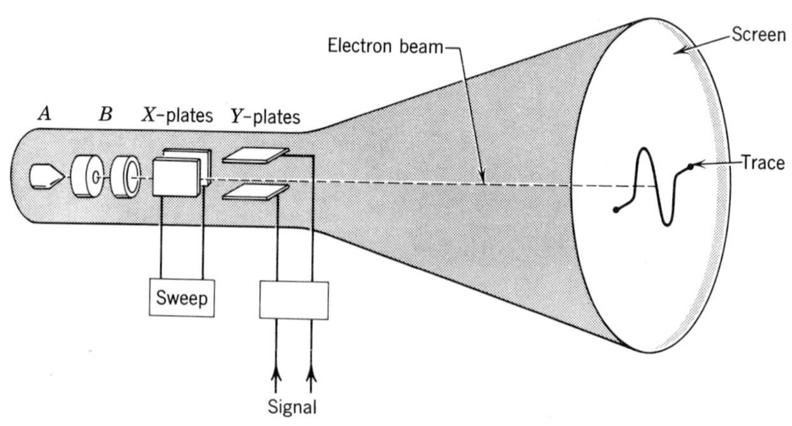

FIGURE 18.3. Oscilloscope.

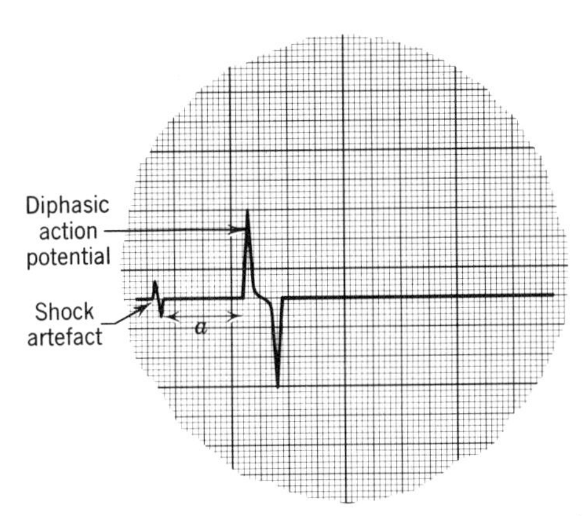

FIGURE 18.4. Oscillograph trace. Distance *a* is proportional to conduction velocity.

The action potential travels more slowly than the electric current and so deflects the beam a fraction of a second later (after it has moved further to the right). This gives a picture as shown in Fig. 18.3. If the speed of sweep of the electron beam is known, the time taken by the action potential to travel from the stimulating electrode to the recording electrode can be determined, and the speed of propagation can be calculated. A useful refinement is to synchronize the sweep of the oscilloscope with the firing of the stimulating electrode so that successive shock artefacts and action potentials appear on the screen in the same place. The persistence of the fluorescence of the screen then gives a steady picture rather than a brief flash.

Properties of the nerve fiber. From the preceding techniques, many of the basic properties of nerve (and muscle) have been worked out on isolated tissue preparations as described earlier.

As is to be expected, nerve shares many properties with muscle. Some features in common include:

i. An All or None response.

ii. The existence of liminal stimulus strengths in individual fibers and of maximal strength in bundles of several fibers.

iii. The strength-duration relationship also holds. A weak stimulus is more effective if it lasts longer.

iv. The production of a local excitatory state by subliminal stimuli and their summation to elicit an action potential if repeated sufficiently rapidly.

v. Nerve also shows a refractory period but it is very much shorter even than that of skeletal muscle (0.4 milliseconds in a nerve compared with 3.0 milliseconds in muscle).

Speed of the nerve impulse. The velocity of the propagation of the action potential may be studied with an isolated nerve fiber, with an isolated nerve, or with a nerve in an intact animal preparation. The speed in single nerve fibers is greater in those of larger diameter. Therefore, if a stimulus is applied to a mixed bundle of nerve fibers, a succession of action potentials is recorded from electrodes further along the bundle because of the different velocities in the fibers of various sizes. On this basis, fibers have been classified according to their conduction velocities (see Table 18.1).

TABLE 18.1 Classification of Nerve Fibers

Classification	Range of Diameters in μ	Where Found Characteristically	Conduction Velocity meters/sec
A (myelinated)	1–22	Somatic	5–12
B (myelinated)	3	Autonomic	3–15
C (nonmyelinated)	0.4–1.2	Somatic and autonomic	0.7–2.3

Myelinated fibers carry impulses faster than nonmyelinated fibers of the same diameter. It was at one time supposed that this was because the myelin had a similar function to the insulation on an electric cable. The ingenious theory of *saltatory conduction,* however, now has wide acceptance in explaining the faster conduction in myelinated fibers. The myelin sheath is not continuous but is interrupted at intervals of a few millimeters by the *nodes of Ranvier,* which are reminiscent of the joints in the stems of grasses (see Figure 18.5). The saltatory conduction theory proposes that when the membrane becomes depolarized at one node, local currents are set up in the axoplasm (intracellular fluid) and in the extracellular fluid. If the next node is sufficiently close, these currents may depolarize the exposed membrane at that point before an action potential can reach it by the "normal" means of propagation. The depolarization at that node would trigger the same event at the next one, and so on. The adjective "saltatory" means "leaping."

The *membrane potential* of a nerve may be determined approximately by crushing one end of an isolated preparation and measuring the potential between this portion and the intact surface elsewhere. As was discussed in the section on muscle, this gives a fair approximation of the membrane potential. A relatively recent development is the use of microelectrodes which can be placed *inside* intact nerve cells for the direct measurement of the poten-

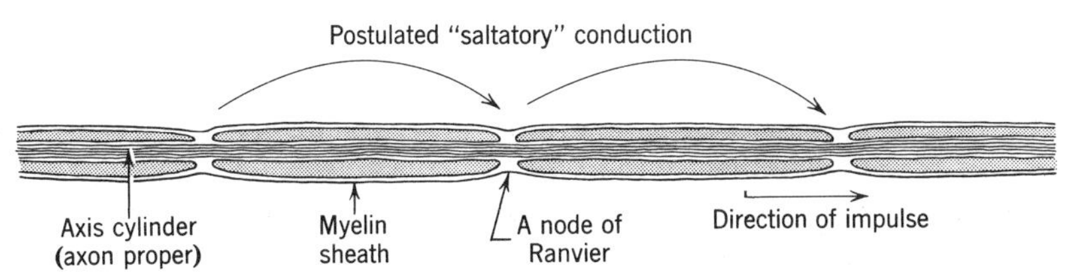

FIGURE 18.5. Diagram of myelinated nerve fiber.

tial. The squid has some axons of far larger diameter than any found in mammals and these *giant axons* have been particularly useful for this kind of work. It has been found that the resting potential in the squid nerve is of the order of -70 millivolts, which is close to that predicted theoretically from the electrolyte composition of the intracellular and extracellular fluids. On depolarization, the resting potential increases to zero and overshoots to about $+15$ millivolts.

Synapses. The junction between the end-tufts of an axon and a dendrite of an adjacent neuron is a *synapse.* Most histologists agree that there is no continuity of the cytoplasm of the two cells and that they are at least separated by their membranes. The mechanism of transmitting a nerve impulse across this "gap" is a topic of intensive research. At present, several important pieces of evidence suggest a neurohumoral transmission like that by acetylcholine in the myoneural junction. For example, careful measurements have shown that there is a slight delay in transmission at the synapse. Furthermore, the repeated stimulation at very high frequencies of two nerve fibers with a connecting synapse eventually causes a fall-off in transmission, evidently due to fatigue. If each fiber's conduction is then tested separately, it is found that they are still functioning well. This is again suggestive of *synaptic* fatigue, analogous to the muscle *end-plate fatigue* due to depletion of acetycholine. It is unlikely that acetylcholine itself is the synaptic transmitter substance, however, since experiments show that it takes unphysiologically high concentrations of it to evoke discharges from central nervous system neurons.

Homeostasis and nerve. As discussed in Chapter 5, the resting potential is affected by the composition of the extracellular fluid; the normal functioning of nerves and muscles depends on homeostasis of the internal environment. In particular, an increase in the potassium ion concentration or a decrease in the calcium ion lowers the resting potential, making the membrane more easily depolarized. That is to say, less stimulus is required to produce an action potential, and we say the cell is *more excitable.* Reciprocal changes decrease the excitability. In the intact body, a general increase in excitability leads to exaggeration of normal reflexes, uncontrolled movements and finally to convulsions. Conversely, decreased excitability is first seen in diminished reflexes, then weakness, and finally complete paralysis.

The effects of decreased oxygen tension in the interstitial fluid are even more dramatic, especially on the cells of the brain. Unconsciousness ensues very rapidly when the blood supply (and therefore the oxygen supply) to the brain is diminished. We recognize this state as *syncope,* or fainting.

Carbon dioxide accumulation has less sudden effects but can be equally serious (see chapter on acid-base balance); a modest rise in the carbon dioxide tension reduces the excitability. A serious fall in the circulating glucose concentration leads to hyperexcitability and convulsions. This is actually employed therapeutically in the *insulin shock treatment* for some mental illnesses.

These few examples show the dependence of nervous tissue on the constancy of the internal environment. We must also recognize that nerves themselves control and regulate many of the homeostatic mechanisms, and this is especially true of the autonomic nervous system. For example, the rate of exchange between plasma and interstitial fluid is largely determined by the blood pressure, and this is controlled by the heart rate and vasomotor tone. Both of these factors are mediated by the autonomic nervous system.

It should be realized that the action potential is the unit of *all* nervous activity, including the functioning of the highest centers of the brain. All of man's hopes, fears, and memories are made up of fantastically complicated patterns of action potentials weaving through the millions of nerve cells of his brain. Every great work of art originated in the delicate interplay of minute potential changes in the cells of a man's brain.

REFERENCES

1. Brazier, Mary A. B., *The electrical activity of the nervous system,* Macmillan, New York, 1958.
2. Katz, Bernhard, "The nerve impulse," *Sci. American* **187** (5), 55, November 1952.
3. Keynes, Richard D., "The nerve impulse and the squid," *Sci. American* **199** (6), 83, December 1958.
4. Shurrager, P. S., "'Spinal' cats can walk," *Sci. American* **183** (5), 20, November 1950.

ORGANIZATION OF THE NERVOUS SYSTEM

19

The nervous system is usually considered in two main categories: the central nervous system (CNS), which includes the brain and spinal cord, and the peripheral nerves. Each of these categories contains both *somatic* and *visceral* nerves. The somatic nerves supply the skeletal musculature and are therefore also referred to as *voluntary* nerves. The visceral nerves supply the internal organs, blood vessels, and some other structures; they may be called *involuntary* nerves. It will be recalled that some nerve fibers convey information from receptors *toward* the CNS, and that these are the *afferent* nerve fibers. The *efferent* fibers carry orders from the CNS to the effectors. Therefore, there are afferent and efferent fibers in both the somatic and visceral nerves.

The visceral efferent nerves are known collectively as the *autonomic nervous system,* and this name refers to the involuntary nature of the control exerted by them on the internal organs. The *visceral afferent* nerves which convey information about the state of the internal organs to the "visceral centers" of the CNS are *not* considered as part of the autonomic nervous system. This arbitrary exclusion is due to the historic development of neurology and it is sometimes a source of confusion.

The central nervous system may be compared with the chain of command in an army. There are many levels of command to which the afferent nerves bring reports and from which the efferent nerves carry orders. Routine, everyday matters are dealt with at lower levels (in the spinal cord), frequently with a message sent higher up to indicate that action has been taken. More serious occurrences are referred to higher levels for decision and sometimes even the very highest level is called in to impose a general policy on the lower levels that normally "go by the book." The instructions from higher levels may actually *contradict* the usual procedure for dealing with a situation. Plans originating at the higher levels may be carried out by orders sent directly to the effectors, by general orders to lower levels which reformulate them in detail, or more commonly, by a mixture of both.

THE CENTRAL NERVOUS SYSTEM

Embryologically and phylogenetically, the central nervous system develops from a *neural tube*. The anterior end of the tube is enlarged and greatly modified to form the brain, and this reaches its greatest complexity in man. The original tube structure is still apparent in the intercommunicating *ventricles* of the brain and the central *canal* in the cord. For simplicity, we will first consider the anatomy and function of the spinal cord and the nerves arising from it.

The spinal cord. It is generally accepted that the remote ancestors of the vertebrates were segmented, much as a modern earthworm is segmented. In the earthworm, each segment contains nerves, blood vessels, muscles, etc. in a pattern that is approximately repeated in successive segments. This sort of arrangement is clearly seen in the embryonic stages of vertebrate development, and some manifestations of it persist even in the adult. Chief among these is the periodicity of the *vertebral column* (or back bone) and the spinal cord which lies within it. At each of the small spaces between the vertebrae, *spinal nerves* arise from each side of the cord and distribute themselves in the tissues in a way that can be partly related to some archaic segmentation of the muscles, etc. There are about as many pairs of spinal nerves as there are vertebrae but at the level of the first lumbar vertebra, the cord divides into all the spinal nerves which emerge from the vertebral column below that point and continue as separate strands.

This structure is aptly called the *cauda equinae,* which means "horse's tail."

The spinal nerves arise at each level as two roots on each side, a *dorsal* and a *ventral root,* (see Fig. 19.1). In man, the dorsal and ventral roots are sometimes called *anterior* and *posterior* respectively, because of our upright posture. The *dorsal roots* contain the afferent or sensory fibers which bring information in to the CNS from the receptors. The fibers are either myelinated or not. Their cell bodies are located outside the cord proper (but still within the vertebral canal) in a swelling called the dorsal root *ganglion.* The *ventral roots* carry the efferent, *motor* fibers, and their cell bodies are inside the spinal cord. The motor fibers are heavily myelinated.

On reaching the foramen through which the spinal nerve emerges from the vertebral column, the dorsal and ventral roots join to form one structure. After emergence, this structure divides again into the *anterior* and *posterior rami* which go to deep and superficial structures respectively. As indicated in Fig. 19.1, the rami each contain fibers from *both* roots, sensory and motor. This is an important point.

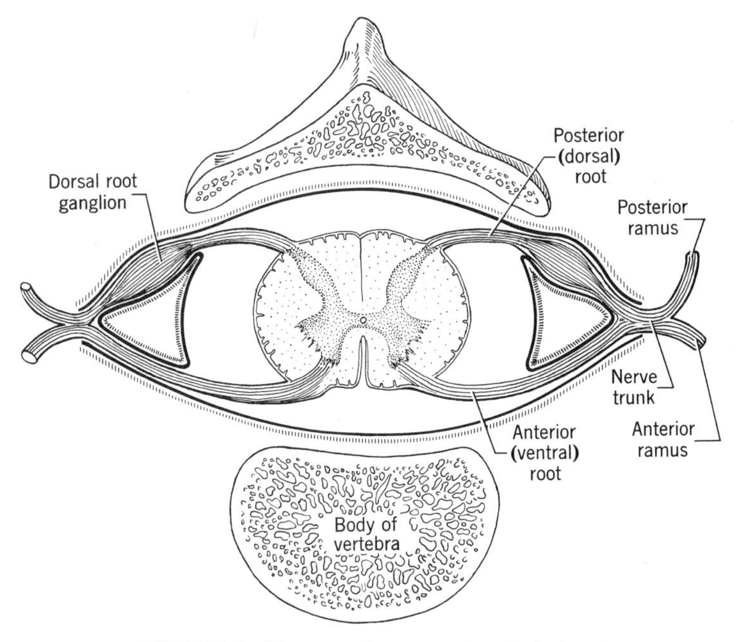

FIGURE 19.1. Diagrammatic cross section of spinal cord.

In cross section, the spinal cord itself is a tube in which the great thickness of the walls all but obscures the central canal. The spinal cord is made up of white matter with a large, H-shaped core of gray matter. The white matter comprises the millions of myelinated nerve fibers conveying information up and down the cord, and the gray matter is largely made of the cell bodies of neurons arising at that and other levels. The gray matter also contains fibers which lie entirely within the CNS, making interconnections between afferent and efferent peripheral nerves. Therefore, in addition to the motor and sensory nerve fibers, we must also consider the third category of *associative fibers.* These are also found in tremendous numbers in the complex interconnections of the brain.

Spinal reflexes. The simplest example of nervous function is a *spinal reflex.* If an animal's limb is pinched or otherwise stimulated, the limb is withdrawn. The basic events are diagramatized in Fig. 19.2. The pain receptors at *A* are stimulated, an impulse travels over the sensory nerve to the spinal cord where it synapses at *B* with an *internuncial* neuron. This in turn synapses with the motor fiber at *C,* and an impulse travels out to stimulate the muscle at *D* which contracts and withdraws the limb. In a very few cases, the sensory fiber synapses directly with the motor fiber, but usually at least one internuncial neuron is interposed. Other associative fibers may convey the impulse to other motor fibers at different levels in the spinal cord (and also to the brain) to bear the information that the noxious stimulus has been applied. This

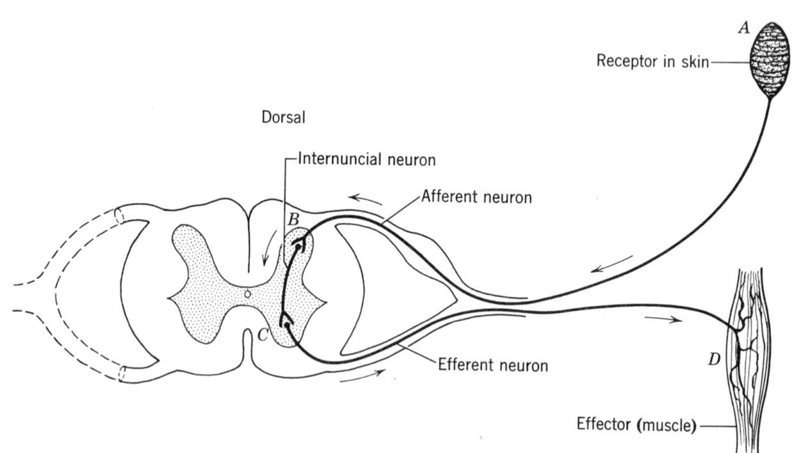

FIGURE 19.2. Diagram illustrating a simple reflex arc.

illustrates the immediate action at the local level with a message despatched to higher centers for information.

Less dramatic than the withdrawal reflex are the multitude of reflexes which maintain the *tone* of the skeletal muscles. The antigravity extensor muscles, to maintain posture, must be in a constant, or *tonic,* state of contraction. The quadriceps femoris of the thigh for example, keeps the knee straight while standing. If its tonic contraction diminishes, the knee starts to bend, but as it does so, the quadriceps is stretched and this stimulates *stretch receptors* located in the muscle and its tendons. The receptors send impulses to the spinal cord evoking a reflex contraction of the quadriceps which restores the original degree of extension. Of course, our upright posture is not maintained by repeatedly correcting a sag at the knees, but rather the stretch receptors send continual volleys of impulses which maintain the tonic contraction.

The nature of this reflex can be seen by applying a sudden stimulus to the stretch receptors. If a seated subject crosses one knee over the other and is tapped below the kneecap, there is a sudden extension of the leg, or *knee jerk.* The tapping of the knee momentarily stretches the tendon at that point and this causes an immediate discharge by the stretch receptors, just as if the knee had been suddenly flexed. This brief volley of impulses to the cord causes a reflex contraction of the quadriceps, as if it were indeed correcting a sudden loss of postural tone. The integrity of the knee jerk reflex is a valuable diagnostic sign because it may be depressed or exaggerated in some neurological diseases.

It should be noted that we have described extremely simple spinal reflexes. Modern studies have shown that the functions controlled mainly at the spinal level are of even greater complexity than was previously believed. For example, the basic contractions and relaxations in walking motion can be seen by appropriate stimulation of an animal with the brain removed. Not only do the spinal reflexes stimulate muscles in the moving limb, but they simultaneously *inhibit* those in the limb on the other side. Of course, the balance and coordination required in locomotion depend on the brain, and without a brain an animal will display only very crude walking movements. All of the reflex activities of the cord are greatly modified by impulses from the brain, and this is especially true in man.

Motor fibers. The motor fibers which actually run out to the muscles have their cell bodies at various levels in the gray matter of the spinal cord. They can be stimulated by spinal reflex mechanisms like those just described, or by impulses coming down other neurons from the brain to initiate true "voluntary" movements. The neurons from the brain are also "motor" neurons, but they must synapse with those arising in the cord before the impulse can actually reach the effector muscles. On this basis therefore, we distinguish between *upper* and *lower* motor neurons. The lower motor neurons arise from the ventral horn of the gray matter and are therefore frequently referred to as *ventral horn cells.*

There are so many pathways by which the lower motor neurons can be stimulated that the term "upper motor neuron" is perhaps too broad to be useful. There are, however, two main groups of fibers which descend the cord to synapse with the lower motor neurons; the *pyramidal* and *extrapyramidal* tracts.

The *pyramidal tract* fibers arise in the *motor area* of the cerebral cortex, or the "voluntary" part of the brain; they descend the cord to synapse directly with the lower motor neurons. They are therefore also known as *corticospinal tracts.* As they descend through the brain, the large bundles of fibers are visible as pyramid-shaped projections on the under surface of the medulla; this is the source of their other name. At this level, the fibers arising from each side of the brain cross over and descend in opposite sides of the spinal cord. These nerve fibers convey impulses voluntarily initiated in the brain and delivered to a specific muscle for its contraction on the other side of the body.

The *extrapyramidal tracts* comprise all the other motor or *descending* tracts. They arise from other sections of the brain, particularly from the vestibular nuclei of the medulla (concerned with balance and posture) and from the reticular formation of the hind brain. Both the vestibulospinal and reticulospinal tracts convey impulses to the lower motor neurons via many internuncial neurons rather than directly, as in the pyramidal tracts. The signals from the brain follow a maze of pathways in a generally downward direction, and in consequence, the impulses not only take longer, but tend to spread out more to other motor neurons. Such impulses exert facilitatory or inhibitory effects in the cord, so that they modify and supplement the effects of pyramidal signals and of the spinal reflexes. In particular, they modify the spinal tonus discussed previously.

It will be noted that whatever the source or pathway of the original stimulating impulse, the actual contraction of a muscle can result only from an impulse from the cord over the lower motor neuron. This has therefore been aptly called the *final common pathway.*

Sensory fibers. Impulses from sensory receptors enter the spinal cord via fibers in the dorsal roots. Many of these synapse with internuncial neurons in the gray matter of the cord, which in turn connect with motor neurons at various levels, thereby completing the spinal reflex arcs. Other secondary neurons receiving sensory activity in the cord ascend to the brain. In addition, there are important sensory fibers which do not synapse on entering the cord, but ascend all the way to the medulla, the lowest part of the brain. It will be appreciated that such fibers from receptors in the toes, for example, are very long, and that there are no motor fibers which compare with them for directness of communication.

The ascending, or sensory fibers can be discussed in terms of the *tracts* in which they run, in a similar manner to the descending fibers.

Spinothalamic tracts contain fibers running from the cord to the two *thalami,* which are important sensory receiving centers in the brain. Fibers from pain and temperature receptors synapse in the cord with the spinothalamic fibers which then cross over (decussate) and ascend to the brain.

Spinocerebellar fibers are similar to the spinothalamic fibers, but they receive impulses from proprioceptive fibers. They ascend to the medulla and thence to the cerebellum, which as we shall see, is largely concerned with the *coordination* of muscular activity. In this connection we should note that other branches of the incoming proprioceptive fibers also connect with motor fibers in the cord (via internuncial neurons); they can elicit spinal reflec corrections to changes in posture, etc. This is therefore a good example of an incoming message which is acted on at a low level of command but which is also relayed to a higher center, so that other activities may be coordinated with the action decided upon at the lower level.

The *dorsal columns* contain the long fibers that enter the cord and ascend as far as the medulla without any intermediate relay. In the medullary nuclei, the fibers synapse with others which ascend higher in the brain, cross the midline and terminate in the thalami. Again, information from proprioceptors is relayed to a higher center, and from the thalami, the information is relayed even higher to the conscious areas of the cerebral cortex.

By experiments on animals and by observing the consequences of injury to the spinal cord in man, the locations of the various ascending and descending tracts in the cord are quite well known. Diagrams showing their arrangement may be found in any text of neuroanatomy.

Cranial nerves. In addition to the spinal nerves arising at each segment of the cord, there are twelve pairs of nerves arising from the brain itself. This is not surprising, as the brain is quite clearly an elaboration of the anterior end of the cord as it existed in the ancestors of the vertebrates. It is traditional for students to memorize the names and functions of these nerves, which are listed in Table 19.1.

TABLE 19.1 The Cranial Nerves

Nerve Pair	Function
1. Olfactory	Smell.
2. Optic	Vision.
3. Occulomotor	Eye movements.
4. Trochlear	Eye movements and sensations.
5. Trigeminal	Sensory from face and head. Motor to jaw, etc.
6. Abducens	Eye movements and sensations.
7. Facial	Sensory and motor to face and scalp.
8. Acoustic	Cochlear division—hearing. Vestibular division—balance, etc.
9. Glossopharyngeal	Sensory to pharynx. Taste for back of tongue. Sensory from carotid body, etc.
10. Vagus	Extensive sensory and motor, visceral and somatic throughout thorax and abdomen.
11. Accessory	Sensory and motor to upper thorax.
12. Hypoglossal	Sensory and motor to tongue.

The functions of the cranial nerves are largely what would be expected by analogy with spinal nerves at various levels, with the exception of the tenth cranial nerve, the *vagus*. This large nerve carries many of the fibers of the craniosacral or *parasympathetic* division of the autonomic nervous system, and it has extensive ramifications. Its very name means "wanderer," and it is the Latin root of the English words "vagabond" or "vagrant." The autonomic fibers in the vagus regulate the heart beat, the contractions of the smooth muscle of the gut, and many other widely separate effectors. It should be noted however, that most of the fibers in the vagus nerve are *afferent,* bringing impulses from visceral receptors to the autonomic centers of the brain stem.

BRAIN

20

In our analogy of the chain of command in an army, the brain itself comprises many levels of command. In general, the lower levels of command deal with automatic functions such as heart rate, respiration, etc.. Higher levels deal with voluntary motion, still higher levels are concerned with complicated, skilled activities, and at the very highest levels we find the processes of reasoning and abstract thought. In an exactly parallel fashion, the incoming sensory impulses that deal with the state of the viscera are dealt with at the lower levels, and the more and more complex "information reports" are referred to higher and higher centers.

There is good evidence for believing that the lower levels of command in the brain are the most primitive or "older" parts of the brain, and that these portions are similar in most of the mammals. The higher mammals show greater size and complexity in the "newer" anterior parts of the brain, culminating in the great size of the forebrain in man. The more primitive parts of the brain still fulfill their original functions in man but they are subject to various degrees of control and *integration* by the higher centers. Therefore, a useful theme in studying the functional anatomy of the brain is the "*Level of Organization.*" This concept should not be extended too far in neurology, but it is adequate at the level of this text.

As is true of most organs, the function of the brain cannot be discussed without constant reference to its structure. Brain anatomy is complicated and can only be properly understood by examination of dissected specimens or at least of good three-dimensional models. The following account, therefore, is only a bare introduction and should be supplemented by laboratory work for a proper understanding of the subject.

GENERAL DESCRIPTION

The human brain looks remarkably like half a walnut, see Fig. 20.1. This appearance is due to the *cerebral hemispheres* (or cerebrum) which are the large and complicated anterior enlargements referred to previously. They tend to hide the more primitive (but equally vital) structures of the *brain stem.* In the most primitive mammals, the cerebral hemispheres are insignificantly small, and even in animals as intelligent as the dog, they are but a fraction of their relative size in man. There is no doubt that the cerebrum is the seat of intelligence.

The parts of the cerebrum are recognized by the *fissures* which separate the principal *lobes,* as shown in Fig. 20.1. In man the surface of the cerebrum is highly convoluted. This is of great interest since it means that the outer part, or *cortex,* has a large surface area. The cortex is the gray matter containing the cell bodies of neurons concerned with intelligence and many of the other higher activities. Therefore, although it is only 3–5 millimeters thick, the mass of the cortex is large. The inner bulk of the cerebral hemispheres is white matter, comprising the myelinated axons of the cortical cells and others. It also contains large masses of gray matter or *nuclei,* which will be discussed.

Below and behind the cerebral hemispheres lies a smaller, but similarly convoluted pair of hemispheres, the *cerebellum* (Latin—little brain).

If we examine a brain from below, as in Fig. 20.2, we can better see the brain stem, which consists of the more primitive parts out of which the "newer" parts (cerebrum and cerebellum) have grown. The brain may be likened to a flower in which the spinal cord is the stalk and the brain stem the calyx. This is particularly apt because in a growing plant the stalk is present first, then the calyx structure appears, and from this the corolla springs later.

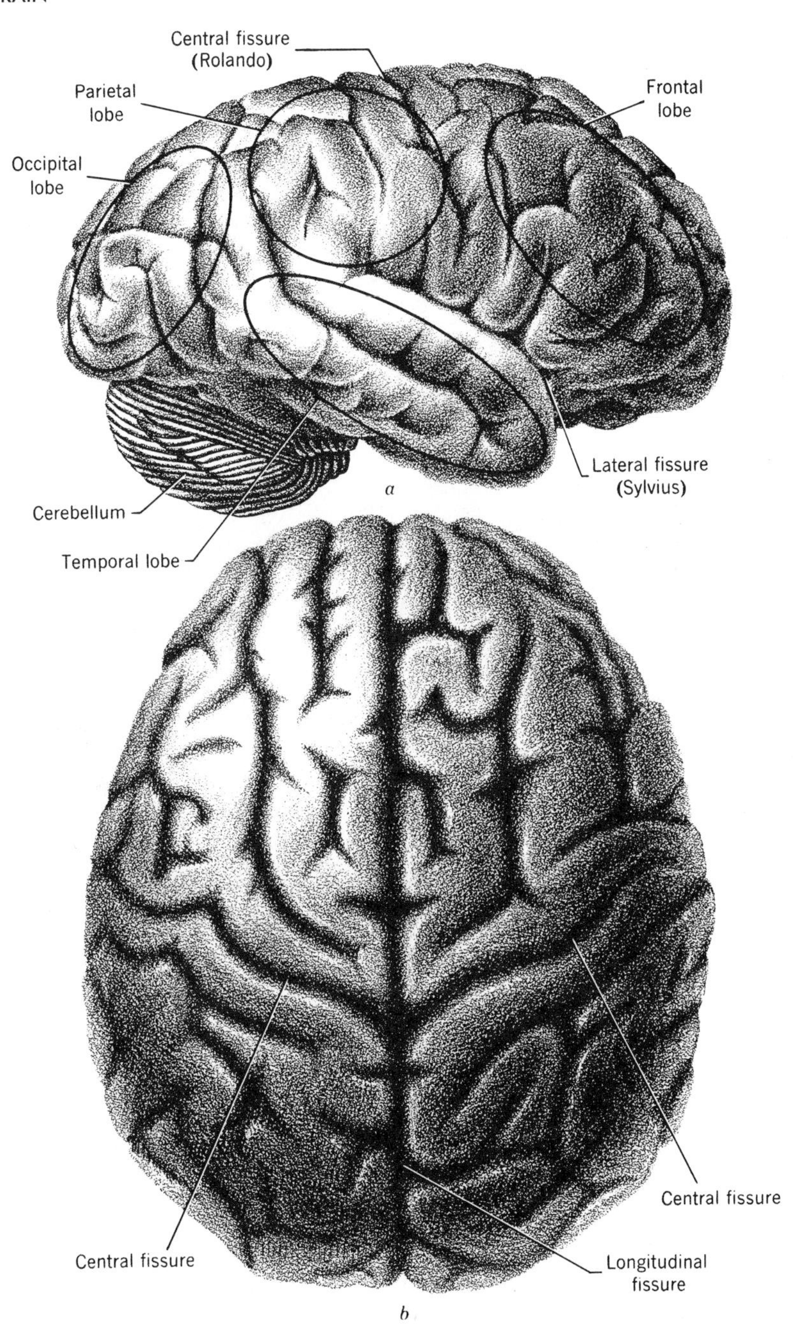

FIGURE 20.1. The human brain; (*a*) from the right, (*b*) from above.

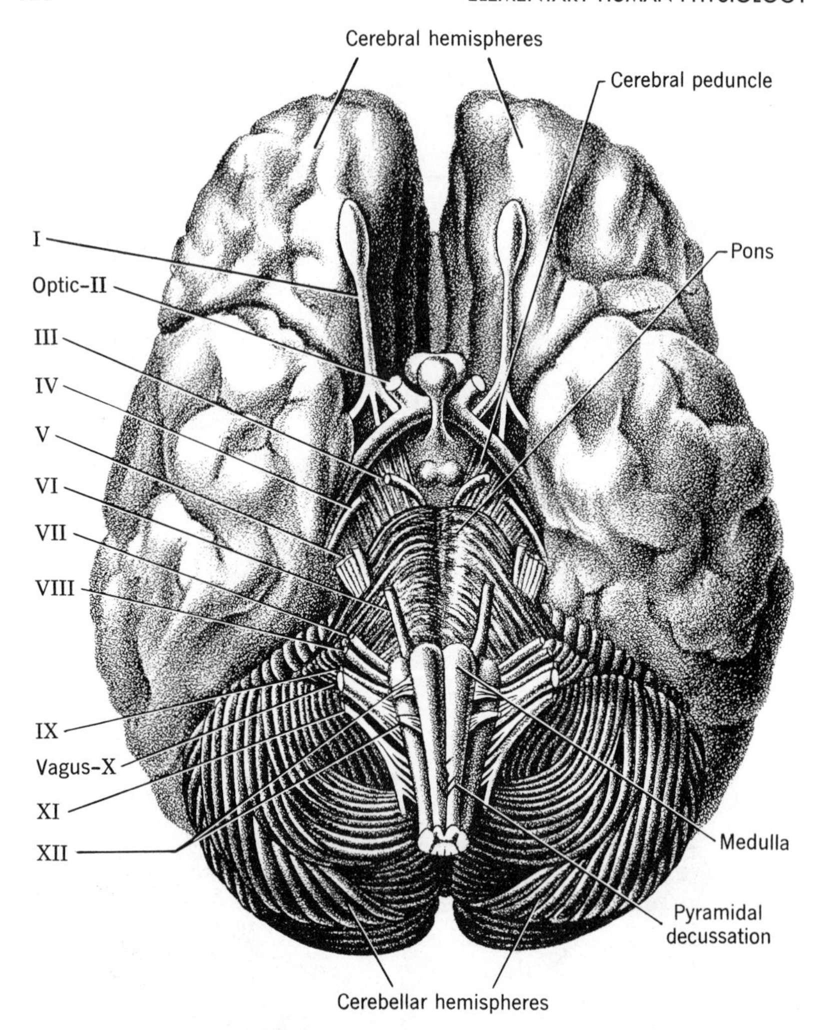

FIGURE 20.2. The human brain from below.

In the view from below, the twelve pairs of cranial nerves can be seen arising from the brain stem; the optic nerves are especially conspicuous since they join and appear to cross over. This structure is the *optic chiasma.* The olfactory nerves run still more anteriorly.

The components of the brain stem are best seen in a midsagittal section as in Fig. 20.3, where they are not obscured by the cerebrum and cerebellum. Joining the cerebral hemispheres to the lower structures are the *cerebral peduncles,* which are like twin stalks.

They are in fact, largely made up of fibers originating in the cerebrum and running into the spinal cord. In front of these is a forward-bulging structure, the *pons*. Below the pons is the *medulla oblongata* from which most of the cranial nerves arise. On the inferior surface of the medulla are two elevations, *the pyramids,* composed of the pyramidal tracts arising in the cerebral cortex and running clear down the cord to synapse with lower motor neurons. At the level of the medulla, these fibers cross over so that fibers arising in the left side of the cortex synapse with neurons serving muscles on the right side of the body.

The originally tubular nature of the CNS is reflected in the system of ventricles in the brain. The *first and second,* or *lateral,* ventricles are in each of the cerebral hemispheres and each connects through a small hole (foramen of Monroe) on its medial surface with the *third* ventricle, which lies in the middle. The third ventricle communicates by the *cerebral aqueduct* through the upper brain stem with the *fourth* ventricle, which lies between the cerebellum

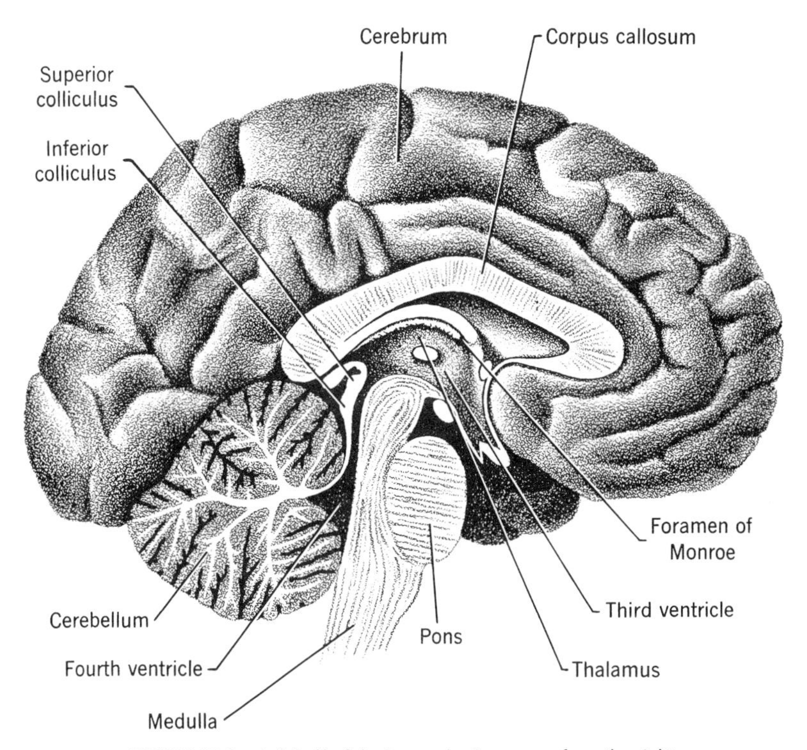

FIGURE 20.3. Left half of the human brain as seen from the right.

and medulla. The central canal runs from the fourth ventricle into the spinal cord. Several important parts of the "old" brain make up the walls of the third and fourth ventricles.

CLASSIFICATION

The classification of the parts of the brain is based on its embryological development, which probably reflects its phylogenetic development. There are three main divisions: the forebrain, midbrain, and hindbrain, and these are subdivided as follows:

Forebrain

1. Cerebral hemispheres.
   (*a*) Cortex
   (*b*) Basal ganglia
   (*c*) Rhinencephalon (Olfactory portion)
2. Diencephalon. This comprises most of the structures making the walls of the third ventricle.
   (*a*) Thalamus and epithalamus
   (*b*) Hypothalamus

N. B. As these structures lie on each side of the third ventricle, they are therefore paired, that is, there are two thalami.

Midbrain

Comprises structures surrounding the cerebral aqueduct
   (*a*) Ventrally–cerebral peduncles.
   (*b*) Dorsally–corpora quadrigemina

Hindbrain

Comprises structures surrounding the fourth ventricle.
1. Metencephalon.
   (*a*) Cerebellum
   (*b*) Pons
2. Myelencephalon–medulla oblongata.

FUNCTION AT VARIOUS BRAIN LEVELS

Hindbrain. The simplest aspect of central nervous system function is seen in the spinal reflexes, although these are not as simple as was once thought. The next degree of complexity may be taken as those reflexes which are mediated by the centers in the *medulla oblongata.* Of all the brain, this structure most clearly displays its origin as an enlarged part of the cord. Visceral afferent fibers bring information from pressoreceptors and chemoreceptors to the cardiac and respiratory centers in the medulla; from these, efferent impulses are sent out to mediate the heart rate and the muscles of respiration. In the control of respiration, the medullary center is particularly sensitive to modification from higher levels of command; for example, from the pneumotaxic center in the pons, or from conscious interference arising from the cerebral cortex (as in breath-holding). The cardio-inhibitor and accelerator centers in the medulla are similarly subject to control by impulses from the anterior and posterior hypothalamus and indeed from the cortex itself. Damage to the medulla is always serious because of its effects on these *vital centers.*

The medulla also contains the *vestibular nuclei.* Sensory impulses from the labyrinth structures within the skull travel over the eighth nerve conveying information about the attitude and movement of the head in space. Other impulses from proprioceptors elsewhere in the body travel to the vestibular nuclei or to the cerebellum, which relays the information. These incoming signals evoke motor impulses to the muscles responsible for maintaining posture and they strengthen the spinal tonus reflexes. The posture maintained through the vestibular reflexes is quite crude and it is greatly modified by impulses arising from higher centers in the cerebrum. An animal in which the connection between the cerebrum and the lower brain stem has been cut shows a characteristic rigidity which is an exaggerated caricature of its normal posture. This state is called *decerebrate rigidity.* If the brain is transected below the vestibular nuclei, the rigidity disappears and the animal is then a *spinal animal.*

In addition to these relatively discrete centers, or nuclei, the upper part of the medulla contains a more diffuse *reticular structure* which also extends up into the pons. This *bulboreticular* area sends impulses which facilitate (and/or inhibit) functions in the spinal cord. It has more recently been appreciated that the reticular formation also sends facilitatory impulses upwards to higher centers, to the thalamus and the cortex for example. It is considered likely that the "alertness" thus induced in the thalamus is characteristic of wakefulness, and that sleep follows decreased facilitation of the higher centers.

The *pons* contains part of the bulboreticular formation, the pneumotaxic center (concerned with respiration, see Chapter 8), and other scattered nuclei. The bulk of the pons however comprises the vast number of fibers passing through it.

The *cerebellum* is a structure concerned with the smooth coordination of muscular activity and we have seen that the spinocerebellar tracts bring information about the tone and position of muscles. The cerebellum similarly receives information from the cerebral cortex relating to the initiation of voluntary muscle movements. The efferent fibers from the cerebellum pass impulses to the spinal cord via the red nuclei, the reticular formation, and the vestibular nuclei. It is believed that motion sickness is partly due to a disorganization of the part of the cerebellum that coordinates posture and balance through the vestibular nuclei of the medulla. Damage to the cerebellum leads to jerkiness of voluntary movement and poor balance. Other centers can coordinate voluntary activity, however, and patients have "learned" to manage in spite of cerebellar lesions. Futhermore, there are several reported cases of individuals in whom the cerebellum was congenitally absent and where this was only discovered at autopsy.

Midbrain. The *cerebral peduncles* are bundles of myelinated fibers joining the cerebral cortex with other levels of the brain and cord.

The *corpora quadrigemina* are of gray matter and are important sensory receiving centers. Impulses from the eyes and cochleae (of the ears) pass first to the thalamus and then to appropriate cortical areas. The thalamus also relays part of this input to the corpora quadrigemina which elicits efferent discharges reflexly moving the eyeballs and head in response to visual or auditory stimuli. These centers are therefore important in the maintenance of posture and for the reflexes which right the body when posture is grossly disturbed.

Forebrain. The *hypothalamus* is the principal center integrating the autonomic functions. The activities of the craniosacral, or parasympathetic, division of the autonomic nervous system are mediated by the anterior portion, and those of the thoracolumbar, or sympathetic division by the posterior portion. The extensive afferent input to the hypothalamus is from all parts of the brain and periphery. The efferent impulses go out over the cranial and spinal nerves that contain autonomic fibers (see Chapter 21). The maintenance of homeostasis is absolutely dependent on the integrity of the hypothalamus since all homeostatic mechanisms are under some degree of autonomic control. Lesions of discrete portions of the hypothalamus lead to loss of temperature regulation, to tremendous changes in appetite, or to other bizarre disturbances.

The *thalamus* is probably the most important sensory receiving center of all. Most of the impulses conveying information about "sensations" in the body are sent directly or by relay to the thalamus. We might say that the thalamus is the seat of crude "awareness" in the brain, from which the sensory impulses are sorted out and relayed, or *projected,* to other parts of the brain. Those that are projected to the cerebral cortex give rise to conscious sensations.

The hypothalamus and thalamus are clearly the "oldest" part of the forebrain and are able to interpret the external and internal environments adequately for existence at a primitive level. Consciousness, as we know it in man, seems to develop with increased size of the cerebral hemispheres and particularly with the increased complexity of the cerebral cortex.

There has been great interest in a layer of tissue around the thalami and related structures which is functionally similar to the cerebral cortex and may indeed be thought of as the highest level of the "old brain." This part of the brain is called the *limbic system.* Experiments have shown that the sensations received by the thalamus (and hypothalamus) are projected to the limbic region, where they are "perceived" in a manner presumably analogous to, but cruder than, the perception by the cerebral cortex. The limbic region in several of the lower mammals is of comparable size to that in man, whereas the cerebral hemispheres and their cortex are insignificantly small even when the differences of brain size are allowed for.

If the limbic region is indeed responsible for the primary interpretation of sensation, we may reasonably assume that the lower mammals perceive sensations and emotions more crudely than man. On the other hand, it would follow that man still has this seat of primitive interpretation of sensation in addition to the great refinement due to the projection to the cerebral cortex. There is good evidence that the very "basic" sensations of hunger, thirst, fear, rage, and sexual drive are developed in the limbic region, although they reach the level of conscious sensation, probably in a modified form, in the cerebral cortex. This is of great psychiatric interest because it would suggest that the raw material of our emotions originates in our "old brain" and that our relatively reasonable and "human" behavior depends on the ascendancy of the cerebral cortex. These matters are under intensive investigation because this is one of the few areas in which psychiatric observations may be correlated with observable physiological phenomena.

CEREBRAL HEMISPHERES

Basal ganglia. These are relay stations for impulses rising in the cerebral cortex that control the "associated" movements of voluntary muscular activity. The classic example is the arm swinging which accompanies normal walking. In the basal ganglia, fibers descending from the premotor areas of the cortex synapse with neurons that descend the cord to connect with the final common pathways. The courses of these impulses are therefore included among the *extrapyramidal tracts,* that is, they do not connect directly with the lower motor neurons.

The basal ganglia themselves are four large masses, two on each side of the thalami. The larger pair are the *caudate nuclei,* the smaller are the *lenticular nuclei,* and the whole structure is often referred to as the *corpus striatum.* Between each lenticular nucleus and the thalamus is a gap through which many fibers pass from the cortex to lower portions of the brain; this is called the *internal capsule.*

Rhinencephalon. This is probably the "oldest" part of the cerebral hemispheres and part of it is concerned with the sense of smell. The afferent impulses from the receptors in the nose are relayed to all parts of the brain, not only to the cortex for conscious interpretation, but to the visceral centers through which many reflex responses are elicited. Although the sense of smell is of minor importance in civilized life, lower animals depend on it to a great extent for interpreting environmental changes. There is good evidence that fishes for example, have only rudimentary vision and hearing but they can detect infinitesimal changes in the chemical content of the water in which they swim. Even though we do not depend on our sense of smell for survival, it is a common experience that an odor can be more evocative of half-forgotten memories than any other sensory stimulus. This would suggest that olfaction is more important in our appraisal of the environment than we might normally think.

Corpus callosum. The tough white matter which joins the two cerebral hemispheres at the depths of the longitudinal fissure is the corpus callosum. This structure is composed of the myriad fibers connecting the two hemispheres for the integration of their activities.

Cerebral cortex. The very fact that it has been necessary to refer to the cortex in discussing almost all the other parts of the brain is an indication of its overwhelming importance. It comprises all of the outer 3–5 millimeters of the cerebral hemispheres, including the parts that dip down into the fissures.

Experiments with animals and observations on human subjects with specific lesions have made it possible to determine the functions of many areas of the cortex. Over the years, neurophysiologists have been able to relate motor and sensory functions of various parts of the body with discrete areas of the cortex with remarkable accuracy. In the sensory areas, the neurons are stimulated by impulses from the thalamus, etc., and by means of this nervous activity we interpret the environmental changes which initially stimulated the receptors. The warmth of a fire for example, stimulates temperature receptors in our skin but it is the relay of the impulses to the cerebral cortex that enables us to *interpret* the stimulus as warmth. Similarly, the impulses from the retina of the eye are first transmitted to the corpora quadrigemina, but only when these are relayed to the *visual cortex* of the occipital lobe is the image on the retina interpreted as a picture of the environment. The destruction of the visual cortex leads to blindness just as surely as does destruction of the eye.

Motor area. The motor area is the thickest part of the cerebral cortex and contains the cells from which most motor impulses originate. The localization of the various functions is shown diagrammatically in Fig. 20.4. If we take the *central fissure* (fissure of Rolando) as a landmark, we note that the motor area lies in a band on each hemisphere just anterior to the fissure and, as indicated by the arrow in the diagram, it extends down into the longitudinal fissure on each side. The many kinds of cells in the motor cortex include those of the pyramidal tracts, and most conspicuous are some especially large ones known as *Betz* cells. These are the cell bodies of pyramidal fibers but not all pyramidal fibers originate from them.

The impulses for voluntary muscular movement come from the motor area of the cortex, and they can be initiated by applying electrodes to the cortex itself. When a small shock is passed, contractions are observed in muscles on the opposite side of the body. By varying the location of such stimuli we have found that the regions of the motor cortex relate to muscles at various levels of the body in a distinct order, but upside down. The size of the motor cortex region for each group of muscles is related more to the complexity of the movements performed than to the bulk of the muscles. Thus, the regions concerned with the fingers or the facial muscles are proportionately larger than those for the muscles of the trunk.

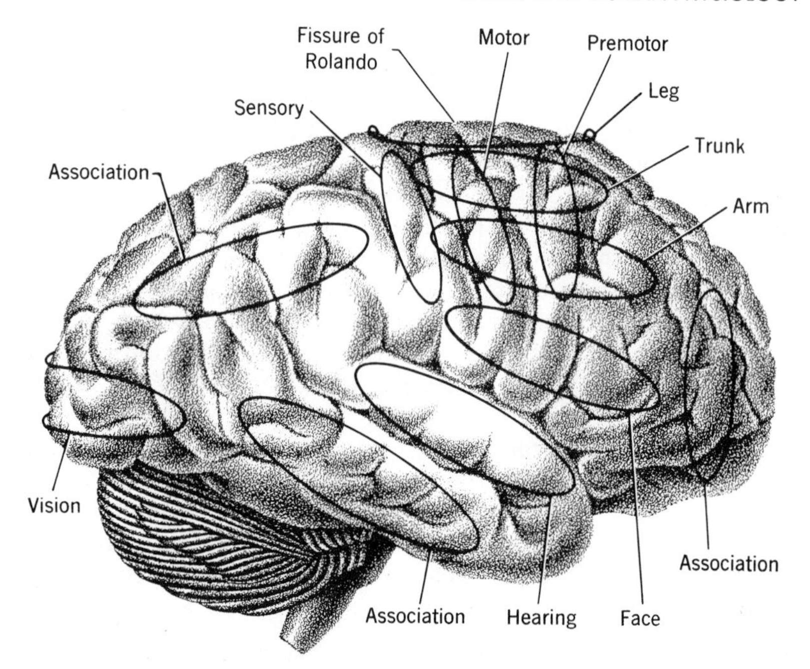

FIGURE 20.4. Localisation of cortical functions.

Supressor strip. Just anterior to the motor area is a strip which is reported by some workers to exert a profound inhibitory effect on the motor systems if stimulated.

Premotor area. Still more anteriorly lies the cortical area from which the impulses for associative movements originate. It will be recalled that these are relayed via the basal ganglia over extra-pyramidal tracts. In addition, there are important connections with the cerebellum, again concerned with coordination. The relationship of the regions of the premotor cortex to the musculature is the same as in the motor cortex.

Sensory area. As discussed previously, the sensory impulses reaching the upper levels of the brain are sorted out by the thalamus and *projected* to various parts of the brain. Those which are consciously interpreted are projected to the *sensory cortex,* which lies just posterior to the central fissure. The pattern of projection is similar to that for the motor regions, with large areas devoted to highly sensitive receptor areas like the hands and tongue.

The cortical area for hearing lies lower on the side of the brain, actually on the *temporal lobe,* and extends into the *Sylvian fissure.* As mentioned previously, the visual cortex is on the posterior portion of the occipital lobe, right at the back of the head.

Association areas. Clearly, even these extensive areas which have been mapped do not account for all of the area of the cortex. Large areas used to be called "*silent areas,*" but this term has been abandoned in favor of "*association areas.*" In these regions activities such as deduction, memory, and planning probably take place.

Physiologists are a long way from being able to account for the neural events in even the simplest intellectual processes, and the processes of imagination or artistic creativity are practically beyond comprehension. It would seem reasonable to suppose, however, that a memory, for example, is made up of a series of definite and unique circuits in the cortical cells. The experience to which the memory relates could well cause the establishment of these circuits in the first place and then the subsequent re-use of them "plays back" the experience as a memory. There are enough potential circuits in the cortex for this theory to be tenable, since the tremendous numbers of neurons can make an astronomical number of combinations.

ELECTROENCEPHALOGRAM

The activity of the cerebral cortex is determined to a great extent by the constant barrage of impulses from the thalamus. This has been mentioned previously in connection with the effect of the reticular formation on the thalamus in maintaining wakefulness. The constant activity of the cortex can be detected as changes in potential if electrodes are applied to the brain, or even to the skull. A record of the potentials is called an *electroencephalogram* or *EEG*. The potentials are conventionally recorded from the occipital, frontal, and parietal regions of the skull. In a normal subject, their form is quite characteristic of the region and the level of wakefulness or mental activity. In particular, the slow basic rhythms of relaxation or sleep are replaced by waves of very high frequency during intense thought or when carrying out intricate muscular activity (see Fig. 20.5). Many disorders of the brain give rise to abnormal "brain waves" so that in skilled hands the EEG is a valuable diagnostic tool. It is often possible to distinguish between several kinds of epilepsy by this means.

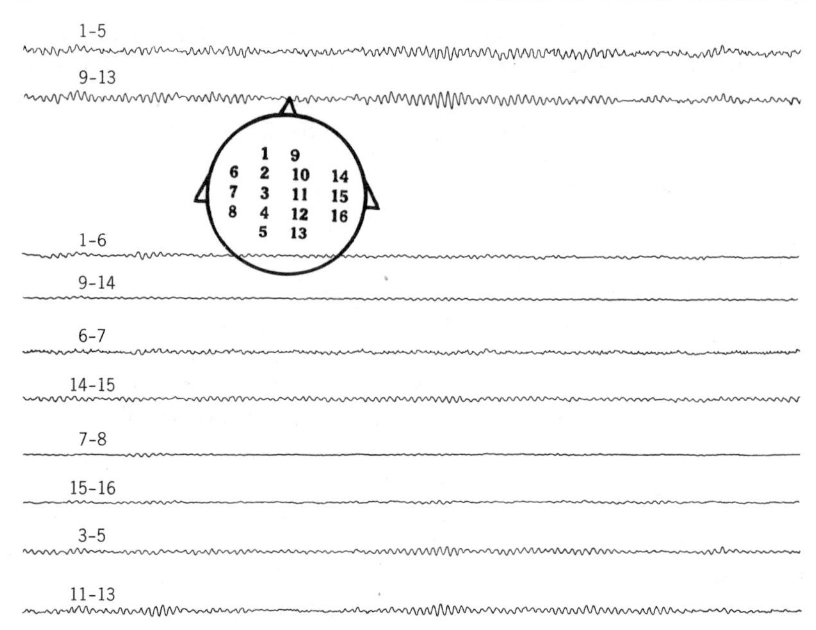

FIGURE 20.5. An EEG. The electrodes are placed on the skull in the positions indicated in the diagram. Each tracing is from a pair of electrodes identified by numbers.

IMPORTANCE OF AFFERENT IMPULSES

Since it has been realized that cortical activity depends so much on projection from the thalamus, there has been great interest in the physiological and psychological impact of *sensory deprivation.* In some preliminary experiments, subjects have been floated in warm, dark, soundproof tanks of water, in which they are virtually weightless and receive no outside stimuli. The psychic stress is severe and the subjects report bizarre patterns of mental activity.

INHIBITION AND EXCITATION AT THE CELLULAR LEVEL

In all the preceding discussion we have repeatedly spoken of impulses being relayed from one system of fibers to another, or frequently to several others. There must obviously be some discrimination at the interconnections so that some impulses are passed on and others are not.

The basic unit of the central nervous system is the *ganglion cell,* similar to the idealized neuron described in an earlier section except for the large number of fibers from other cells which terminate on its surface. Contact is made through the *terminal knobs* of the connecting fibers. It is important to note that most fibers branch near their ends and may contact many different ganglion cells. A single ganglion cell may have up to a hundred terminal knobs through which it receives impulses from many parts of the nervous system. Impulses over some fibers excite the ganglion cell and sufficient excitation of this kind causes the ganglion cell itself to depolarize. Impulses over other fibers are *inhibitory* and they cancel out the effects of the excitatory fibers. Thus, the "decision" as to whether a CNS cell discharges or not depends on the sum of the impulses affecting it. Even from this brief description we can see how this mechanism can account for discrimination in the relaying of nerve impulses in the CNS and how this relates to the "levels of command" described before. Similar systems are employed in man-made electronic circuits which can make "decisions."

REFERENCES

1. Aird, Robert B., "Barriers in the brain," *Sci. American* **194** (2) 101, February 1956.
2. Beecher, Henry K., "Anesthesia," *Sci. American* **196** (1), 70, January 1957.
3. Eccles, John C., "The physiology of imagination," *Sci. American* **199** (3), 135, September 1958.
4. French, J. D., "The reticular formation," *Sci. American* **196** (5), 54, May 1957.
5. Gerard, Ralph W., "What is memory?" *Sci. American* **189** (3), 118, September 1953.
6. Gray, George W., "The great ravelled knot," *Sci. American* **179**, (4), 26, October 1948.
7. Kleitman, Nathaniel, "Sleep," *Sci. American* **187** (5), 34, November 1952.
8. Olds, James, "Pleasure centers in the brain," *Sci. American* **195** (4), 105, October 1956.
9. Snider, Ray S., "The cerebellum," *Sci. American* **199** (2), 84, August 1958.
10. Walter, W. Grey, "The electrical activity of the brain," *Sci. American* **190** (6), 54, June 1954.

AUTONOMIC NERVOUS SYSTEM

21

If we consider the entire animal in relationship to its environment, we see that it is aware of external changes through the external receptors (eyes, ears, etc.), and that it adapts to the external environment through the external effectors (skeletal muscles). Similarly a set of internal receptors signal information to the "vital centers" of the brain about the state of the internal environment. Impulses are sent to the internal effectors which regulate the internal environment. Those parts of the nervous system relating to the internal environment may be called the *visceral nervous system.*

In the introductory section on the organization of the nervous system, it was mentioned that the *visceral efferent* nerves are considered as a separate system—the *autonomic nervous system.* From the functional point of view it seems rather illogical that only the *efferent* nerve fibers should be considered part of this system, because clearly, the afferent fibers from the receptors to the central nervous system complete the reflex arcs and are essential to autonomic control. Nevertheless, the original classification, which was made on anatomical grounds, still persists. The student will occasionally see reference to *autonomic sensory* fibers, but according to the strict definition, this is incorrect. The "correctness" of a definition depends on usage, and the main concern here is to forestall any possible confusion due to conflicting statements about the extent of the autonomic nervous system.

The anatomical peculiarities of the visceral efferent nerves have important effects on the physiology and pharmacology of this part of the nervous system. One important characteristic is that instead of there being one neuron to transmit an impulse from the spinal cord to a muscle effector (as in a lower motor neuron), an autonomic pathway has a relay ganglion between the cord and the effector. There are therefore *two* components to a visceral efferent pathway, a *preganglionic fiber* (which is myelinated), and a *postganglionic fiber* (which is nonmyelinated). Many fibers pass to and from each ganglion and each of these structures is able to discriminate in its relaying. The transmission of impulses in the ganglia is always by acetylcholine release, but the postganglionic fibers stimulate the visceral effectors by either nor-epinephrine or acetylcholine release.

The autonomic ganglia are of three kinds. First, the *lateral,* or *vertebral* ganglia lie in a chain down each side of the spinal cord and fairly close to it. This chain is frequently called the *sympathetic trunk.* Second, the *collateral,* or *prevertebral,* ganglia lie further from the cord within the abdomen and thorax. Third, the *terminal* ganglia lie actually within the visceral organs the fibers serve, and it is apparent that these postganglionic fibers are very short.

The cell bodies of the preganglionic fibers are within the spinal cord (as are those of lower motor neurons), but the courses of their axons are different at various levels. The axons rising from the thoracic and lumbar segments of the cord pass to lateral ganglia or to collateral ganglia. These thoracolumbar fibers are part of the *sympathetic division* of the autonomic system. Fibers arising from the cranial and sacral portions of the cord go to terminal ganglia within the viscera. These are part of the *parasympathetic* division (Fig. 21.1).

THE SYMPATHETIC DIVISION

The preganglionic fibers of the sympathetic division emerge from the cord in the ventral roots (which also contain the lower motor neurons, see Fig. 21.2). They branch off as *white rami communicantes* and lead to the lateral ganglia down each side of the cord. Here, some fibers synapse with their postganglionic fibers which emerge as the *grey rami communicantes* and rejoin the

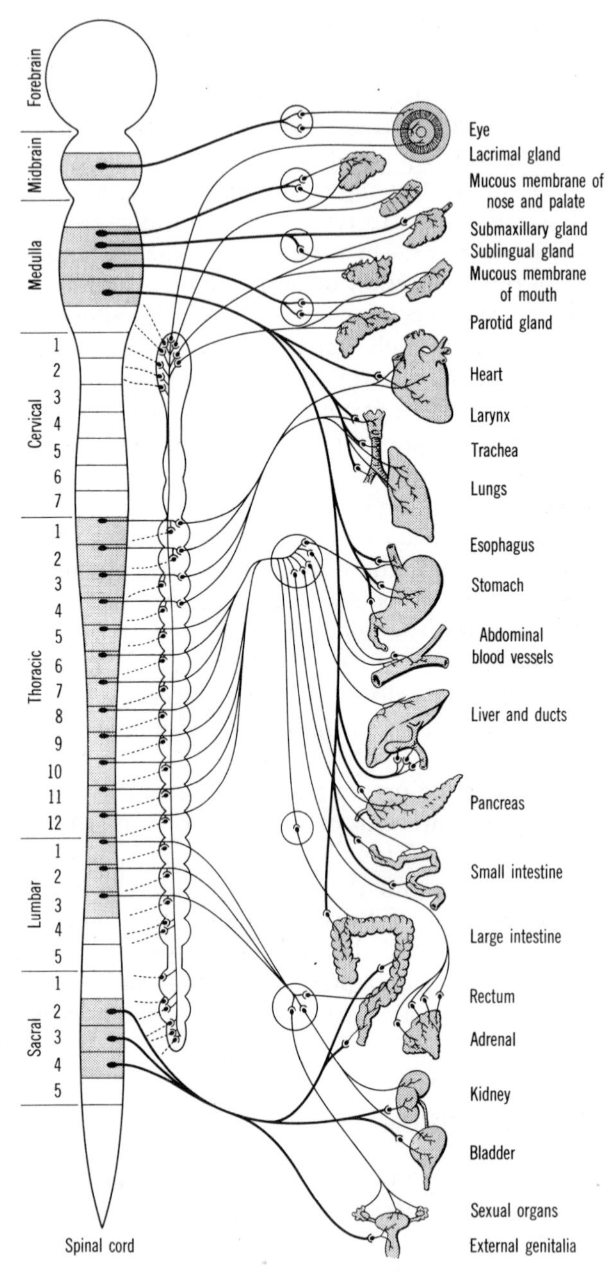

FIGURE 21.1. Diagram of the autonomic nervous system. The parasympathetic branches, arising from the brain and sacral vertebrae, are indicated by heavy lines; the sympathetic branches, arising from the thoracic and lumbar vertebrae, are lighter.

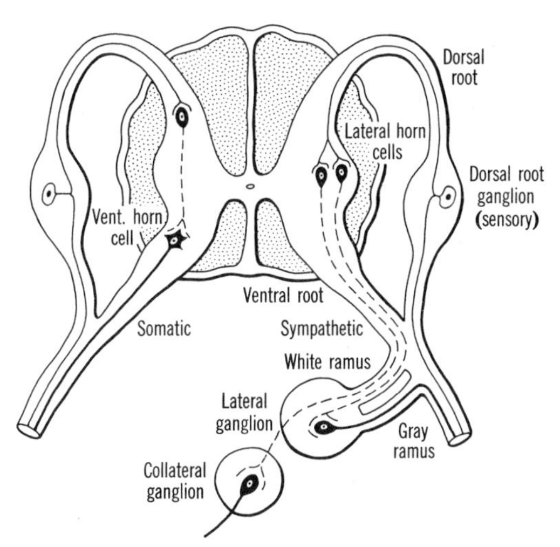

FIGURE 21.2. Diagram to illustrate somatic (at the left) and autonomic (at the right) nerves. The preganglionic fibers are represented by broken lines; the postganglionic by heavy black lines. Redrawn by permission from Bard, *Medical Physiology*, Mosby, 1956.

spinal nerves for distribution to vasomotor and pilomotor muscles. Other fibers pass uninterrupted through the lateral ganglia to the collateral ganglia, which are near the bifurcations of important blood vessels in the abdomen. From these ganglia, postganglionic fibers called the *splanchnic nerves* are distributed to the abdominal viscera.

The arrangement of the lateral ganglia is related to the "segmentation" of the spinal column, but, as shown in Fig. 21.1, those of the neck and upper back (cervical and upper lumbar) are fused to make three larger ganglia. Each lateral ganglion is joined to those above and below it; an arrangement partly responsible for the widespread response to sympathetic discharges. Furthermore, there are more than ten times as many postganglionic fibers as preganglionic. Therefore, a visceral efferent impulse arising from the hypothalamus has a more diffuse effect than a comparable somatic motor impulse.

The sympathetic outflow is usually considered in two functional categories:

i. For the maintenance of various *tonic* states, as with peripheral vasomotor tone, and the cardio-acceleratory impulses (which partly balance the stronger parasympathetic inhibition).

ii. To evoke a general response to meet any increased demands on the animal, typically those involved with exercise, fear, rage, etc. Because the sympathetic division mediates an *integrated* response in increasing the heart rate, shunting of blood from the viscera, sweating, etc., there is a tendency to regard a sympathetic discharge as a simple *mass response* to an emergency. This is an oversimplification, because clearly the response varies with the situation. The nature of the integrated response is determined in the hypothalamus and appropriate impulses sent over fibers within the cord to the preganglionic fibers. The specificity of conditions in which sweating is induced, for example, is sufficient evidence that the sympathetic response is not always a simple "emergency" reaction.

Transmitter substances. In both the sympathetic and parasympathetic divisions, the transmission of the impulse from the preganglionic to the postganglionic fiber is by acetylcholine release, analogous to that at a somatic neuromuscular junction. The sympathetic stimulation of effector organs, such as the smooth muscle of visceral blood vessels, is by nor-epinephrine release at the ends of postganglionic fibers. The postganglionic fibers of the parasympathetic division release acetylcholine. In some effector organs, nor-epinephrine is excitatory and acetylcholine is inhibitory, in others, *vice versa,* and in a few effectors, they are both excitatory.

An important area of pharmacology is concerned with drugs that will enhance or inhibit the effects of the humoral transmitters of the autonomic nervous system, both in the ganglia and at the effector organs. For example, an injection of nor-epinephrine has, as might be expected, a similar effect to sympathetic stimulation.

The *adrenal medulla* is of particular interest in this respect. As was mentioned in the chapter on endocrinology, this gland may be well regarded as part of the sympathetic division of the autonomic nervous system. It is innervated by preganglionic fibers only, and their acetylcholine release evokes the gland's release of epinephrine and nor-epinephrine. It is as if the gland were a mass of postganglionic fiber tissue. The catecholamines released from the adrenal medulla into the blood stream can reach all tissues and therefore have a general effect. Therefore, this aspect of the sympathetic response has no specificity.

THE PARASYMPATHETIC DIVISION

The autonomic fibers arising from the cranial and sacral portions of the CNS are classified together as the parasympathetic division.

As indicated, the ganglia lie in the organs served, so that the postganglionic fibers are relatively short and serve only the tissues of that organ. The parasympathetic innervation is therefore much more specific and without the spreading of response characteristic of the sympathetic. The autonomic fibers of the cranial nerves control the pupillary size, salivation, cardiac inhibition, and gastro-intestinal secretion. The sacral autonomic fibers control the smooth muscles of the bladder, colon, and rectum.

From Fig. 21.1, it will be noted that the *vagus* nerve, although it is one of the cranial nerves, extends to many of the abdominal viscera. It contains about 80% of the total parasympathetic fibers. This is not to say that the vagus *nerve* as seen in the animal is made up mostly of parasympathetic fibers, because it also contains visceral afferent, somatic efferent, and somatic afferent nerve fibers. Only about 20% of its fibers are parasympathetic.

Dual innervation. Many visceral effectors are innervated by both divisions of the autonomic nervous system, being excited by the one and inhibited by the other. The heart is of course the prime example; others are listed in Table 21.1. Some visceral effectors receive dual innervation which is *not* antagonistic, but rather *synergistic;* the salivary glands secrete in response to impulses from both divisions.

Single innervation. Table 21.1 also lists those effectors innervated by the sympathetic division alone. There are no effectors with solely a parasympathetic innervation.

The secretion of sweat, the erection of hairs, and the contraction of the spleen are under sympathetic control only. The smooth muscle in the blood vessels of skeletal muscle is also innervated by the sympathetic alone, but whereas the normal tonic state is one of moderate vasoconstriction, there are some sympathetic *vasodilator* fibers that function in muscular exercise.

CENTRAL REPRESENTATION OF THE AUTONOMIC NERVOUS SYSTEM

Medulla oblongata. In our discussion of the control of the blood pressure, we have seen that the centers controlling heart rate and vasomotor tone are located in the medulla oblongata. These are prime examples of autonomic mechanisms. Visceral afferent fibers bring impulses from the baroreceptors which modify the output from the centers over the visceral efferent fibers. As noted in the discussion of blood pressure regulation, the brain stem centers are

TABLE 21.1 Some Functions of the Autonomic Nervous System

Organ	Parasympathetic Fibers	Sympathetic Fibers
1. Iris	Contraction of circular fibers	Contraction of radial fibers
2. Ciliary muscle of eye	Contraction accommodates for near vision	Relaxation
3. Lacrimal gland	Secretion and vasodilation	Vasoconstriction
4. Cerebral blood vessels	Dilation	Constriction
5. Salivary glands	Secretion	Secretion
6. Bronchi	Constriction	Dilation
7. Heart	Inhibition	Acceleration
8. Coronary arteries	Constriction	Dilation
9. Stomach	Increased motility and secretion	Inhibition of motility and secretion
10. Intestine	Increased motility	Inhibition of motility
11. Adrenal medulla		Secretion
12. Blood vessels of abdominal and pelvic viscera		Constriction
13. Sweat glands		Secretion
14. Blood vessels of skin		Constriction
15. Pilo-erector muscles		Contraction

subject to control from higher centers in the brain. The control of respiration is a rather special case, because although it is clearly an autonomic function, it is actually carried out by *somatic* neuromuscular mechanisms. The control of the size of the pupil in the eye is a special case in the reverse direction; it is controlled by the autonomic nervous system, but the effective stimulus is the amount of light falling on the retina, which is clearly an *external* stimulus and unrelated to the internal environment.

It should be noted that exploration of the brain stem with modern techniques has shown that the cardiac and respiratory "centers" are not the discrete nuclei of nervous tissue that the name implies; they are more diffuse structures in a reticular formation. In our present context, however, the import is the same.

Hypothalamus and thalamus. Following transection of the brain of an experimental animal between the pons and the thalamus, the automatic functions controlling heart rate, vasomotor tone, respiration, and the urinary bladder remain relatively intact. In this state however, the animal is unable to react to environmental changes and cannot regulate its body temperature. This is because the autonomic functions which permit *adaptation* depend on the thalamus, which is the principal sensory receiving center. It relays sensory impulses to higher centers, including the cortex, and also to the hypothalamus which in turn contributes largely to the sympathetic and parasympathetic outflow regulating the visceral organs.

The posterior portion of the hypothalamus sends out impulses over sympathetic fibers and the anterior portion over parasympathetic fibers. If the environment is cold and the body temperature tends to fall, sympathetic impulses from the hypothalamus promote vasoconstriction in the skin and the other mechanisms tending to conserve heat. Conversely, warm conditions evoke parasympathetic discharges promoting sweating and skin vasodilation.

The hypothalamus is subject to modification from even higher centers (via the thalamus or not), as is illustrated by the sympathetic response evoked by emotions.

SPECIAL SENSES

22

INTRODUCTION

Most of the organs regulating or distributing the internal environment are subject to feedback controls in which nervous mechanisms play a part. A vital link in each feedback system is the *internal receptor,* which is sensitive to changes in a particular component of the internal environment. Some examples we have discussed are:

1. The baroreceptors in the carotid sinus and in the arch of the aorta are specialized nerve endings which increase their rate of impulse formation when they are distorted by increased arterial pressure. Their impulse frequency is part of the information supplied to the cardiovascular centers of the medulla which regulate the circulation.
2. The chemoreceptors in the carotid body are sensitive to changing oxygen content of the blood. Signals from these specialized nerve endings inform the respiratory center in the medulla if the oxygen tension falls.
3. The stretch receptors in the lungs are stimulated when the lungs are adequately inflated at each respiration; their signals inhibit the inspiratory center and permit passive expiration to follow.
4. The cells of the respiratory center

are sensitive to p*H* and to changing carbon dioxide tension in the internal environment. They adjust the ventilation rate accordingly.

In each of these receptors, an impulse is initiated by the depolarization of the membrane as described on page 289. In experimental work we commonly use a small electric shock as the stimulus for inducing the depolarization, but the same effect can be achieved by mechanical or even chemical stimulation of the nerve. The specialized receptor nerve cells are extremely sensitive to stimuli of specific kinds; the chemoreceptors are sensitive to changes in oxygen tension and the stretch receptors are sensitive to mechanical distortion. In these receptors however, and in the external receptors to be discussed in this section, a sufficiently powerful stimulus of *any kind* will cause the emission of impulses.

The *special senses* are the receptor systems through which the animal is "aware" of its *external* environment. The conditions in the external environment constantly stimulate the sensory receptors, which in turn send impulses to the central nervous system. These impulses provide the information required by the higher centers of the brain for the integration of the muscles, etc. in adapting to changes in the external environment, just as the receptors previously discussed provide information about the internal environment.

The special receptors in the eye are the light-sensitive nerve cells of the *retina;* those of the ear are the vibration-sensitive cells of the *organ of Corti.* Closely associated with the auditory apparatus of the inner ear are receptors sensitive to the position of the head with respect to gravity and to changes in movement. The receptors for taste and smell are located in the tongue and nasal passages, respectively.

THE SOMESTHETIC SENSATIONS

In addition to the special senses, the body has a great many receptors at the surface and within the muscles. These receptors keep the brain informed of the immediate effect of the external environment on the tissues. They are the receptors to touch, heat, cold, pain, and *proprioception,* or muscle sense. It is through the impulses from these receptors that we are aware of temperature changes, the contact of foreign objects with the skin, and the position of the limbs.

VISION

Elementary optics. Light, like radio signals, is a form of electromagnetic radiation. It is propagated in a form which may be equally well considered as either waves or corpuscles, but for our purposes it is best thought of as a wave form. The wavelength of light determines its color; sunlight is a mixture of the colors of the *spectrum* as seen in a rainbow, and which in the visible range, extend from red to blue. The wavelength of ultraviolet light is even shorter than that of blue light, and because the retina is not sensitive to it, it is invisible. Similarly, infrared light has a wavelength longer than the visible red and is also invisible.

Light does not need a medium in which to be propagated, as is quite apparent in starlight which travels unbelievable distances across the interstellar emptiness. The speed of light is so great that for ordinary earth-bound distances, its propagation can be regarded as almost instantaneous. It has been measured however, and the staggering result is 300,000 kilometers, or 186,000 miles per second. This is the speed in a vacuum; it is slower through air, glass, water and other materials. If we think of a beam of light as a series of wave fronts, as in Fig. 22.1, their *frequency* remains the same even when their rate of travel is slowed. Therefore, the

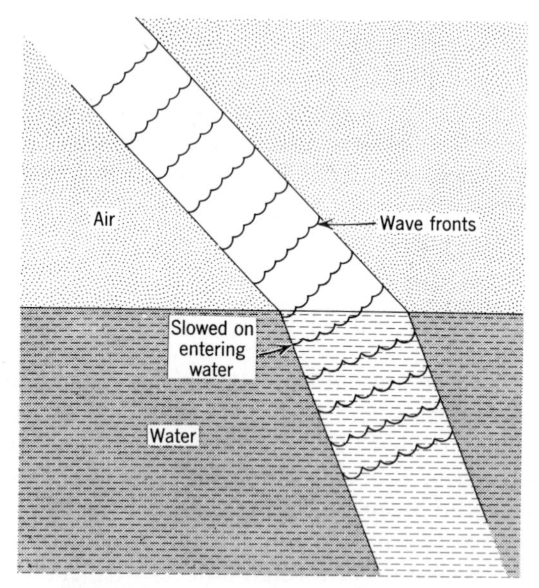

FIGURE 22.1. Diagram showing refraction of light.

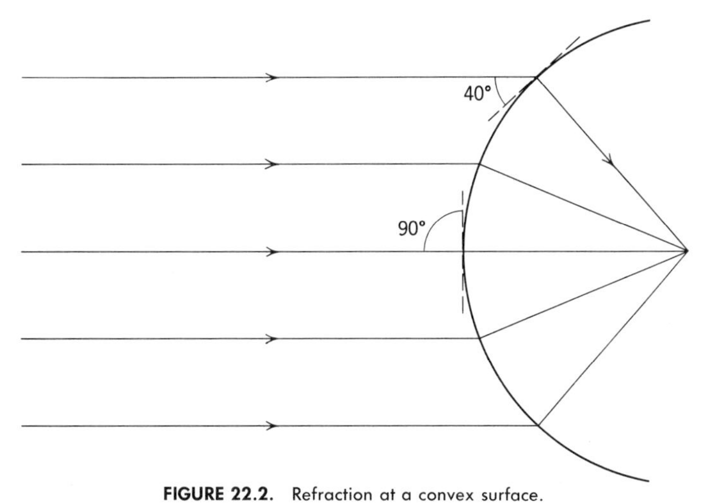

FIGURE 22.2. Refraction at a convex surface.

slowing down on entering an optically denser medium shortens the wavelength, as is also shown in Fig. 22.1.

When a beam of light strikes the surface of a material at an angle (as on entering water, for example) the waves in the bight of the angle are slowed and shortened before the waves on the other side of the beam. In consequence the beam is bent. The extent of bending depends on the *angle of incidence* of the ray of light and on the relative speeds of light in air and water (in this example). The smaller the angle of incidence, the earlier is the slowing of light on one side of the beam relative to the other. The angle of the refracted light therefore, bears a definite trigonometrical relationship to the angle of incidence. When the light enters the surface at right angles, the slowing is synchronous in all the wave fronts and there is no bending.

The relative slowing of light in a medium is a function of its *refractive* index, which (for light of a given color) is just as characteristic as density or hardness. The refractive index of air is taken as 1.000.

Lenses

When parallel light rays strike a uniformly curved surface, the angle of incidence is 90° at the center and decreases in each direction toward the periphery. The angle of refraction therefore, increases toward the periphery with the result that the parallel rays are bent toward a single point (Fig. 22.2). The properties of *convex lenses* are dependent on this effect.

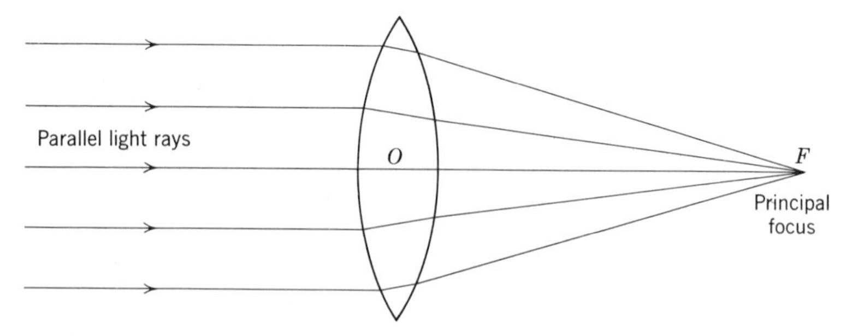

FIGURE 22.3. A biconvex lens bends parallel light rays so that they meet at the principal focus of the lens. The focal length, OF, is characteristic of the lens.

Figure 22.3 illustrates a *biconvex* lens which bends parallel rays of light so that they meet at the point *F*, the *principal focus* of the lens. The distance *OF* is the *focal length* and is a characteristic of the lens dependent on the curvature of the faces. This kind of convergence of light can be seen when a magnifying glass is used to focus rays of sunlight to a point. Only the light rays from an extremely distant source such as the sun can be considered as parallel. All other light from a single point must be thought of as composed of divergent rays.

In Fig. 22.4, we can see that rays from the point source, *S*, can be focused by a biconvex lens to a point, *C*, a *conjugate focus*. The conjugate focus is not a fixed property of the lens, because it varies with the distance of the lens from the source. It will be noted that in both Fig. 22.3 and Fig. 22.4, the line representing a ray of light passing from the source through the center of the lens is not bent; it is at right angles to each of the curved surfaces.

The "strength" of a lens is determined by its focusing power. A lens which will focus parallel light rays to a point 100 centimeters

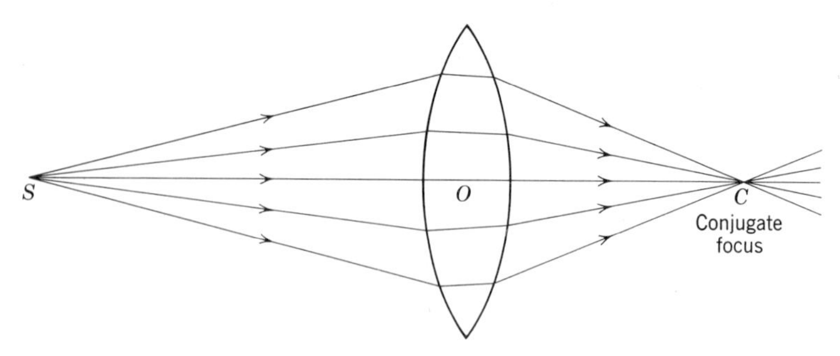

FIGURE 22.4. Light from a point source, S, is focused at the conjugate focus, C, The distance OC depends on SO, as well as on the lens characteristics.

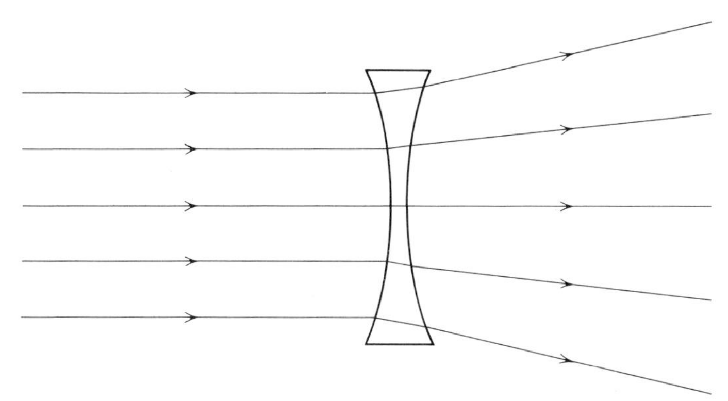

FIGURE 22.5. Concave lenses scatter parallel light rays.

behind it is arbitrarily given the strength of 1 *diopter.* Lenses with shorter focal distances are stronger, i.e., they have greater *refractive power.* A lens with a focal length of 50 centimeters has a refractive power of $100/50 = 2$ diopters; a lens with a focal distance of 10 centimeters has a refractive power of 10 diopters.

Lenses with *concave* surfaces do not focus parallel light rays to a point, but scatter them (Fig. 22.5). This property is useful in correcting some of the focusing defects in the eye.

The usefulness of optical instruments usually lies in their ability to focus images rather than single points. Figure 22.6 shows how a convex lens projects the rays of light from a candle (AB) as an inverted image (CD) on a screen placed at a suitable distance behind the lens. This is the principle of a slide or movie projector, or in the reverse direction, the principle of a camera. It will be recalled that slides must be placed in a projector upside down in order to get the image on the screen the right way up, and that the image actually projected on the film in a camera is inverted. We can focus the image by moving the lens back and forth until it is at the right point between the object and the screen.

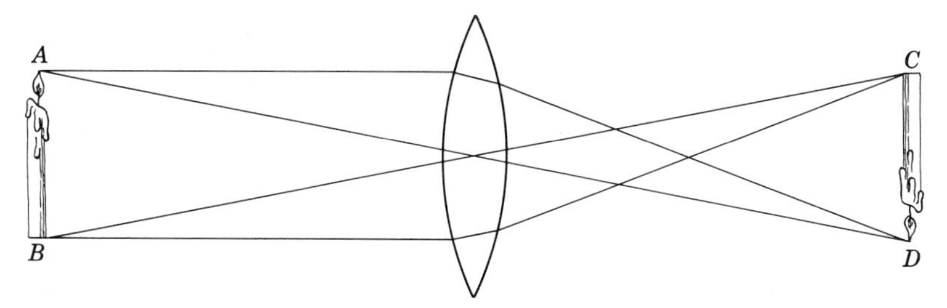

FIGURE 22.6. Image projection by a convex lens.

The eye. Each eye consists of a hollow sphere, *the eyeball,* held by muscles in the eye sockets of the skull, the *orbits.* From the back of each eyeball, the optic nerve connects through the *optic chiasma* with the brain.

The front of each eyeball is covered by the transparent *cornea,* but the rest of the outer coat consists of the tough, white *sclera.* Under the sclera is another coat, the vascular *choroid layer.* At the back of the eyeball there is yet a third and inner layer, the *retina,* which is made up of the visual receptor cells and their associated tissues. Behind the cornea, in the anterior portion of the eye, is a structure like the aperture diaphragm of a camera, the *iris.* It contains pigment granules which give color to the eye —gray, blue, brown, etc. Its central aperture is the *pupil.* Immediately behind the iris is the *lens,* held by its *suspensory ligaments.*

Reference to Fig. 22.7 will show that there are three chambers in the eye. The *anterior chamber* lies between the cornea and the iris; the *posterior chamber* is very small and lies between the iris and the lens. Like the anterior chamber, it is filled with a transparent fluid, the *aqueous humor.* The largest chamber of the eye, between the lens and the retina, is filled with a jelly-like medium called the *vitreous humor.*

The muscles of the eye can be considered in two main categories, *intrinsic* and *extrinsic.* The intrinsic muscles control the lens and the iris, whereas the extrinsic muscles move the eyeball within the orbit and open and close the eyelids.

The *eyelids* serve a protective role and each contains dense connective tissue which strengthens it as a mechanical shield for the eye. The inner surface of the eyelid is part of the *conjunctiva, a* mucus membrane which is also reflected on part of the eyeball and is concerned with the lubrication of the surfaces of the eyeballs. The *blink reflex* is elicited by the contact of a foreign body with the cornea; the movement of the eyelid helps to dislodge dust, etc., so that it can be washed away. This reflex is still present in light anesthesia and so is used as a rough guide to depth of anesthesia.

The *lacrimal gland* is within the orbit, lateral to each eye. It secretes *tears* which run through ducts into the sac of the conjunctiva, between the eyelids and eyeball. The tears wash the cornea and keep it free of dust.

Accommodation

The lens focuses light rays from the surroundings on the retina. Unlike the focusing in a camera, this is not accomplished by moving the position of the lens but rather by changing its refractive

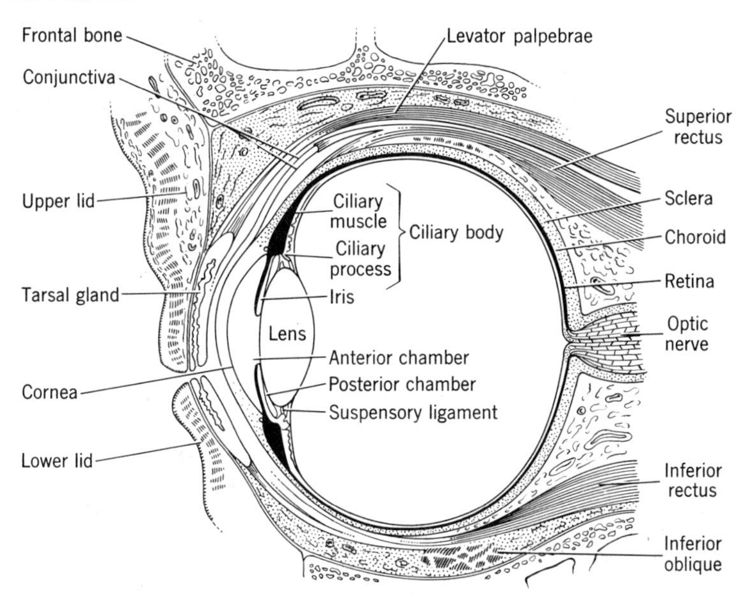

FIGURE 22.7. Diagrammatic cross section of the eye.

power. This is possible because the lens is elastic and not a brittle, glass-like material. The elasticity is due to the springy protein forming the outer coat of the lens, the interior is a viscous fluid. At rest, the suspensory ligaments keep the lens stretched to its minimum curvature, which is suitable for focusing distant objects. To focus on closer objects, the ciliary muscles pull the suspensory ligaments in such a way that the tension on the lens is *reduced.* This lessened tension permits the lens to become more spherical, because its own elasticity causes it to assume that shape. The degree of contraction by the ciliary muscles determines the extent of the increase of lens curvature, and therefore the extent to which the eye focuses more closely. The focusing process in the eye is called *accommodation.*

By accommodation, an optical image is focused on the retina, which contains dense ranks of light-sensitive cells. The patterns of light and dark (and color) in the image stimulate the retinal receptor cells and the consequent impulses travel over fibers of the optic nerve to the brain, where they are interpreted as a subjective image.

The Iris

The focusing of parallel light rays to an exact point depends on the perfection of the spherical surfaces of a biconvex lens. If the curvature of the surfaces varies from point to point, the refraction is not uniform and a blurred image results, a defect especially noticeable in cheap cameras. The distortion of the image is said to be due to *spherical aberration.* Furthermore, light of different wavelengths is differently refracted, and so the point of focus for one part of the spectrum is different from that of another. This cause of blurring is called *chromatic aberration.* Distortions due to spherical and chromatic aberration can be minimized if the peripheral refraction of a lens is not used and the light is allowed to pass through only a small part in the center.

If a lens is set up at a fixed distance from a screen, an object must be, theoretically, at one exact distance from the lens to be in sharp focus. In practice however, the image will seem equally sharp over a range of distances. The Figure 22.8 shows that with a small aperture, an image is in good focus through a larger range than with a wide aperture. Indeed, photographs taken with a *pinhole camera* have everything in the foreground, middle, and far distance in equally sharp focus. Because of these advantages of a small aperture, particularly the increased depth of focus, a photographer "stops down" the lens of his camera to the smallest diaphragm aperture diameter consistent with the existing light conditions.

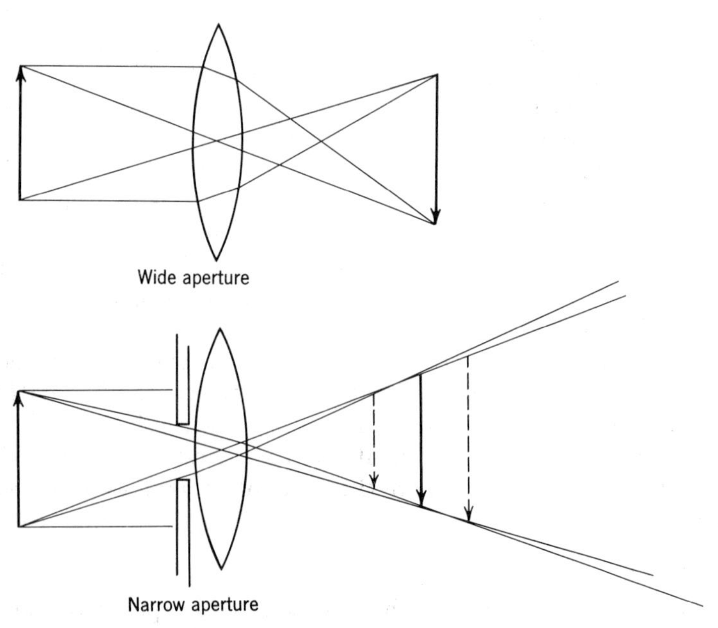

FIGURE 22.8. The optical advantage of a small aperture.

As might be expected from its elasticity, the lens of the eye is not a particularly "good" lens with respect to the geometric perfection of its surfaces. The distortions of the retinal image from chromatic and spherical aberration are considerable, but are normally minimized by the constriction of the *iris*. Its intrinsic muscles are reflexly synchronized with the ciliary muscles of the lens, so that the pupil size is reduced in accommodating to close vision. This not only cuts out distortion due to the aberrations, but improves the range of clear vision. A separate function is the constriction of the iris in a bright light; this protects the retinal receptors from excessive stimulation. Therefore, in addition to participating in the *accommodation reflex,* the iris has a *light reflex;* these two reflexes can be observed separately. When looking at a close object, a subject's pupils become small and then on looking at a distant object they enlarge. If a light is then flashed in the eye, they become small again, independent of the accommodation. Neurological disorders commonly associated with syphilis cause a loss of the light-adaptation reflex, leaving the accommodation reflex intact; this symptom is called the Argyll-Robertson pupil.

It follows from this discussion that individuals with poor powers of accommodation benefit from a good light. The smaller pupillary size in a bright light improves the range of clear vision. It is commonly observed that parents scold their children for reading in what they themselves regard as a poor light, but which is entirely adequate for the flexible accommodation of the children's eyes.

Myopia and Hypermetropia

When the normal eye is relaxed and the lens is at its minimum curvature, parallel rays are sharply focused on the retina. In this respect the normal eye is *emmetropic*. Defects in the lens or in the suspensory ligaments or muscles can cause a blurred image, because the point of clear focus lies either behind or in front of the retina. These conditions are illustrated in Fig. 22.9. As indicated, the condition where the point of clear focus falls short of the retina is called *myopia,* or near sight; the other condition where it is beyond the retina, is *hypermetropia,* or far sight. The lens of the myopic eye may be thought of as having too great a refracting power, which can be corrected by placing a *concave,* or divergent, lens in front of the lens of the eye, giving a system with less total refractive power. Similarly, the hypermetropic lens has too weak a refractive power and this can be corrected by placing a convex lens in front of it, increasing the net refractive power of the system.

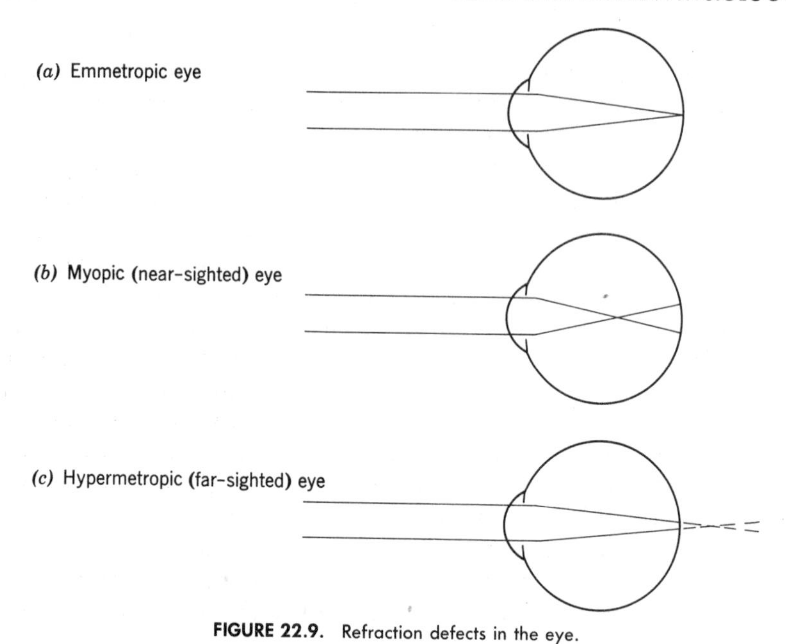

FIGURE 22.9. Refraction defects in the eye.

Short-sighted people have a small advantage over the long-sighted in that there is one range of visual distance in which they get a sharp image, even if it is very close to the eye. There is no point at which a far-sighted individual can achieve a sharp image.

Astigmatism

If a lens is not symmetrically curved, but rather has one quadrant with a greater or lesser refractive power than the rest of the lens, one part of the retinal image is always out of focus relative to the rest. This condition is called *astigmatism*. It is usually manifested in one or more planes which can be diagnosed by a simple test using Fig. 22.10; the instructions are in the legend to the figure. The astigmatic subject is more conscious of "eye-strain" rather than of frankly blurred vision, and it leads to a feeling of fatigue and irritability. The condition is corrected by using a lens reciprocally asymmetrical to the lens of the eye.

Presbyopia

As an individual grows older, the elasticity of the lens decreases, and therefore the powers of accommodation are reduced. Children can change the refractive power of their lenses from 25 diopters up to 40 or more, giving a range of accommodation of 15 or more diopters. By the age of 45 or 50, the accommodation range is reduced to 2 diopters, which makes unaided close vision very difficult. This optical change with age is called *presbyopia.*

The loss of accommodation in middle age theoretically demands a whole series of corrective lenses for each visual distance. This is unnecessary in fact, however, because a good depth of focus can be achieved with only one corrective lens for distant vision; and most close work, such as reading, is conducted at a fairly constant distance from the lens. Therefore, presbyopic subjects are usually well served by bifocal spectacles, in which the main part of the lens is corrective for distant vision, and a smaller portion is corrective for near work. Automobile mechanics and others who customarily work at yet another distance in addition to their reading distance (at arms length, for example) can profitably wear trifocal spectacles.

Another lens change with advancing age is that it becomes progressively yellow and less clear. This has the effect of filtering out some of the colors, and the older person is generally less color-sensitive than younger people. Elderly people who have had a lens removed because of a cataract can be fitted with suitable spectacles to correct for the absence of the eye lens. They are frequently delighted with their awareness of the leaf colors in the fall, espe-

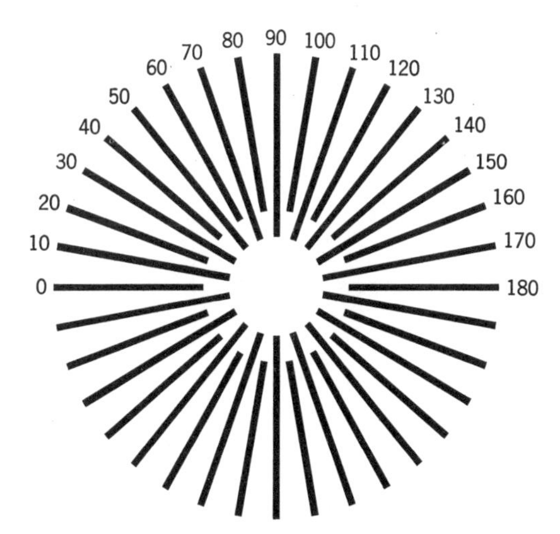

FIGURE 22.10. Simple test for astigmatism. With a normal eye, all the lines in this radiating pattern look equally dark and distinct. With an astigmatic eye, some lines appear darker, and others at right angles to them appear paler and less distinct.

cially as they have a direct comparison with their color awareness through the other eye with its yellow lens.

The Retina

The retina is made up of several layers of cells (see Fig. 22.11). The deepest layer is the one lying over the choroid layer, called the *pigment layer.* Above this, next closest to the vitreous humor, are two kinds of light sensitive cells, the rods and cones. The next layer is of intermediate nerve cells called the *bipolar* cells, which again make connection with the next layer of *ganglionic* cells. From each of these, fibers run inward across the retina to the *optic disk,* where they converge and penetrate to the back of the eyeball as the *optic nerve.* We see then the rather surprising arrangement where the light-sensitive cells are actually underneath layers of relay cells. On examining the retina with an *ophthalmoscope* we see a generally pinkish ground with blood capillaries branching across it. Somewhat off-center is the optic disk, which appears as a pale indentation. As there are no light-sensitive cells at this part of the retina, it is a blind spot. The blind spot can be dem-

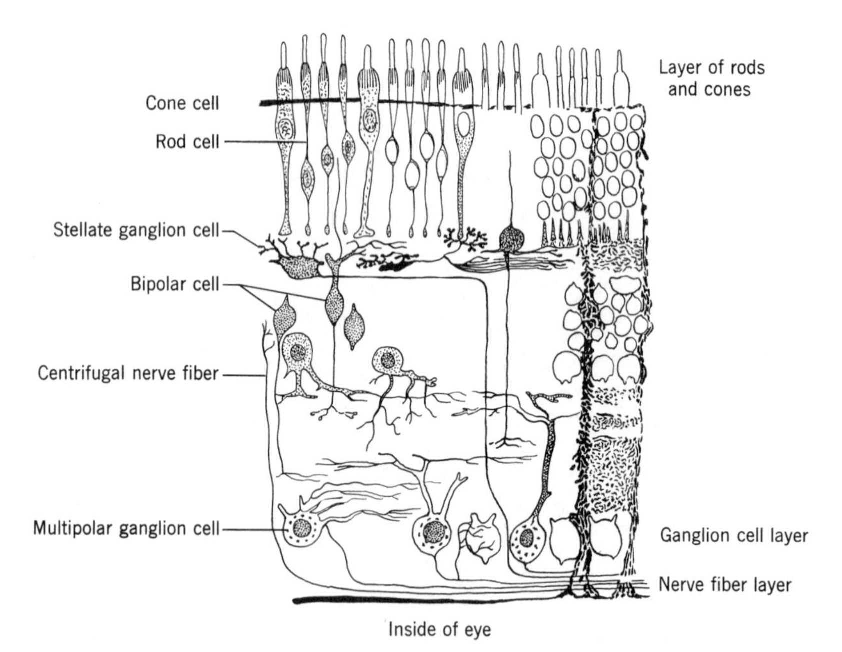

FIGURE 22.11. Diagram of the human retina. Redrawn by permission from Bremer and Weatherford, *Textbook of Histology,* Blakiston, 1946.

FIGURE 22.12. Close the left eye and hold the figure about twelve inches in front of the right. Look steadily at the white disk and move the page slowly toward the eye until the cross disappears. When this occurs the image of the cross has fallen on the optic disk, the blind spot.

onstrated by using the cross and spot in Fig. 22.12 and the instructions in the legend.

At approximately the geometric center of the retina is a yellowish spot, the *macula,* which has a central depression, the *fovea centralis.* This is the region of the most acute vision.

Rods and cones. The two kinds of light-sensitive cells are called rods and cones respectively, because of their shapes. They have different functions which are being very intensively investigated. The functional differences are indicated by the fact that the fovea contains cones at a very high density and has no rods at all. The visual acuity and color perception are the greatest when the image is focused here, but only in a good light.

Rods and cones are found together in other parts of the retina and in general, the proportion of rods increases towards the periphery. Vision in a poor light is better if the image is focused more towards the periphery of the retina, but color perception is then poor. Sailors have been aware for centuries that it is useless trying to make out details of an object in the dark by peering straight at it. They have learned the technique of looking at it slightly sideways, to utilize what we now know to be the better twilight vision in the periphery of the retina. This illustrates the *duplicity* of vision; the response of the cones to bright light is the basis of *photopic* vision and that of the rods is for *scotopic* vision.

The rods contain a pigment called *rhodopsin,* or *visual purple,* which is synthesized from vitamin A. It is bleached by exposure to light and it is presumed that the photochemical changes in the rhodopsin are part of the mechanism stimulating the rods in a poor light. A bright light will bleach all the visual purple immediately, and the individual cannot see in the dark during the period of several minutes while it is regenerating. This is why it takes a few minutes to adjust to the light in a dim room after walking in from bright sunlight and why our night vision is temporarily impaired if we are dazzled by a flashlight or the headlights of an oncoming automobile. One of the symptoms of vitamin A deficiency is night blindness, but there is no evidence that a diet particularly rich in vitamin A will improve night vision. During World War II, the British press gave considerable publicity to a reportedly large intake of carrots by night-fighter pilots, presumably in the naive belief that this would distract the enemy from realizing that air-to-air radar was being used for night interception.

Color vision. Sunlight is made up of all the colors of the spectrum, and "white" artificial lights have similar compositions. When white light falls on the surface of an object, some wavelengths are reflected and others are absorbed. The wavelength of the reflected light determines the "color" of the object. If the material absorbs all except red wavelengths, the reflected red light gives the object a red color. If *all* the light is absorbed, none is reflected and the material appears black; conversely, if *all* the light is reflected, the material appears white. A white object placed in a red light appears red and in a blue light appears blue because it is still reflecting all the light striking it. Colored lights may be achieved by several methods, including:

i. White light can be shone through a prism to produce a spectrum; a screen with a slit can be arranged so that only a monochromatic light is allowed to pass.

ii. A white light can be placed behind a filter which absorbs light of almost all wavelengths and only refracts light of one color.

iii. Some elements when heated to incandescence emit light over a narrow range of wavelengths, again giving a monochromatic light.

Color vision is almost certainly due to the specificity of the receptor cells to light at different wavelengths, but as yet there is no irrefutable biochemical or physiological evidence for any mechanism. Of the several theories of color vision, the Young-Helmholtz theory probably has the widest currency. This theory postulates that there are three kinds of light receptors (cones), each sensitive to one of the three *primary colors.* Any color can be obtained by appropriate mixtures of these three primary colors, and so conversely it is entirely reasonable to suppose that the differential stimulation of three types of primary color receptors in the retina produces the subjective impression of any color. *Color blindness* can then be accounted for by varying degrees of incompetence or absence of one of the primary color receptor cell types. Red-green color blindness is the most common and can vary from a complete inability to distinguish between these two colors all the way to a minor defect which shows up only by very special tests. Color blindness is hereditary and sex-linked to the male.

Binocular vision and fusion. The two images, one on the retina of each eye, are perceived as one. This *fusion* is achieved partly in the occipital cortex by processes which are learned in infancy and partly by the decussation of the optic nerves themselves.

Figure 22.13 is a diagram showing how fibers from the different parts of the retina lead into the brain. If we arbitrarily divide each retina into a nasal and a temporal half, the part of the visual field projected on the temporal side of the retina of the right eye will be projected on the nasal half of the left retina. The partial crossing of the optic nerves insures that the pattern of impulses due to the projection of the right half of the visual field on the two retinas is conveyed to the right half of the brain. Fibers from the temporal (or lateral) side of each retina do not cross and therefore enter the brain on their own side. Fibers from the nasal (medial) half of each retina cross over and enter the brain on the opposite side.

Even with this ingenious arrangement, it is necessary to adjust each eye so that the two images fall on neurologically related areas of the retinas. As would be expected, the neural pathways for achieving fusion are very complex and the basis is present at birth. Even newborn infants show considerable coordination in their eye movements, but the fine adjustments for fusion are probably perfected during infancy. Some drugs, notably alcohol, can disturb the coordination of the eyes sufficiently to cause double vision, or *diplopia.*

Judging distances. Our judgment of distances is largely based on our experience of the relative sizes of objects. Thus, if one automobile appears smaller than another, and our experience tells us that the two automobiles are likely to be of the same size, the smaller is perceived as further away. If our experience does not include familiarity with small European cars, a small car at the same distance from the eye as a large car might be perceived as further away. For distances of less than a few hundred yards, our binocular vision assists our judgment of distances. In order to fixate on a single point, each eye must be turned slightly inward, and the degree of turning will depend on the distance of the point. The extent of binocular adjustment necessary to fixate on an object is conveyed to the visual centers which integrate this information in their unconscious judging of distance. This is in fact the principle of the optical range finder, because if two telescopes

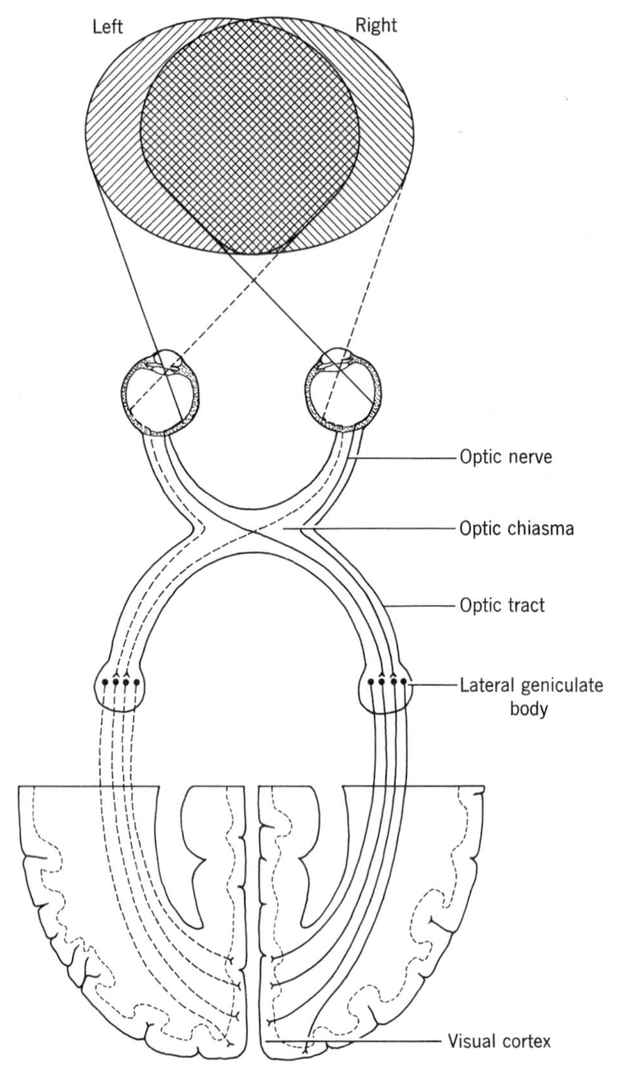

FIGURE 22.13. The visual pathways. Redrawn by permission from Youmans, *Fundamentals of Human Physiology*, Yearbook, 1957.

are a fixed distance apart, the angle of convergence to point at the same object will bear a trigonometrical relationship to the distance. The value of the two eyes as an optical range finder is limited by the small interpupillary distance.

An individual with only one eye has some disadvantage in judging distances, but this is usually apparent only when looking at unfamiliar objects. People with two eyes are in the same predicament when looking at very distant, unfamiliar objects. Tourists who are unfamiliar with mountain landscapes notoriously underestimate the distance of a range of mountains.

Visual acuity. The ability to distinguish details in the images projected on the retina can be thought of simply as the ability to distinguish between two points that are brought closer and closer together. In general, when two points projected on the retina are far enough apart to stimulate separate cones, they are perceived separately. Experiments have shown that even if the separate cones are immediately adjacent, the two points are still perceived separately. The cones are at their densest at the fovea, and this is the region of greatest visual acuity in the retina.

It is more difficult to distinguish details of an object at a distance than when it is nearby, because the image of a distant object is smaller, and a larger proportion of the image falls on each single cone. Furthermore, the sharpness of the image on the retina also limits visual acuity. No matter how discriminatory the response of the cones, details of a blurred image can never be clear. People with "poor eyesight" due entirely to defective accommodating power of the lens, can wear suitable correcting lenses and then have visual acuity as good as those with "good eyesight."

There are clearly psychological aspects to perception beyond the neural control of the eye movements and accommodation. Sailors, for example, through training and experience, can see another ship at enormous distances, whereas an untrained person with equally good visual acuity can overlook a vessel in relatively close proximity.

Vision and balance. The maintenance of posture and balance depends largely on the proprioceptors in the muscles and on the balance and movement receptors in the ears which will be discussed later. The eyes, however, also participate in the maintenance of balance, as can be shown by the slightly greater difficulty of balancing on one leg with the eyes closed than with the eyes open. In fact, minor defects in the balance reflexes can often be manifested only if the subject closes his eyes. Pilots flying in cloud where they have no horizon or other outside reference for their eyes frequently get the impression that they are either turning or changing altitude, when they are in fact flying straight and level. Their sensations can be so acute that they begin to disbelieve their instruments which tell them that they are flying straight and level, and in order to correct their illusory movement, they take action which can lead to their losing control of the airplane.

The ophthalmoscope. A difficulty in examining the interior of someone's eyeball is that the head of the observer tends to get in the way of whatever light is available. The invention of the ophthalmoscope solved this problem, as can be seen by Figure 22.14. The illumination is provided by a bulb in the handle of the instrument, and its light rays are reflected at right angles by a half-silvered mirror into the subject's eyes. The observer is then able to look through the half-silvered mirror into the illuminated interior of the eyeball. In modern instruments, the mirror is replaced by a prism, which occupies only half the observer's line of sight; the principle, however, is the same. The optics of the instrument are rather more complicated than the simple illumination system, because the front of the subject's eye is not a simple window, but is a lens, just like the lens of the observer's eye. Therefore, for examination of the retina of the subject's eye, the ophthalmoscope must include a concave lens to correct the refraction.

The ophthalmoscope is used for the simple examination of the eye ground, especially for the ruptured capillaries associated with hypertension. The appearance of the optic disk, the beginning of the optic nerve, is also of interest. If there is a rise of intracranial pressure, the disk is distorted, giving an appearance called *choked disk.* If there is a rise of intra-ocular pressure, the disk is pushed outwards (towards the brain) giving rise to a *cupped disk.*

The ophthalmoscope may also be used to determine defects in the lens of the subject. The sharpness of the image of the retina

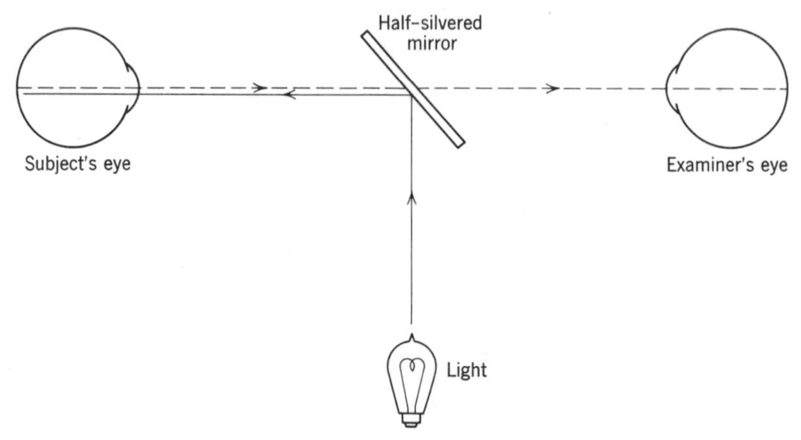

FIGURE 22.14. The principle of the ophthalmoscope. The solid line represents light from the source which is reflected on to the subject's retina. Light from the subject's retina is shown by the broken line.

of the subject on the retina of the observer depends on the refractive powers of the lenses of the two eyes and of the lens of the ophthalmoscope. In the head of the instrument is a "turret" containing a series of lenses which can focus the image of the subject's retina on that of the observer. When the eyes of both the observer and the subject are emmetropic, a lens with a refractive power of -4 diopters gives a clear image of the subject's retina. If the image is not clear, the observer can rotate the turret, trying successive lenses, until a clear image is obtained. The refractive power of the lens used to achieve this effect is therefore an indication of the kind of correction the subject's eyes need for emmetropy. For convenience, the lens of -4 diopters, suited to two emmetropic eyes, is labeled as 0, and the other lenses are labeled by the differences of their refractive powers from this reference lens (in diopters $+$ or $-$).

For successful employment of the ophthalmoscope, the subject must cooperate by focusing on a distant point as asked by the observer, so that the accommodation of his eye is not continually changing. If any extended ophthalmological examination is planned, the subject's pupils are previously dilated by the use of the drug, *atropine.*

HEARING

Physics of sound. Sound waves, unlike light waves, are not a form of electromagnetic radiation; they are, rather, actual vibrations of the medium in which they are propagated. Ordinarily this medium is the air, but sound can also be propagated through water and solid objects. Sound travels relatively slowly (1,087 feet per second in air). This is shown by the common observation that the roll of thunder associated with a distant lightning flash is heard some considerable time after the flash is seen. Even when watching a man drive a stake at a distance of a couple of hundred yards, the noise of the hammer striking the stake is oddly out of phase with the visual impression.

Sound wavelengths are very much longer than light wavelengths, and since they vary with the medium through which the sound is traveling, they are usually expressed as *frequencies.* At any given velocity of sound, the wavelength is a function of frequency and *vice versa.* High frequency sounds (with short wavelengths) are what we recognize as high sounds, and low frequencies are low sounds. As with the retinal receptors, the auditory receptors are sensitive to only part of the whole range of vibrations which may be regarded as sound in the strictly physical sense. Some people can detect notes of higher frequency than can others, and it is well known that dogs can hear notes beyond the human auditory range. Very low frequency sounds can cause visible vibrations in nearby objects, including the nonauditory tissues of the human body. This is apparent when we are close to an organ playing a very deep note or in the passage of a mass flight of several hundred airplanes.

A pure musical note is made up of sound of a single frequency. The notes of a piano are achieved because each string is of the exact length to vibrate at a particular frequency; the notes of an organ are correspondingly due to the length and diameter of the pipe of each note. In string instruments, such as the violin, the length of string vibrating can be varied by sliding the finger, thereby making it possible to produce notes of intermediate frequencies rather than at the discrete "intervals" of piano notes. The intervals between the notes of an octave are quite arbitrary in the purely physical sense, but they are nevertheless the intervals to which our western ears are accustomed. Music of different intervals, such as Oriental music, sounds peculiar to us.

Chords are combinations of pure notes, but almost all the *noises* we hear in everyday life are made up of bewildering complexes of frequencies which can be analysed only with elaborate equipment.

For a wild animal, the survival value of an awareness of the sounds of the external environment is so obvious that it need not be elaborated. In man, the difficulties of life for the completely deaf are sufficient indication of the value of hearing, not only for learning, but also for the warmer interpersonal relationships so important to human happiness.

The ear. The structures of the ear are an elaborate arrangement for transmitting sound waves from the air to the actual receptors, which lie deep in a *bony labyrinth* hollowed out of the *temporal bone* of the skull. Functionally and anatomically, the ear may be considered in three parts; the external ear, the middle ear, and the internal ear. This is shown diagrammatically in Fig. 22.15.

The *external ear* consists of the *pinna* (which is what we usually think of as the ear), and the *auditory meatus.* In man, the pinna is relatively flat against the skull so that its value in trapping sound is slight. Animals such as the cat and dog have more funnel-shaped ears, which, furthermore, can be directed toward the sus-

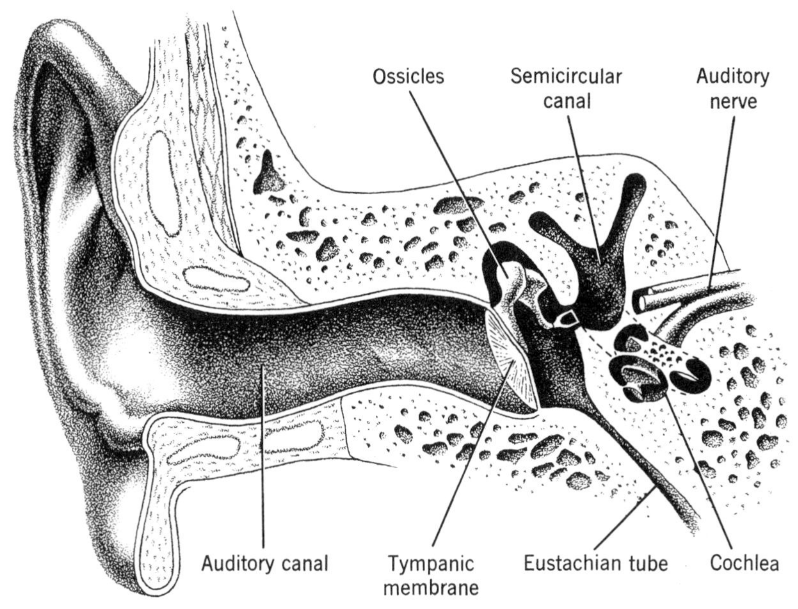

FIGURE 22.15. Parts of the ear.

pected source of sound. Thus, the pinna is undoubtedly a useful structure for these animals.

The auditory meatus is the short tube leading inward from the pinna; the inner end is closed by the *tympanum* (eardrum). Sound waves entering from the outside strike this membrane and cause it to vibrate at their own frequency.

The *middle ear* is a cavity beyond the tympanum. It contains three small bones (ossicles), whose arrangement is shown in Fig. 22.16. They are the *malleus,* the *incus* and the *stapes* (these names are Latin for hammer, anvil, and stirrup, respectively). They are joined together as shown in the diagram, but not rigidly. The handle of the malleus is attached to the tympanum so that it also vibrates when sound waves strike the tympanum. The vibrations are transmitted through the head of the malleus to the incus, which in turn moves the stapes back and forth. The foot part of the stapes is inserted like a piston in the *oval window,* which communicates with the inner ear. This piston-like movement transmits vibrations at the same frequency as the original sound waves to the fluid of the inner ear.

The lever-like arrangement of the ossicles is such that the *amplitude* of the vibrations transmitted to the inner ear is only

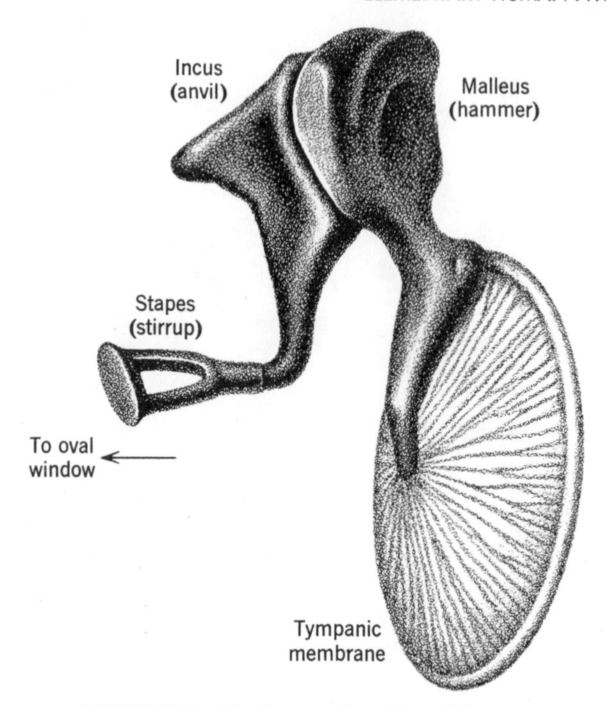

FIGURE 22.16. The three ossicles of the middle ear.

two-thirds of the amplitude of the vibrations in the tympanum. The *force* of the vibration in the stapes, however, is correspondingly increased.

A small muscle, the *tensor typmani,* is attached to the handle of the malleus and its tonic contraction draws it inward, thereby pulling the tympanum into a slightly conical shape which has accoustical advantages. Another small muscle, the *stapedius,* is inserted on the stapes and tends to pull it out from the oval window. A loud noise causes both of these muscles to contract reflexly, and further reduces the amplitude of the vibrations of the ossicles. This helps protect the even more delicate structures of the inner ear.

The cavity of the middle ear is connected to the nasopharynx by the *eustachian tube.* Through this tube, the middle ear is kept at atmospheric pressure; otherwise, changes in the external atmosphere would cause the eardrum to bulge inward or outward. The eustachian tube is normally closed but can be opened by a small pressure differential, or by swallowing. It can be temporarily or chronically plugged in upper respiratory infections, thereby preventing pressure equalization and causing acute discomfort through the bulging of the eardrum with modest changes in altitude (and atmospheric pressure). Furthermore, infection from the nasopharynx can spread into the middle ear via the eustachian tube and from there into spaces in the mastoid bone. The type of ear infection is very common, and previous to the development of antibiotics it frequently reached the stage where surgical drainage or removal of part of the mastoid bone was necessary in order to prevent the infection from spreading to the brain and associated tissues.

The *inner ear.* The structures of the inner ear are contained in a bony labyrinth—a complex of passages within the temporal bone itself (see figure 22.15). The bony labyrinth in turn contains an even more complex arrangement of membranes which is better understood from three-dimensional models than from a verbal description and two-dimensional diagrams. The inner ear actually contains three sets of receptors, only one of which is related to hearing; the other receptors are concerned with the sense of movement and with the sense of orientation with respect to gravity. The *cochlea* contains the auditory receptors, and the *semicircular canals* and the *utricle* are concerned with the other functions to be described later.

The cochlea is shaped rather like a snail's shell spiralling around a central pillar of bone. Its spiral cavity is divided by membranes which are shown in cross section in Fig. 22.17. Each of these three cavities enclosed by the membranes (the *scala vestibuli,* the *scala tympani,* and the *scala media*) runs the full length of the spiral cochlea. The oval window (in which the stapes is inserted) opens into the scala vestibuli. This chamber communicates through a small hole at the apex of the cochlea with the scala tympani. The two chambers are filled with *perilymph* (which is actually cerebrospinal fluid). When sound strikes the tympanum, the movement of the stapes sets up vibrations in the perilymph of the scala vestibuli.

The scala vestibuli and scala tympani are partly separated by a bony shelf and partly by the membranes enclosing the scala media. Therefore, vibrations in the perilymph of the scala vestibuli strike the *membrana vestibuli* and cause similar vibrations to pass through the endolymph of the scala media to the perilymph of the scala tympani on the far side. Here the vibrations are dissipated by moving the membrane covering the *round window,* which connects the scala tympani with the cavity of the middle ear.

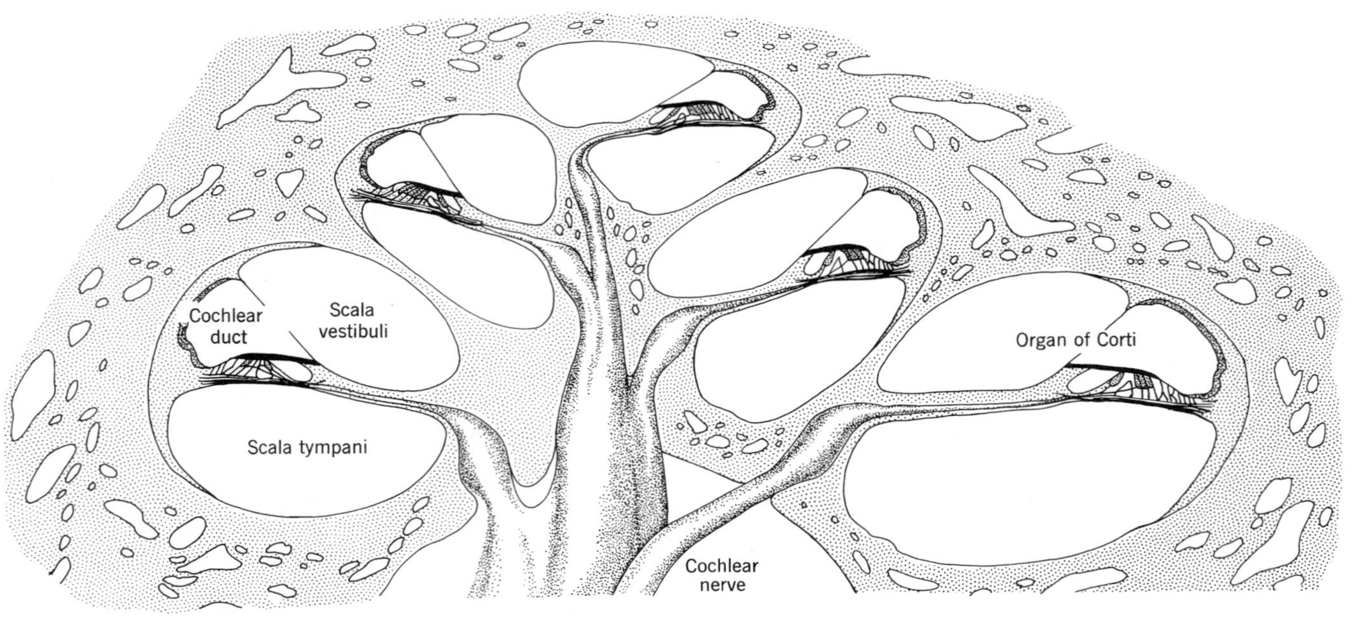

FIGURE 22.17. Cross section of the tube of the cochlea. (From von Békésy, G., *Sci. Am.* **197** (2) 66, 1957.)

In their passage through the endolymph of the scala media, the vibrations stimulate hair cells in the *organs of Corti,* which are the auditory receptors. Impulses from the receptor cells pass over the auditory nerve (VIIIth cranial) to the brain, where they are interpreted.

Because we are able to distinguish between sounds of different frequencies, the receptors must discriminate in some way. Helmholtz's "harp theory" postulates that the fibers in the *basilar membrane* vary in length systematically along the cochlea. Therefore, each part resonates with vibrations of a particular frequency in the endolymph. This would stimulate the hair cells in that one specific part of the organ of Corti, and the impulses received over fibers from those cells are interpreted as caused by a particular frequency. This ingenious theory is by no means universally accepted.

Deafness. From this discussion it is clear that there can be three kinds of deafness: conduction deafness, perception deafness, and central deafness. *Conduction deafness* is due to defects in the conducting mechanisms. A surprisingly common cause is the accumulation of wax in the auditory meatus, and this can be simply removed by syringing. Damage to the ear drum by explosions or by the entry of sharp objects is common, but a simple puncture of the tympanum usually heals with little impairment of hearing.

The calcification of the points of contact of the ossicles "freezes" them so they cannot act as levers to transmit the vibration of the tympanum to the perilymph. A delicate operation to "mobilize" calcified ossicles has been developed and is giving good results.

Vibrations can be conducted to the inner ear through the bones of the skull as well as via the ossicles, although much less efficiently. This can be demonstrated in an interesting fashion by jamming the blunt end of a phonograph needle into the end of a pencil or similar stick. If the needle is then applied to a turning record and the pencil clamped between the teeth, the music is clearly audible. Hearing aids work on a similar principle as they apply amplified sound to the bone behind the pinna.

Perception deafness follows damage to the receptor cells of the organ of Corti. The damage can be general, as from an infection, or more commonly it can be selective, with the deafness restricted to certain frequency ranges. *Boilermakers' deafness* is of this nature. Airplane pilots and artillerymen also show deafness at particular frequencies and multiples of these frequencies. It should be noted that the occurrence of deafness at specific frequencies is good evidence in favor of the Helmholtz "harp theory" of auditory perception.

Central deafness is due to damage to the auditory cortex, so that the impulses over the auditory nerve remain unperceived. This kind of deafness can follow psychic disturbance without actual trauma to the auditory cortex, and many bizarre kinds of deafness (and reversals) are manifested by hysteric patients.

Sense of movement and position. In addition to the auditory equipment, the bony labyrinth comprises the *utricle* and the three *semicircular canals.* These contain receptors that are sensitive to changes of the position of the head and to changes of speed and direction of movement.

The function of the utricle is probably more easily understood than that of the semicircular canals. The receptor organ, the *macula,* has cells with long hair-like extensions which project into an overlying gelatinous mass. This contains bone-like fragments called *otoliths* (Latin; ear stones). As the head is inclined forward, backward, or sideways, the effect of gravity causes the otoliths to press on different sensory hairs and thus to stimulate different receptor cells within the organ. The consequent pattern of impulses over the *vestibular nerve* (part of the VIIIth cranial) is delivered to the vestibular nuclei of the medulla and to the cerebellum. On the basis of this information, the reflex "righting" movements and other muscular responses to compensate for changes with respect to gravity are initiated.

The receptors for the semicircular canals are located in the *ampullae,* the enlarged portions near the lower junction of the canals (see Fig. 22.18). The receptors within the ampullae are called *cristae,* and they are very similar to the maculae except that there are no otoliths. Instead, the sensory hairs are bent by the swirling of the endolymph in the canals whenever the speed or direction of movement of the head is changed. A steady movement, as in a smooth automobile ride for example, conveys no sense of motion if the eyes are shut. We are however, well aware of sudden braking, acceleration, or turning of corners.

Figure 22.18 shows that the semicircular canals are arranged at right angles to each other. Movement in any direction in space can be resolved into components in these three planes, and the degree of endolymph swirling in each canal depends on the rate and direction of the acceleration (or deceleration) of the head. The patterns of impulses from the cristae are interpreted within the brain, and appropriate muscle reflexes are initiated.

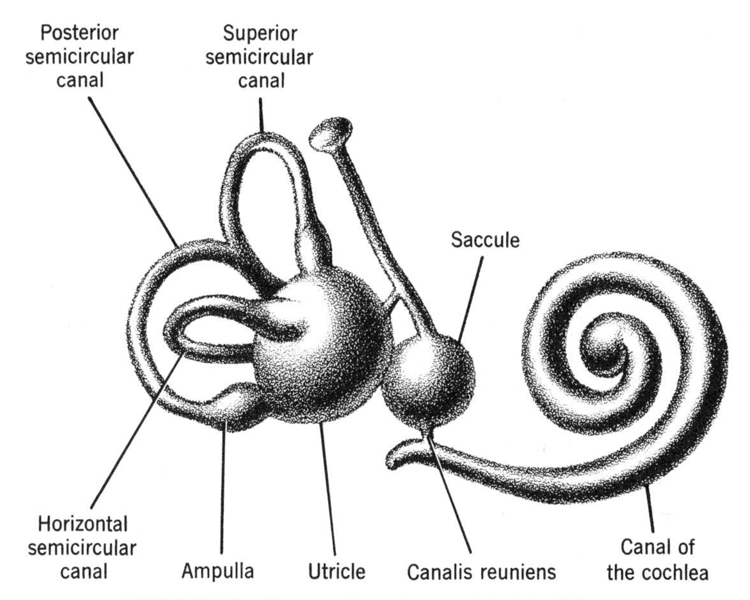

FIGURE 22.18. Diagram of membranous labyrinth of the ear.

An interesting and conveniently demonstrated vestibular reflex is *nystagmus.* When a subject is seated in a revolving chair, the eyes move in the opposite direction to the spin so that they fixate on one point. As that point moves out of the field of vision, the eyes turn very rapidly in the direction of spin, select another fixation point and follow that until it also disappears. If the spinning is stopped suddenly, the endolymph continues to swirl and the subject not only has the sensation of continued spinning (and cannot stand up), but the reflex eye movements persist. The eyeballs move as if tracking a moving object and then quickly return to the original position and track once more. The movement is reminiscent of that of a typewriter carriage. Some teaching laboratories are equipped with strongly built rotating chairs (Bárány chairs) and equipment for recording the eye movements while spinning. With due consideration for the safety of the subject however, the phenomenon can be shown using a revolving office chair or piano stool.

Motion sickness, as on the ocean, results from an overstimulation of the position and movement receptors by the unfamiliar pitching and tossing of the ship or airplane. *Ménière's syndrome* is a condition in which the patient is subject to periodic attacks of vertigo and nausea, just as if he were being violently seasick. The vertigo is usually accompanied by ringing in the ears (*titinnus*). The cause of the condition is still obscure but may well be related to ischemia of the labyrinth due to more general vasomotor disturbances. It is certainly one of the most miserable of the nonlethal diseases to which mankind is subjected.

TASTE

The taste receptors, or taste buds, are of modified epithelial tissue and are located on discrete areas of the tongue. They are mainly found on the sides of *papillae,* nipple-like projections of various sizes from the surface of the tongue (Fig. 22.19).

The sensation of taste when eating a delicately flavored meal is actually more complex than can be accounted for by the stimulation of the taste buds alone. Our sense of smell is an extremely important contributor to the overall sensations while eating. There are four primary taste sensations; sweet, salty, sour, and bitter. It is presumed on this basis that there are four kinds of taste buds, but this has not been demonstrated beyond all doubt. Foodstuffs seldom yield only one sensation, but rather a combination of two or more of the primary sensations.

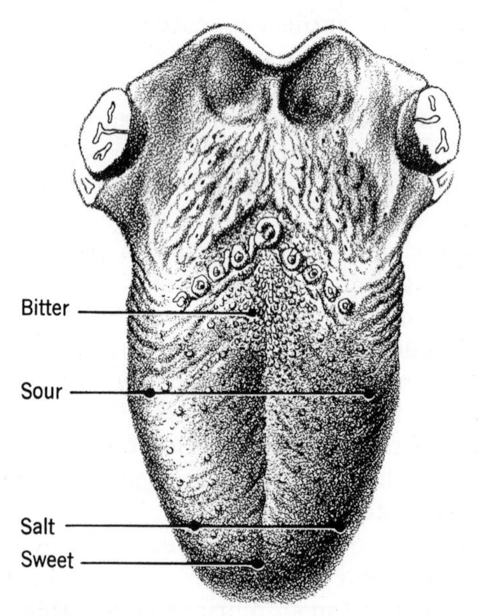

FIGURE 22.19. Location of the taste buds on the tongue.

The taste buds at the tip of the tongue are those the most sensitive to sweet and salty tastes. The salty taste is also perceived by taste buds along the lateral margins of the tongue. The sour taste seems to be detected by taste buds all over the tongue and indeed by the few taste buds in the sides of the mouth and pharynx. At the back of the tongue, the *circumvallate papillae* are large, clearly-visible projections forming a rough V. The taste buds on these papillae are particularly sensitive to the bitter taste.

The sweet taste is evoked by the monosaccharides and disaccharides described in an earlier chapter, but also by several compounds chemically unrelated to the sugars. Notable among these are the poisonous lead acetate (also known as sugar of lead) and the complex synthetic compound known as *saccharin.* Saccharin is used as a sugar substitute in beverages by diabetics and those wishing to restrict their caloric intake, but some doubts have been raised concerning its harmlessness.

The salty taste is primarily evoked by sodium chloride, and to a lesser extent by the chlorides and bromides of the other light metals (lithium, potassium, etc). It is interesting to note that the stimulus is probably due to the anions, because a solution of partly-ionized sodium acetate does not evoke the sensation of saltiness.

The sour taste is probably the most specific of all. The taste buds respond to all acids, that is, compounds dissociating to yield hydrogen ion. Finally, the bitter taste is perceived on contact with solutions of alkaloids, such as quinine.

Curiously enough, the taste receptors respond to high concentrations of materials suddenly injected into the blood stream, as well as when placed on the tongue. People who receive thiamine injections usually detect the yeast-like flavor of this compound. Another compound, sodium dehydrocholate, has a useful application of this connection. After it is injected in the arm, there is a short delay and then the subject becomes conscious of its characteristic bitter taste. If the time interval is carefully measured with a stop watch, it is a useful index of the circulation time and therefore of the cardiovascular status of the subject.

SMELL

The receptors for smell are found in the *olfactory membrane* located in the upper passages of the nose (Fig. 22.20). Their sensitivity is greater than that of the taste receptors by a factor of

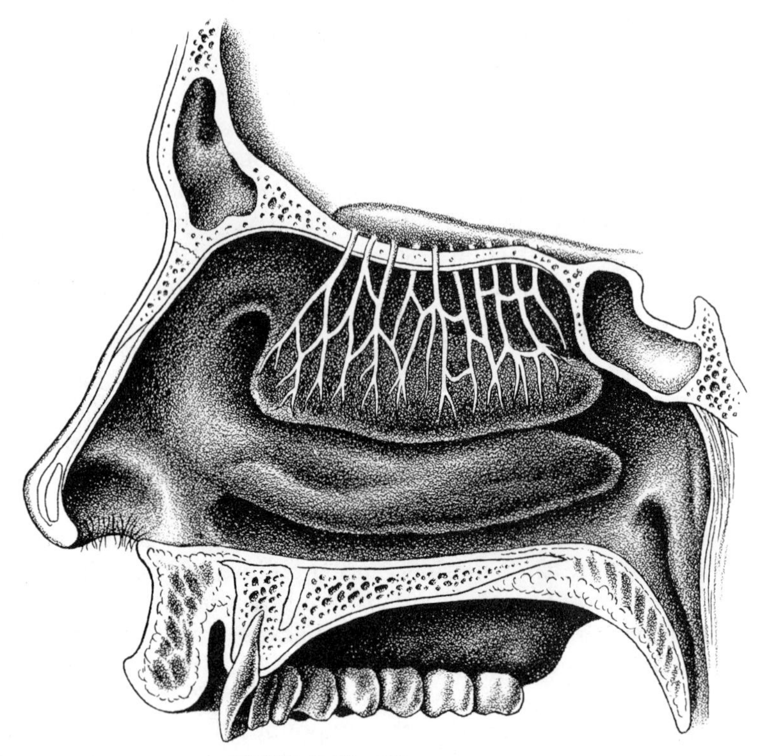

FIGURE 22.20. Olfactory receptors.

several thousand. This is one of the reasons why the sense of olfaction is so mysterious. The limiting concentrations at which different compounds can be detected by our sense of smell is roughly correlated with their volatility; the more volatile substances being the more readily detected. It might be reasonably supposed, therefore, that the molecules of the vapors of substances dissolve in the mucus on the olfactory membrane. Experiments with animals and with human cadavers have shown, however, that the air drawn in through the nose during a normal inspiration does not actually pass directly over the olfactory membrane, and that this is probably too simple an explanation. It has been suggested that the olfactory receptors are sensitive to electromagnetic radiation in the infrared region from the molecules of the stimulating substances.

In characterizing the sensory response to light and sound, we do at least have the advantage that the stimulus in each case can be properly characterized by frequency, etc. Taste, and to an even greater extent smell, can only be described in strictly subjective terms. The complexity of the sensation of smell has made it very difficult to classify odors, although several determined efforts have been made in the past. Because taste and smell are both "chemical senses," they are commonly lumped together. Neurologically however, they are quite different. The impulses from the receptors in the olfactory membrane are the only sensory impulses which do not pass first to the thalamus in the brain. Instead, they go to the olfactory bulb, a forward projection of the brain itself, whence they are transmitted to the amygdaloid and pyriform nuclei. From there, the impulses seem to be projected to all levels of the brain. This is extremely interesting from an evolutionary standpoint and suggests that olfaction is the most primitive of all sensations.

THE SOMESTHETIC SENSATIONS

The receptors for the somesthetic sensations are small and much more scattered than those in the retina or the organ of Corti. Some of them are illustrated in Fig. 22.21. It will be seen that they are histologically quite distinct, but the assignment of various functions to these structures is not yet completely certain.

Pain. The receptors for pain seem to be simply branched, free nerve-endings. They are found in all the tissues of the body except the brain, which is paradoxically insensitive to pain. Despite the length to which we are prepared to go to avoid pain, it is clearly a very valuable sensation in that it warns us of contacts damaging to the tissues. A patient's description of the nature and location of a pain is important to the physician in attempting to diagnose an illness.

Touch. Probably the lightest contact of which we are conscious is one that merely bends the hairs on the skin slightly. Many hairs have nerve endings twined around their roots and these are stimulated by the distortion as the hair is bent. The resulting signal is interpreted by the brain as a light contact in that region of the body. Meissner's corpuscles are other receptors in the skin that are sensitive to light touch. A firmer contact, amounting to a *pressure,* stimulates also the *Pacinian corpuscles* which lie much more deeply under the skin.

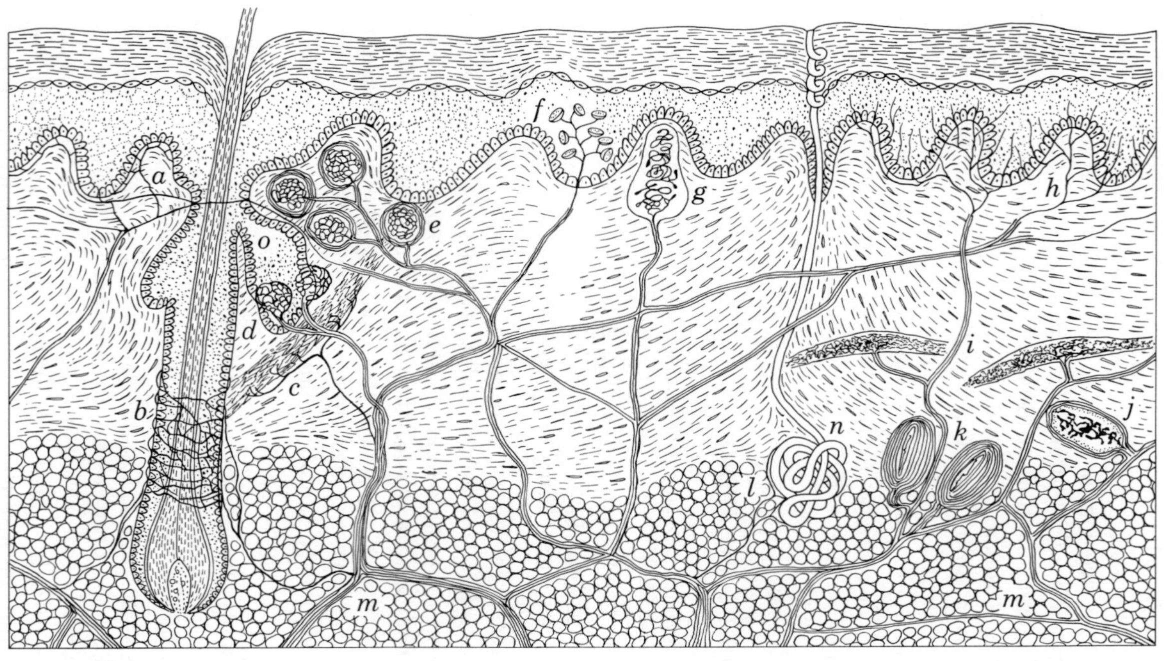

FIGURE 22.21. An idealized section of skin showing several types of receptors. *a* indicates free nerve endings; *b*, nerve endings around a hair follicle; c, sympathetic nerve fibers suppling a small muscle; *d*, Ruffini's endings; e, Krause's end bulbs; *f*, Merkel's disks; *g*, Meissner's corpuscles; *h*, free endings; *i*, Ruffini's endings; *j*, Golgi-Maxxoni endings; *k*, Pacinian corpuscles; *l*, sympathetic fibers innervating a sweat gland; *m*, nerve trunks; *n*, sweat gland; o, sebaceous gland. The function of each type of ending is not known. (From Livingston, W. K., *Sci. Am.*, **188** (3), 59, 1953)

The distribution of the touch receptors is not uniform all over the body. They are very close together on the finger tips and the lips, for example, as is evidenced by the great tactile sensitivity of those regions. The distribution of the touch receptors can be charted by use of *two-point discrimination* tests. If two points in contact with the skin are far enough apart to stimulate touch receptors which are separated by at least one unstimulated touch receptor, we can distinguish between the two points. If the points are close enough together so that they stimulate the same touch receptor or two immediately adjacent ones, we are able to perceive the contact of one point only. The points of a pair of dividers can be applied to the skin of a blindfolded subject at various locations. It will be found that on the lips or finger tips, the subject is able to discriminate two-point contact when the dividers are open only one or two millimeters, but in the lumbar region he may interpret as only one point the contact of two points as much as a centimeter apart.

Temperature receptors. The perception of heat and cold are through separate receptors. *Ruffini's end organs* are those sensitive to heat. Those for cold are *Krause's end bulbs.* Extremely hot objects stimulate the pain receptors in addition to the Ruffini end-organs.

The proprioceptors. The brain is informed of the degree of contraction or relaxation of the skeletal muscles by impulses from receptors within the muscles, in the tendons, and in the fascia around the muscles. The contraction of a muscle causes the stretching of the *muscle spindles* which are actually wrapped around the muscle fibrils. At the same time, the *Golgi tendon organs* are also distorted by the increased tension in the tendon and their impulses contribute to the barrage of information. *Ruffini end organs* in the fascia of the muscle are also contributory, as are also the pressure sensations on the soles of the feet when standing or in other areas when sitting.

Dermatomes. We have mentioned earlier that the nerves entering or leaving the spinal cord at each level generally serve segments of the body related to that level of the spinal cord. The *dermatomes* are the areas relating to each spinal segment (Fig. 22.22). For simplicity the diagram shows clear-cut borders between each dermatome, but of course there is much merging and overlap.

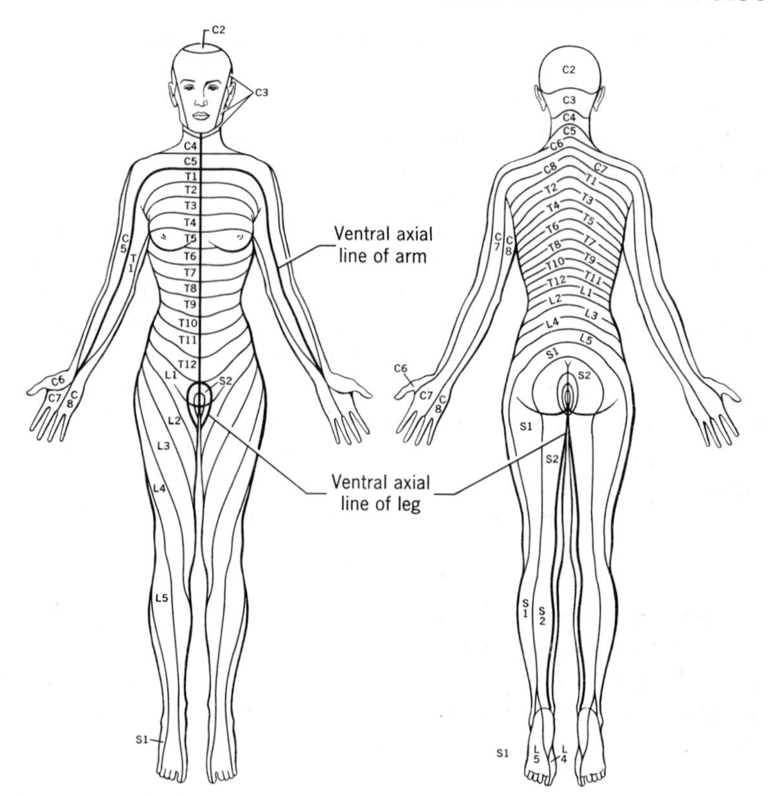

FIGURE 22.22. The dermatomes. Reproduced by permission from Keegan and Garrett, *Anatomical Record,* **102,** 411, 1948.

REFERENCES

1. Von Békésy, Georg, "The ear," *Sci. American* **197** (2), 66, August 1957.
2. Chapanis, Alphonse, "Color blindness," *Sci. American* **184** (3), 48, March 1951.
3. Haagen-Smit, A. J., "Smell and taste," *Sci. American* (3), 28, March 1952.
4. Kalmus, Hans, "The chemical senses," *Sci. American* **198** (4), 97, April 1958.
5. Land, Edwin H., "Experiments in color vision," *Sci. American* **200** (5), 84, May 1959.
6. Livingston, W. K., "What is pain?" *Sci. American* **188** (3), 59, March 1953.
7. Milne, Lorus J. and Margery J., "Electrical events in vision," *Sci. American* **195** (6), 113, December 1956.
8. Sherrington, Sir Charles, "Sherrington on the eye," reprint, *Sci. American* **186** (5), 30, May 1952.
9. Sperry, R. W., "The eye and the brain," *Sci. American* **194** (5), 48, May 1956.

PART 5

REPRODUCTION

INTRODUCTION

The simplest form of reproduction is the binary fission of a cell to yield two daughter cells, each identical with the original (see p. 22). Microorganisms such as amoeba commonly reproduce by this asexual method, although they have sexual methods of reproduction which are analogous to those in higher animals. The regeneration of tissues in the mammalian body, typically the intestinal mucosa, for example, is by the production of daughter cells by mitotic division (see page 22). Similarly, *growth* takes place in part by the multiplication of cells. The higher animals reproduce their kind by sexual reproduction exclusively.

As was briefly described in an earlier chapter on cells and tissues, the essential characteristic of sexual reproduction is the formation of gametes by *meiotic* division (reduction division). Each gamete contains half the chromosome content characteristic of the cells of the parent, and the union of the two gametes brings the chromosome content back to the full complement, with contributions from each parent. Therefore, the offspring from sexual reproduction are not identical copies of the parents as in the binary fission of cells, but can show endless variations of their hereditary characteristics. This kind of reproduction is common to both plants and animals. In animals the gametes are the sperm and egg cells, which are produced by the male and female sexes respectively. They originate in special germinal epithelia of the reproductive system where they develop through a series of stages which will be described. In mating, the sperm have access to the eggs and can fertilize them. Once an egg has been fertilized, it enters into rapid cell division of a mitotic nature, developing into an embryo and later, into a separate, free-living individual.

In some simple marine organisms, the gametes are released into the sea water, where the sperm, which are independently motile, seek out and fertilize the eggs. In some fishes, the male approaches the female closely and expels the sperm at the same time that the female expels the ova. In mammals and birds, the male deposits the sperm within the genital tract of the female and the fertilization takes place in the *oviduct.* In birds, the fertilized egg is "laid" and the embryo develops inside its shell. In the mammals the fertilized egg is retained in the *uterus* of the female, where it undergoes extensive development in a completely protected environment.

LIBIDO

The sex drive, or *libido,* is one of the behavioral characteristics controlled by the hormones secreted by the gonads, and these are in turn under the control of the gonadatrophic hormones secreted by the adenohypophysis. We have indicated that the behavioral repercussions of sexuality extend far beyond the mere seeking of a mate and copulation. In some mammals and birds there are elaborate courting procedures wryly reminiscent of human courting and sexual competition.

The sexuality in the human has several differences from that in lower mammals. Most mammals grow rapidly and continuously, reaching puberty and sexual activity relatively early in their lives. Human children, on the other hand, grow rapidly between birth and seven or eight years of age and then their development slows down considerably until the age of 11 or 12, when puberty begins. Even after puberty, when the individual is physiologically capable of reproducing, cultural factors tend to delay this by several years. This long human childhood is clearly important for the learning processes by which the individual benefits from the cumulative experience of previous generations.

Freudian psychology is based on the theory that during this

childhood period, the child develops a sexuality centered on himself (and his relationship with his parents), which determines the structure of his adult personality. The Freudian theory of the predominantly sexual motivation in children and adults is now widely accepted, if perhaps in modified form, but at one time it was vigorously opposed. A good part of the opposition arose no doubt from an unwillingness to accept the concept of sexuality in "innocent" little children.

REPRODUCTION IN THE HUMAN MALE

23

The male genital equipment is illustrated in Fig. 23.1. The sperm-producing organs are the *testes,* which are located outside the body in the bag-like *scrotum.* The germinal epithelium is in the highly convoluted *seminiferous tubules* which make up the mass of each testis. The sperm, after their production in the seminiferous tubules, pass along them into the *vasa recta,* from there into the intricately coiled *epididymis.* They then ascend from each side to the abdominal cavity via a tube, the *vas deferens.* Here, the tube enlarges to form a storage section, the *ampulla;* adjacent to this are two *seminal vesicles* which secrete a fluid for the nutrition of the sperm. From the ampulla, the duct passes through the *prostate gland* into the *urethra,* a tube which passes out of the abdominal cavity down the middle of the *penis,* the other portion of the external male genitalia. The urinary bladder is located just above the prostate gland and the urethra is also the route for excretion of urine. The penis is an approximately tubular organ plentifully supplied with large venous sinuses which can fill with blood under the appropriate psychic or mechanical stimulation in sexual excitement.

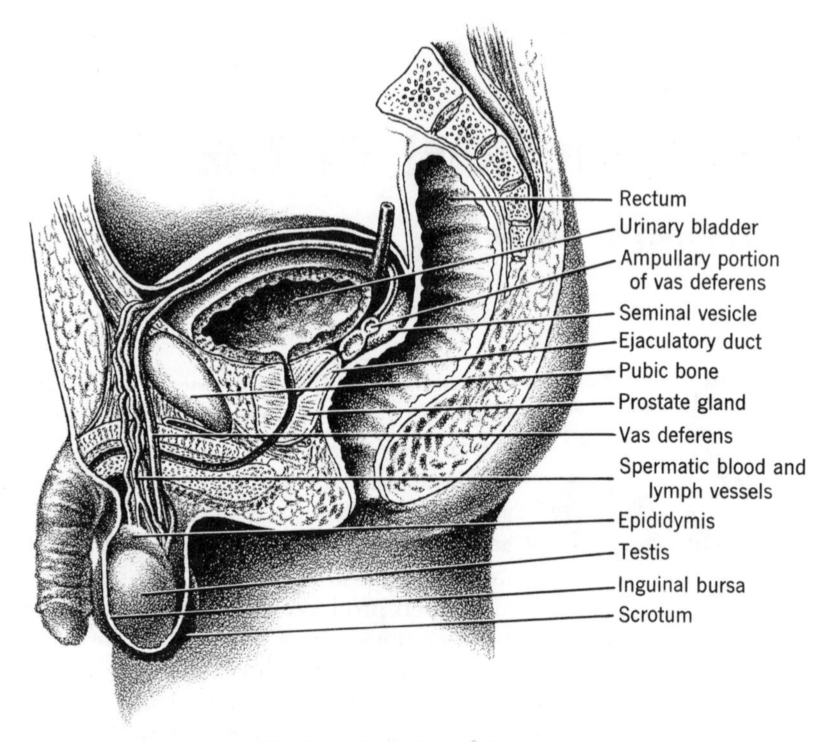

FIGURE 23.1. The male reproductive system.

HORMONAL FUNCTION OF THE TESTES

In the spaces between adjacent seminiferous tubules there are the *interstitial cells;* these are stimulated by the ICSH (interstitial cell stimulating hormone) of the adenohypophysis and produce the male hormones, the *androgens.* Through some biological "clock" which we do not yet understand, the gonadotrophic secretion starts at the age of 10 or 11, initiating the proliferation of the interstitial cells and their production of androgens. It will be recalled from the chapter on endocrinology, that the male hormones are responsible for the libido and for the secondary sexual characteristics, that is, the masculine appearance of the male. If the testes are absent, due to accident or congenital defects, or if there is hyposecretion of the gonadotrophins at the time of normal puberty, there is no reproductive potential, no libido, and the secondary characteristics do not develop. Such an individual is a *eunuch.* The loss of the testes after puberty leads to some recession of the secondary sexual characteristics but not to the marked extent that is seen in prepubertal castrates.

SPERMATOGENIC FUNCTIONS OF THE TESTES

The epithelium lining the walls of the seminiferous tubules contains continuously proliferating cells called *spermatogonia.* These develop to a form called *primary spermatocytes,* which undergo meiotic division to yield two *secondary spermatocytes* each. These immediately divide again, but mitotically, to yield two *spermatids* each. The meiotic division yielding the secondary spermatocytes means that they and the spermatids have only half the chromosome content of the parent cells. While still in the seminiferous tubule, the spermatids develop the tail and other structural characteristics of the mature sperm shown in Fig. 23.2. The newly formed sperm pass into the epididymis where they remain for some time and undergo further maturation processes, not fully understood. The mature sperm are stored in the vas deferens and the ampulla.

There is almost overwhelming evidence that spermatogenesis proceeds best at temperatures slightly below that of the normal

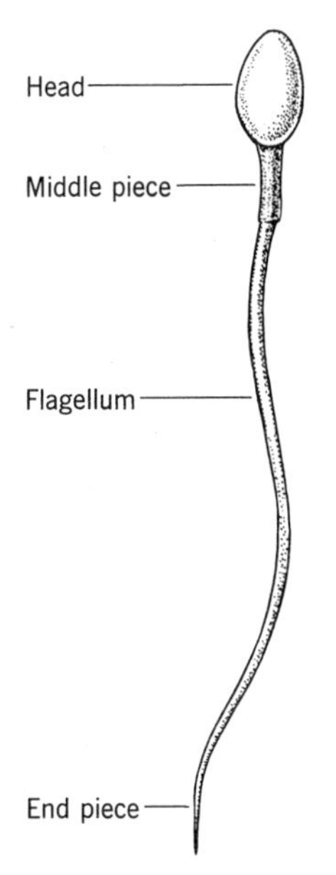

FIGURE 23.2. A spermatozoon.

body. The location of the testes in the scrotum outside the body cavity does in fact keep their temperature a few degrees lower. It has been suggested that frequent hot baths cause a temporary depression of fertility. In cold weather the skin of the scrotum shrinks and the testes are drawn up close to the body; conversely in hot weather, the skin of the scrotum is loose and flaccid and the testes are well separated from the body.

In the embryo, the testes develop within the body cavity, but before birth they descend the *inguinal canals* to their location in the scrotum. Sometimes one or both testes do not descend, giving rise to the condition called *cryptorchidism.* An undescended testis develops its endocrine functions under gonadatrophic control at puberty but does not produce sperm, presumably because of the higher temperature in the abdomen. Most cases of cryptorchidism can be corrected by a relatively simple surgical procedure provided it is undertaken well in advance of puberty.

THE SEX ACT IN THE MALE

The sex act is known as *copulation* or *coitus;* in the more sociological literature it is frequently referred to as *sexual intercourse.* In the male it begins with the *erection* of the penis, due to the engorgement of the venous sinuses. The hemodynamic changes are integrated by parasympathetic impulses from the sacral region of the cord. The actual process of erection is therefore involuntary, and can occur during sleep or while otherwise preoccupied. The mechanical pressure of a full bladder on the prostate gland for example, can cause erection. Ordinarily, the autonomic impulses causing erection originate from the hypothalamus which is influenced by higher centers. The state of sexual excitement is one of the "primitive" sensations like hunger or thirst, in which it is believed the *limbic lobe,* or "old cortex," plays an important role. There is certainly no doubt that under appropriate circumstances, sexual excitation is just as distinctly manifested as is hunger and thirst.

In male farm animals, the trigger for sexual excitement is the proximity of a female in "heat." In the human male, sexual stimulation is much more complex, although the proximity of a responsive female is still probably the stimulus *par excellence.* The stimuli for sexual excitement and consequent erection can be entirely psychic, such as erotic thoughts or dreams, or even some intrinsically uninteresting occurrence with erotic associations.

Even in coitus itself, it is clear that psychic considerations relating to the partner and the circumstances are important determinants of the degree of excitation.

Erection of the penis is a mechanically essential prerequisite for the next stage of the sex act, which is the insertion of the penis into the vagina of the female. The enlarged distal portion of the penis, the *glans,* is plentifully supplied with nerve endings. The friction of the lubricated walls of the vagina on these receptors causes afferent impulses to the thalamus and higher centers, giving rise to extremely pleasurable sensations. The cumulative effect is to provoke a further autonomic discharge which causes the *orgasm.* In the male this is accompanied by, and indeed almost identified with, *ejaculation.* Contractions beginning in the vas deferens and spreading throughout the duct, the seminal vesicles and the prostate ejaculate the *semen.* The semen is the complete fluid comprising the sperm and the secretions of the seminal vesicles and the prostate gland. After the semen is ejaculated into the vagina of the female, the sperm can swim independently to achieve the fertilization of an ovum.

Like erection, the ejaculatory reflexes are entirely involuntary and can be evoked in an unconscious male. The degree of excitement due to the psychic factors referred to probably determines the speed with which orgasm is reached.

THE PROSTATE GLAND

The alkaline fluid secreted by this gland serves to neutralize partially the acid conditions in the vagina, which would otherwise inhibit the motility of the sperm.

Infections of the prostate gland are extremely common and the irritation can lead to an almost continual and painful erection. Cancer of the prostate gland is a common cause of death in men, and its occurrence is probably related to the incidence of chronic infection in earlier life. Even without malignant growth, hypertrophy of the prostate gland is common in older men. The swelling causes constriction of the urethra where it passes from the bladder through the gland, making urination and the complete emptying of the bladder difficult. Sometimes the occlusion is complete and urination can only be accomplished by passing a catheter. The prostatic portion of the urethra can be enlarged surgically, but more commonly the prostate is removed entirely, partly as insurance against possible future neoplastic growth.

REPRODUCTION IN THE HUMAN FEMALE

24

The female genital apparatus is shown in Fig. 24.1. The *ovaries,* which produce the eggs, are analogous to the testes in the male, but are located within the abdominal cavity. Arising near each is a *fallopian tube,* which descends and connects with a muscular sac, the *uterus.* In most mammals, the uterus is shaped like a capital Y with long arms, and the fallopian tubes are continuous with the *horns* of the uterus. In the human female, the uterus does not have so distinctly a bifurcated form and the lower part of the Y is the uterus proper. This lower portion of the uterus terminates with a narrow constriction, the *cervix* which projects into the *vagina.*

The vagina is a muscular tube lined with mucus membrane which terminates in the *vestibule,* into which the urethra also discharges. The external genitalia of the female comprise the *labia major* and the *labia minor,* two pairs of fleshy lips on either side of the vestibule. The *clitoris,* a small analog of the male penis, similarly composed of erectile tissue, is located near the anterior junction of the labia minor. A useful inclusive term for the external genital region is the *perineum.*

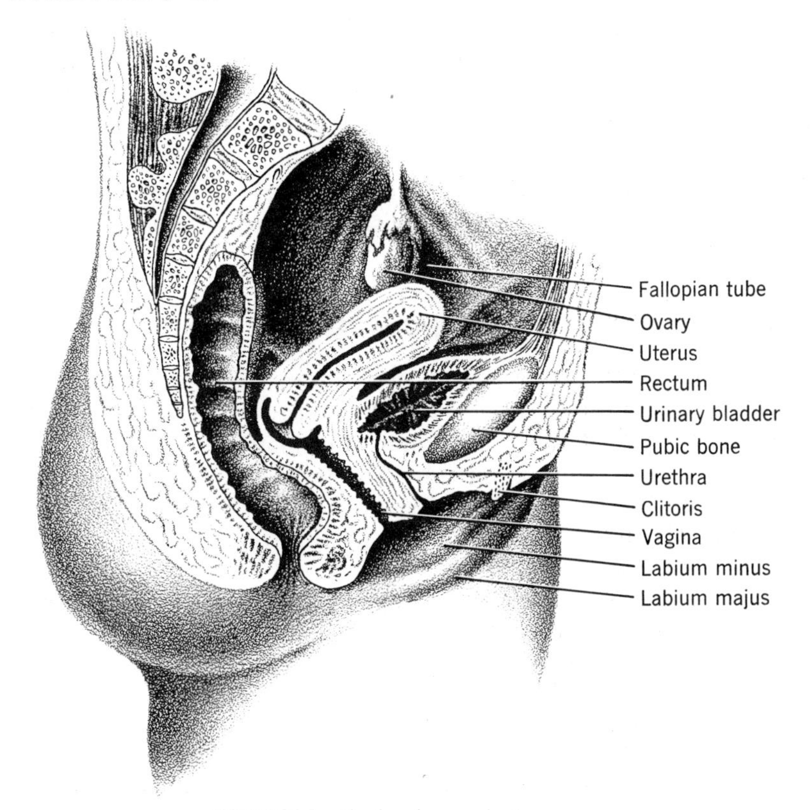

FIGURE 24.1. The female reproductive organs.

THE ENDOCRINE FUNCTION OF THE OVARIES

The estrogens produced by the ovaries are analogous to the androgens produced by the testes. Their production begins before puberty under the stimulus of the gonadatrophic hormones from the adenohypophysis, which promote the development of the primary and secondary sexual characteristics. Under the influence of the estrogens, the size of the vagina and uterus increases and the breasts develop their mature shape. At puberty there is also a relative widening of the female pelvis and a deposition of subcutaneous fat giving the characteristically rounded feminine contours to the body. Hyposecretion of estrogens due to ovarian or adenohypophyseal disorders causes a girl to grow with a masculine body configuration and distribution of body hair.

The other hormone secreted by the ovaries is progesterone. This hormone is concerned with the proliferation of the uterine lining in preparation for receiving a fertilized ovuum. In the pregnant female, progesterone is important in maintaining the pregnancy and in causing the cellular changes in the breasts leading up to the actual secretion of milk. A further account of the functions of this hormone is given in connection with our discussion of pregnancy and lactation.

OOGENETIC FUNCTIONS OF THE OVARIES

In the ovaries, the germinal epithelium is the outer covering rather than an inner lining, as in the testes. The epithelial cells proliferate to yield many thousands of primordial ova, all of which can potentially develop into ova for fertilization. Ova are not released in the same prodigious quantities as are spermatoza, and each month only one or two ova actually ripen and descend the Fallopian tubes.

A ripening ovum on the ovary becomes surrounded by a mass of granulosa cells, making a *follicle* which grows until it is more than a centimeter in diameter. At *ovulation,* the follicle ruptures and expels the ovum. Just before this occurs the ovum undergoes meiotic division within the follicle, but unlike the production of sperm, only one of the resulting cells survives. The other, called the first *polar body,* is expelled. The surviving cell then contains half the chromosomal content of the parent; it undergoes a subsequent mitotic division and one of the daughter cells is expelled as the *second polar body.*

Ovulation occurs midway between the menstrual cycles. Most women are unaware of its occurrence, but the rupture of the follicle causes a small hemorrhage which can irritate the adjacent peritoneum enough to cause a transient discomfort known as *mittleschmerz.* After expulsion of the ovum, the granulosa cells of the ruptured follicule do not degenerate, but rather proliferate greatly and develop into the *corpus luteum.* This is a temporary endocrine gland secreting progesterone, which further promotes the preparation of the uterine lining for the implantation of the fertilized ovum. Its functions extend well into pregnancy. If no fertilization takes place, the corpus luteum stops secreting its hormones after about 12 days, whereupon menstruation occurs.

THE SEX ACT IN THE FEMALE

After the spermatozoa are ejaculated into the vagina, they can swim through the cervix into the uterus and ascend the fallopian tubes. If coitus has been roughly coincident with ovulation, a descending ovum can be fertilized. The fertilized ovum continues to descend the fallopian tube and becomes implanted in the uterine wall, where it undergoes development which will be described.

All of these processes can take place without sexual excitement or even willing participation of the female. In the human female, the libido oscillates with the menstrual cycle, but there is no *heat period* analogous to that seen in other mammals. In most species, the female only displays sexual interest at the time of ovulation. The psychic determinants of sexual excitement in the human female are even more subtle than in the male, the erotic stimuli are usually less definable and are more related to "atmosphere" than to specific acts or associations. Under the appropriate emotional conditions however, the mechanical stimulation of the external genitalia heightens sexual excitement. The friction against the vaginal walls during coitus induces pleasurable sensations comparable to those in the male and which similarly culminate in an orgasm. The clitoris becomes erected in sexual excitement and usually the friction against it during coitus is also an important source of the stimulation leading to an orgasm. The female orgasm, like that of the male, is a release from the pleasurable tension of mounting excitement but it is not, of course, accompanied by an ejaculation. It has been suggested that the rhythmic contractions of the vagina and uterus during an orgasm tend to move the semen inwards, thereby making fertilization more likely. The importance of this is doubtful.

It has often been stated that many women do not derive satisfaction from their sex life in marriage because they have been brought up to regard sex as evil and to be avoided. It is presumed that their ingrained attitudes inhibit their pleasurable participation in intercourse. This is undoubtedly true, but it has probably been discussed and written about so much that, in the college educated population at least, new problems have arisen because young women become too preoccupied with their "success" or "failure" in achieving an orgasm. Several well-meaning writers have given meticulous advice on the mechanical aspects of satisfactory intercourse, but it must be emphasized that under the appropriate emotional circumstances, the mechanical arrangements have a way of taking care of themselves. A suitable emotional atmosphere is not achieved when either partner is preoccupied with technicalities.

THE MENSTRUAL CYCLE

In nearly all mammals there is a distinct rhythmicity to ovulation. In many wild mammals and in the domestic sheep, ovulation only occurs during one period of the year. This is presumably the consequence of natural selection, because the offspring conceived at that time are born under favorable conditions in the spring. At the other extreme, the ovulation cycle in the rat is only a few days; in the rabbit it occurs as a reflex triggered by coitus.

Ovulation occurs at monthly intervals in the human, but the cyclic changes in libido are less marked than in other mammals. A further important difference is the occurrence of menstruation, which, it should be emphasized, is *not* due to the rupture of the follicle. Ovulation occurs about midway between the menstrual periods, and although there may be a small hemorrhage at this time, it amounts to only a barely visible drop or two of blood. Menstruation is the consequence of the monthly preparation of the walls of the uterus to receive a fertilized ovum. The proliferated tissue is sloughed off at the end of the month if fertilization has not occurred, and this is the *menstrual flow.*

Menstruation begins in puberty, soon after the ovaries assume their adult function under the influence of the gonadatrophins. This inception is called the *menarche.* The average age for its occurrence is 13 years, but it may frequently appear as early as 10 or 11 years of age and can be delayed as late as 15 or 16. Menstruation then continues, except for periods of pregnancy and lactation, until the *menopause,* which marks the end of the reproductive life of a woman. The age of the menopause is much more variable, but the normal range is between 45 and 50. Contrary to popular belief, the cessation of reproductive capacity and the menopause does not necessarily mark the end of libido in the human. Its continuance however, probably depends on the emotional conditioning of the sex life of the previous 20 years. Cases have been reported of women whose libidos have increased after the menopause, when the inhibitions due to fear of pregnancy are removed.

The *endometrium,* the lining of the walls of the uterus, is relatively thin immediately after menstruation. During the first two weeks of the cycle (up to the time of ovulation), the estrogens secreted by the ovaries promote the proliferation of the endometrial cells and the growth of blood vessels into the thickened tissue. After ovulation, the progesterone secreted by the corpus luteum promotes even more massive development until it is about a quarter of an inch thick. At this time the tissue contains considerable deposits of fat and glycogen which provide a nutrient reserve for the early stages of the intra-uterine life of the embryo. If there is no implantation, the end of the cycle (as marked by menstruation) is brought about by the abrupt cessation of secretion from the ovary, probably due to reduced gonadotrophic secretion by the adenohypophysis. The reasons for the almost exact rhythmicity are not understood at all. The sudden falling off in the estrogen secretion causes vasoconstriction in the blood supply to the thickened endometrium, and the cells literally die for want of oxygen and nutrients. As in all tissues under such conditions, necrosis sets in and the thickened endometrium sloughs off with considerable hemorrhage. The actual loss of blood in menstruation is variable, but ordinarily may be placed between 50 and 100 milliliters, a surprisingly small amount.

Hyposecretion of the ovarian hormones can lead to irregular menses; hypersecretion is possibly one cause of the excessive bleeding which some women undergo. The hormonal changes just preceeding and initiating menstruation seem to have repercussions throughout the endocrine system. In particular, there is an increased secretion of the mineralo-corticoids which causes a retention of salt and consequent edema. This is seen as the swelling of the hands and ankles and a gain in weight. *Pre-menstrual tension* is also apparent in the increased emotional lability and irritability in many cases. In primitive societies and in some not so primitive, the menstruating female is regarded as "unclean"; she is often forbidden to prepare food or to participate in rituals. It has been facetiously suggested that these rules originated because the menstruating woman is too irritable to do anything properly.

The changing status of women in the modern world has altered even their own attitude toward menstruation. In the "natural" state the adult female is either pregnant or lactating, and menstruation is a relatively rare occurrence. Modern women refer to the menses colloquially as the "curse," but in mediaeval times it was called the "benefit."

FERTILIZATION

When the ovum begins to descend the fallopian tube, it is still surrounded by some of the granulosa cells from the original follicle. These are dispersed by an enzyme secreted by the sperm. Although only one sperm is necessary to fertilize an ovum, it is unlikely that one sperm would have enough of the enzyme to disperse the granulosa cells, and this is one apparent value of the enormous number of sperm in each ejaculation. Once a sperm reaches the ovum, it immediately penetrates into its cytoplasm. This leads to mysterious changes in the membrane surrounding the ovum, rendering it completely entry-proof to all subsequently approaching sperm. The nuclear material in the head of the sperm fuses with the nucleus of the ovum, restoring the full chromosome content characteristic of the species. The cell then begins mitotic division which is continued through the many complex stages of the development of the embryo and fetus.

PREGNANCY

The fertilized ovum descends the fallopian tube and enters the uterus, where after a day or two it becomes *implanted* in the thickened endometrium which has been prepared for it. At this, the *trophoblastic* stage of development, the embryo derives its nutrition by literally digesting the cells of the endometrium. The dissolution of the surrounding tissues implants it all the more deeply in the wall of the uterus.

During this early stage of pregnancy, many women suffer from recurrent nausea, especially in the morning. It has been suggested that the digestion of tissues by the trophoblast in the uterus produces toxic side products which cause the nausea and malaise. Similar cellular digestive processes occur after severe burns or in gangrene, and these conditions are also accompanied by nausea.

During the trophoblastic stage of nutrition, blood vessels grow from the embryo into the endometrium, and comparable vessels grow from the uterine wall in the other direction. The blood vessels and associated tissues develop into the *placenta,* illustrated in Fig. 24.2, which comprises two interdigitated vascular surfaces for the exchange of nutrients and oxygen in one direction, and of carbon dioxide and other end products of metabolism in the other direction. As the placenta develops, the embryo derives more and more of its nutrition by this route, and by about the tenth week of pregnancy, the local digestive processes are dispensed with.

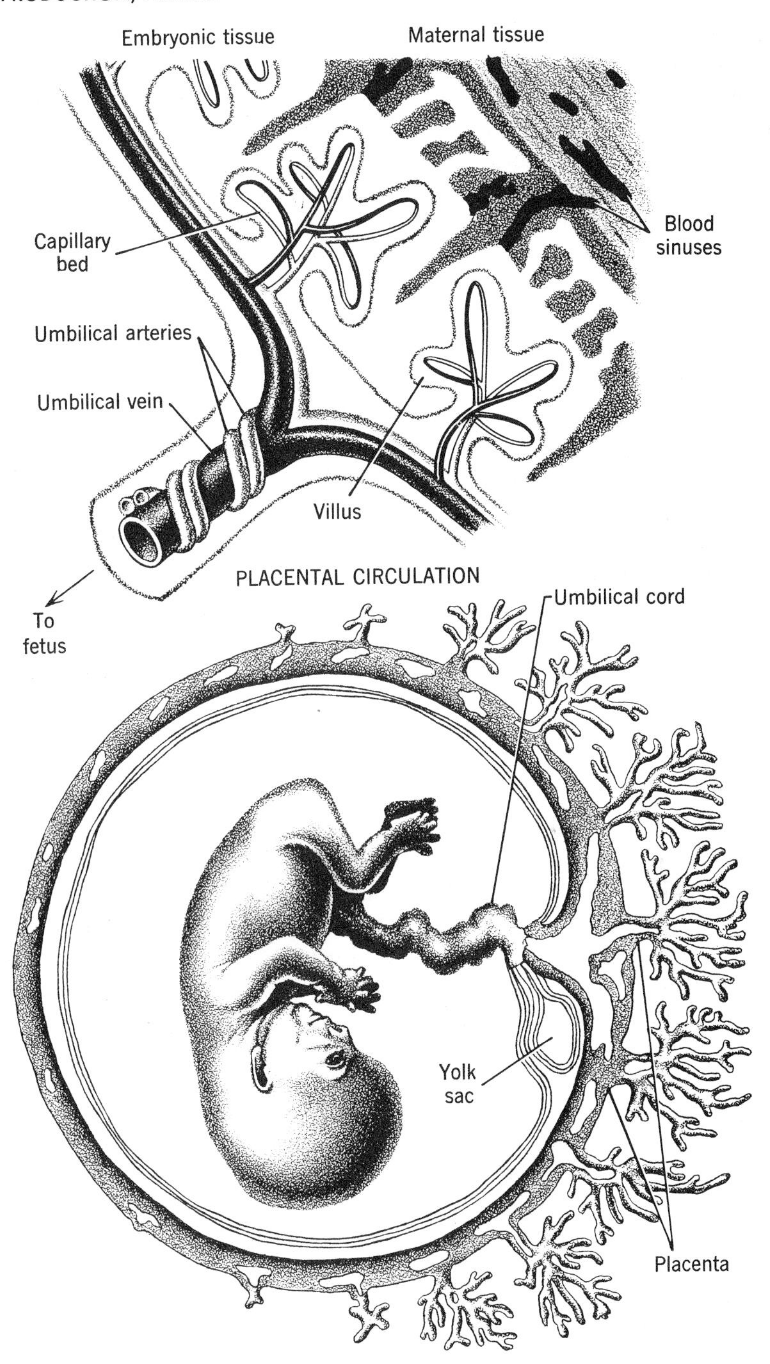

FIGURE 24.2. Diagram of the placenta.

It must be emphasized that there is no direct connection between the blood system of the mother and that of the embryo. In the placenta, the two systems are brought into close proximity and the gases and nutrients are exchanged by diffusion. The placenta grows during pregnancy, but not synchronously with the embryo; it is relatively larger during early pregnancy and then does not grow much during the last six weeks when the fetus grows a great deal. During intra-uterine life, the fetus is attached to the placenta by the *umbilical cord,* which contains the main blood vessels. It should be noted, however, that the cord and half the placenta is of tissues derived from the fetus. At birth, or *parturition,* the fetus is expelled first, usually remaining attached by the umbilical cord to the placenta, which is still in the uterus. The cord can be tied and cut at this stage. Later contractions of the uterus expel the placenta, which is commonly called the *afterbirth.*

Endocrine functions in pregnancy. We have referred to the formation of the corpus luteum as an accessory endocrine organ at the site of the ruptured follicle. The estrogens and progesterone secreted by this body promote the endometrial thickening. If there is no fertilization, changes in the gonadotrophic secretion by the adeno-hypophysis cause the involution of the corpus luteum, a cessation of its secretion, and a consequent menstruation. In the event of the implantation of a *fertilized* ovum, however, the trophoblastic phase is accompanied by a secretion of gonadatrophic hormones by some of the cells of the endometrium itself; these are the *chorionic gonadatrophins.* Their secretion maintains the corpus luteum as an active endocrine organ and therefore maintains the thickened uterine lining for the continuation of pregnancy. Should the chorionic gonatrophins fail to maintain the corpus luteum, the normal sequence of menstruation follows, which we can recognize as an early *miscarriage.*

As the placenta develops, it begins to secrete progesterone and estrogens itself, and by the third or fourth month the placental secretion alone is adequate to maintain pregnancy. The corpus luteum then involutes and is no longer a source of the hormones. Miscarriages between the tenth and fourteenth week of pregnancy can be due to an incomplete overlap between the secretions of the developing placenta and those of the involuting corpus luteum.

In later pregnancy, the placental progesterone promotes the further development of the mammary gland in preparation for lactation.

PARTURITION

The birth of the baby is preceded by considerable hormonal changes in the mother, particularly in a decreased proportion of progesterone secreted to that of estrogen. Although the causes of these changes are not understood, their effect is clearly an increase in the contractility and irritability of the uterus. This is exacerbated by the greatly increased size and activity of the fetus, culminating in the initiation of powerful uterine contractions which expel the fetus (Fig. 24.3).

The baby cannot pass out of the uterus into the vagina unless the cervix is first dilated. The contractions of the uterus force the baby's head against the cervix, and this seems to be the main factor in dilating it. The dilation takes several hours, and especially with the first child, it is probably the limiting factor in the speed at which parturition can be accomplished.

Labor pains are due to the strong contractions of the uterus. The sensations are strange and far from comfortable, but women with relaxed emotional attitudes to childbirth report that they are

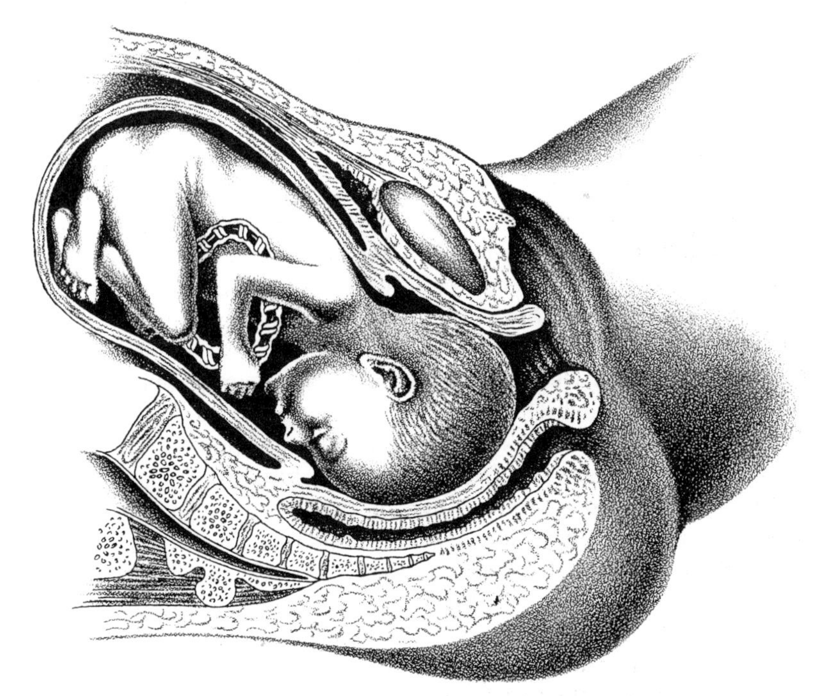

FIGURE 24.3. Parturition.

nothing like the "agonies" described in the obstetrical folklore which is all too current. If the mother is tense and "resistant" to the uterine contractions, the pain is apt to be more severe. Like any other form of hard work, uterine contractions are fatiguing, and toward the end of a long labor the mother may be too tired to provide voluntary muscular assistance; this usually arises only in a first delivery. When the cervix is fully dilated, the uterine contractions usually diminish in strength and frequency for a while, giving the mother a much-needed breathing spell before the shorter spell of powerful contractions which actually expel the fetus. A woman who is not scared by the involuntary contractions at this stage can learn to synchronize voluntary muscular efforts which assist the delivery. Sometimes during this last stage, the contractions are so extreme that the blood supply to the muscles is impeded and this without doubt contributes to the painfulness of the experience.

Further sources of discomfort during and after parturition are the tears of varying severity which can occur in the uterus and other parts of the birth canal. During the very last stage of expulsion of the baby, the distension of the vestibule by the baby's head can cause a tear in the perineal skin. A tear is more painful and more slowly repaired than a clean cut, therefore alert obstetricians frequently make a small cut in the skin at this stage of delivery in anticipation of an imminent tear. This is called an *episiotomy.*

Lactation begins shortly after parturition, and results from the suddenly decreased supply of estrogen consequent to the expulsion of the placenta. The lactogenic hormone secretion by the adenohypophysis is then no longer inhibited by the high estrogen level, and milk synthesis proceeds. If a continued lactation is not in the mother's best interest, it can be inhibited by injecting estrogens.

There are considerable advantages to a mother's feeding a baby with her own milk. The immediate advantage to the mother is that the stimulation of the nipples by the infant causes reflex contractions of the uterus which help to restore its normal size and position. The emotional value is, as always, much harder to evaluate. Many women report that their bond with a suckling infant is the most completely rewarding experience of their lives. For the infant, however, the psychiatric implications of its relations with its mother when fed at the breast rather than with a bottle are clearly beyond the scope of this text. Physiologically, the advantages of breast feeding are clear-cut, at least in the first few days. The milk first secreted after parturition is called *colostrum.* Its nutritional content is somewhat different from the later milk, but more important is its content of antibodies, which can be absorbed across the infant's intestinal walls and confer a passive immunity.

Human milk contains more sugar and less protein than does cow's milk, and many ingenious processes are used to make cow's milk resemble human milk more closely for infant nutrition. In some quarters, goat's milk is highly regarded because of the finely divided state of the fat particles. In primitive societies, and in earlier times, the failure of a mother's milk supply was almost certainly fatal to the infant, not only because of the difficulty of providing suitable milk from other sources, but because of the gastro-intestinal infections which almost inevitably followed the poor sanitation in the domestic milk supply. These hazards have been largely obviated by sterile techniques for the home preparation of babies' foods. Until a short while ago these improvements were contributory to an attitude in which the feeding of a baby at the breast was regarded as almost indecent. This attitude is fortunately no longer current, but in some circles it is still taken for granted that a mother does not wish to nurse her baby unless she indicates her preference to do so very strongly.

As was mentioned in an earlier chapter on nutrition, the nutritional demands on the mother during pregnancy are considerable, but those during lactation are far greater. The enormous output of nutrients, of carbohydrate, protein, fat and calcium, necessitate a large food intake of high quality by the mother.

The normal menstrual cycle is frequently suspended during the first six months of lactation, but this is by no means the rule. The likelihood of conception during lactation is perhaps slightly diminished and this may be one reason for the extremely prolonged nursing periods which are common in some societies. A full lactation period for a human mother may be regarded as 5 or 6 months with 9 months as the maximum, but if the child continues to remove the milk, lactation may continue for two or even three years.

REFERENCES

1. Allen, Robert D., "The moment of fertilization," *Sci. American* **201** (1), 124, July 1959.
2. Farris, Edmond J., "Male fertility," *Sci. American* **182** (3), 16, May 1950.
3. Moog, Florence, "Up from the embryo," *Sci. American* **182** (2), 52, February 1950.
4. Pincus, Gregory, "Fertilization in mammals," *Sci. American* **184** (3), 44, March 1951.
5. Reynolds, Samuel R. M., "The umbilical cord," *Sci. American* **187** (1), 70, July 1952.
6. Reynolds, Samuel R. M., "Obstetrical labor," *Sci. American* **182** (3) March 1950.
7. Stone, Abraham, "The control of fertility," *Sci. American* **190** (4), 31, April 1954.

INDEX